Chromatography in Petroleum Analysis

CHROMATOGRAPHIC SCIENCE

A Series of Monographs

Editor: JACK CAZES
Waters Associates, Inc.
Milford, Massachusetts

Volume 1: Dynamics of Chromatography
J. Calvin Giddings

Volume 2: Gas Chromatographic Analysis of Drugs and Pesticides
Benjamin J. Gudzinowicz

Volume 3: Principles of Adsorption Chromatography: The Separation of Nonionic Organic Compounds
Lloyd R. Snyder

Volume 4: Multicomponent Chromatography: Theory of Interference **(Out of Print)**
Friedrich Helfferich and Gerhard Klein

Volume 5: Quantitative Analysis by Gas Chromatography
Josef Novák

Volume 6: High-Speed Liquid Chromatography
Peter M. Rajcsanyi and Elisabeth Rajcsanyi

Volume 7: Fundamentals of Integrated GC-MS (in three parts)
Benjamin J. Gudzinowicz, Michael J. Gudzinowicz, and Horace F. Martin

Volume 8: Liquid Chromatography of Polymers and Related Materials
Jack Cazes

Volume 9: GLC and HPLC Determination of Therapeutic Agents (in three parts)
Part 1 edited by Kiyoshi Tsuji and Walter Morozowich
Parts 2 and 3 edited by Kiyoshi Tsuji

Volume 10: Biological/Biomedical Applications of Liquid Chromatography
Edited by Gerald L. Hawk

Volume 11: Chromatography in Petroleum Analysis
Edited by Klaus H. Altgelt and T. H. Gouw

Other Volumes in Preparation

Chromatography in Petroleum Analysis

edited by

Klaus H. Altgelt

T. H. Gouw

Chevron Research Company
Richmond, California

Marcel Dekker, Inc. New York and Basel

Library of Congress Cataloging in Publication Data
Main entry under title:

Chromatography in petroleum analysis.

(Chromatographic science ; v. 11)
Includes indexes.
1. Petroleum--Analysis. 2. Chromatographic analysis. I. Altgelt, Klaus H., [Date]
II. Gouw, T. H.
TP691.C57 665'.5'028 79-12150
ISBN 0-8247-6790-X

MARCEL DEKKER, INC.
270 Madison Avenue, New York, New York 10016

Current printing (last digit):
10 9 8 7 6 5 4 3 2 1

PRINTED IN THE UNITED STATES OF AMERICA

FOREWORD

The year was 1958, the place Amsterdam. It was cold and blustery when the Second International Symposium on Gas Chromatography convened its opening meeting. The stolid Dutch scientists, justly proud that their country had been chosen as the site of this important meeting, were not allowing the frivolities of daily life to intrude upon the seriousness of the business at hand. Everybody was seated in time for the opening remarks.

Half an hour later, a side door was abruptly opened. A man in a raincoat, obviously unfamiliar with the layout of the building and the momentous significance of the gathering, blundered in. A shock, almost physical in its intensity, went through the audience. Even though he slipped into a vacant chair mercifully close to the door, there were mutterings, looks of consternation, and a general air of disapproval at his temerity.

The discussion period that followed after that was uneventful until the man rose to pose a query. The chairman acknowledged his presence and asked him, as was the custom, to identify himself. He said he was A. J. P. Martin.

Never have I seen such a short statement produce such a dramatic change in so short a time.

Perchance not with the same suddenness and intensity, but certainly in just a few short decades, has come a similar drastic change in our opinion of the technique that owes so much to the scientist who stood that morning in the meeting room. One of us remembers, in his undergraduate years, an organic chemistry professor making scathing remarks on the future of gas chromatography. (He didn't like his colleague who was working on the technique.) Now, if the question were posed, "Which analytical instrumental technique is the most valuable and widespread in the petroleum industry?", we suspect that the majority of the practicing scientists in the field would vote for chromatography in general and gas chromatography in particular.

It is for this reason that we think the time has come to compile a collective volume to cover the various aspects of chromatography in the petroleum industry.

This book has changed title twice since its inception. First, it was to be "Chromatography in the Petroleum Industry." This, of course, would have slighted the important contributions by other laboratories, notably that of the Bureau of Mines. It was quickly abandoned, although it avoided a problem that we were not able to overcome later. The second title was "Chromatography in the Petroleum Field." By "field" we meant field of interest or field of work, but it sounded too much like a geological field related to drilling for petroleum. Thus, we adopted a third, the present title. It sounded good, but it is still a compromise in view of the intended scope of our book.

We used the concept of petroleum analysis in its widest possible range. For instance, we included the analysis of fuel and lubrication additives, an important and challenging application of chromatography. Another, perhaps even more significant and essential extension is that to other fossil fuels, such as shale oil and coal. In the literal sense of the word, petroleum (=rock oil), these fuels would be included anyway; but traditionally, they have not been. They are not only closely related to crude oil, but the methods of their chromatographic analysis are much the same though generally more demanding. A further widening of the scope suggested by our present title is to petroleum products and processing streams. These span the whole range from petroleum gases to asphalts.

The editors are extremely pleased with their contributors. If, in a few cases, there is an overlap of reported work, it was left that way because of the different emphasis or context used by the authors. Altogether, we hope to have achieved a fairly uniform treatment of the entire field of chromatography in the petroleum industry and in nonindustrial laboratories involved with petroleum and other fossil fuels.

A few general remarks on gas chromatography are probably in order. The literature on it suffers from the fact that too many papers of no, or very little, significance or practical value have been published. This discourages the average reader from finding most of the really valuable papers. On the other hand, it seems that there is also practical folklore of great value to the practicing chromatographer. Much of this has not been published or has appeared only in obscure journals or internal reports. We do not claim that we have managed to separate all the chaff from the wheat, but at least we tried.

One important area is not adequately covered in this book: the work of the ASTM Committee D-2, Research and Development Division IV on Hydrocarbon Analysis. However, the ASTM manuals are published with imposing regularity each year; so it should be fairly easy to determine the current status of any new method by consulting the more recent issues.

The editors are also taking this opportunity to thank their respective spouses for their patience and understanding, and the management of Chevron Research Company for permission to compile and to edit this volume.

T. H. Gouw
K. H. Altgelt

LIST OF CONTRIBUTORS

E. R. ADLARD, Shell Research Ltd., Thornton Research Centre, Chester, England

EDGAR W. ALBAUGH, Chemicals and Minerals, Gulf Research and Development Company, Pittsburgh, Pennsylvania

KLAUS H. ALTGELT, Fuels and Asphalts Research Department, Chevron Research Company, Richmond, California

R. AMOS, Analytical Department, Esso Petroleum Company Ltd., Esso Research Centre, Abingdon, England

M. P. TIMOTHY BRADLEY,* Research and Engineering, The Standard Oil Company of Ohio, Warrensville Heights, Ohio

RALPH A. BROWN, Analytical and Information Division, Exxon Research and Engineering Company, Linden, New Jersey

RONALD D. BUTLER, Fuels Branch, Air Force Aero Propulsion Laboratory, Wright-Patterson Air Force Base, Ohio

DAVID L. CAMIN, Research and Development Department, Suntech Inc., Marcus Hook, Pennsylvania

EMILIO J. GALLEGOS, Research Services Department, Chevron Research Company, Richmond, California

T. H. GOUW, Chevron Research Company, Richmond, California

ELIZABETH A. HARVEY, Research Services Department, Chevron Research Company, Richmond, California

R. E. JENTOFT, Chevron Research Company, Richmond, California

*Present affiliation: Autolab Division, Spectra-Physics, Inc., Santa Clara, California.

DOUGLAS M. JEWELL, Gulf Research and Development Company, Pittsburgh, Pennsylvania

DEWITT R. LATHAM, U.S. Department of Energy, Laramie Energy Center, Laramie, Wyoming

DENNIS L. SAUNDERS, Research Department, Union Oil Co., Brea, California

THOMAS D. SEARL, Analytical and Information Division, Exxon Research and Engineering Company, Linden, New Jersey

MILAN L. SELUCKY, Hydrocarbon Research Center, Department of Chemistry, University of Alberta, Edmonton, Alberta, Canada

LLOYD R. SNYDER, Technicon Instruments Corp., Tarrytown, New York

JOSEPH C. SUATONI, Analytical Technology Department, Gulf Research and Development Company, Harmarville, Pennsylvania

FRED C. TRUSELL, Marathon Oil Company, Denver Research Center, Littleton, Colorado

IRVILLE M. WHITTEMORE, Research Services Department, Chevron Research Company, Richmond, California

CONTENTS

Part I HISTORY

Chapter 1

HISTORY OF CHROMATOGRAPHY IN PETROLEUM ANALYSIS

David L. Camin
Research and Development Department
Suntech Inc.
Marcus Hook, Pennsylvania

I. INTRODUCTION

The application of chromatographic techniques in petroleum analysis predates by several years the discovery of chromatography by Tswett in 1903. In the interim years between Tswett's discovery and the rediscovery of the technique, approximately some 40 years later, chromatographic techniques were being developed and utilized by the petroleum industry, although not recognized as such. Over the past 74 years, nearly all of the many forms of chromatography, i.e., gas-liquid, liquid-liquid, gas-solid, and liquid-solid, have found significant applications ranging from routine analyses to control of refining operations, research separations, and analytical research.

II. EARLY HISTORY

The early history of fractionation of crude petroleum by "filtration" and/or "adsorption" predates by several years and parallels the completely independent discovery and development of chromatography by Tswett [1]. Weil [2-4], in his extensive treatment of the history of petroleum chromatography, tells us that shortly before the turn of the century David Talbot Day, Chief of the Division of Mineral Resources of the American Geological Survey, applied the process of filtration to separate Pennsylvania green petroleum. Day was familiar with the "filtration hypothesis" which theorized that primary Ohio oil was transformed into secondary Pennsylvania oil as it migrated upward through the earth's strata. His experiments demonstrated that crude oil forced upward through a column of powdered limestone was changed in color and composition. The results of this work were published in a short paper in 1897 [5]. Day continued his experimental investigations and in August 1900 reported his results at the First International Petroleum Conference in Paris [6].

While originally using powdered limestone, Day and his co-workers continued their work using British fuller's earth as a column packing. Engler and Albrecht [7] proposed using Florida fuller's earth instead of the British fuller's earth in hopes of achieving a more complete separation.

Whether it was simply optimism or shrewd technological insight, Day fully expected to be able to isolate individual hydrocarbons from complex mixtures. Unfortunately, the accomplishment of this feat was beyond the capabilities of that time. It would be many years before Day's expectations would be realized by the application of high-resolution gas chromatography.

The work of Engler and Albrecht was of great importance and considered by many to be superior to Day's. In the chromatographic system they devised, additional pressure was supplied by gravity, as opposed to Day's general reliance on capillarity, thus obtaining faster flow rates and reduced separation times. Fractions of effluents were collected at fixed intervals in receivers. These fractions were then characterized by classical methods as to specific weight, viscosity, boiling point, and chemical properties. It is now obvious that Engler and Albrecht's isolation technique was, in fact, the first example of a "flowing chromatogram." Ubbelohde [8] was also actively working in the area at this time. He was the first to recognize that filtration was not a capillary diffusion process. He stated, quite correctly, that changes in concentration were affected by adsorption, while capillarity was the propelling force.

Day, Engler, and other early workers were restricted in their efforts by unsatisfactory adsorbents. The outbreak of World War I in 1914 marked the end of an era of activity in petroleum chromatography. Neither Day nor Engler seemed to have followed up his work after the war, although Day, in

1922 in his "Handbook of Petroleum," [9] devoted several paragraphs to adsorption and published a table showing fractionation of crude petroleum in single tubes.

III. THE PERIOD FROM 1914 TO 1935

The period from 1914 to 1935 provided the technique with two developments of significance. The first of these is the work of Gurwitsch [10-18], spanning the period from 1912 to 1926. Although Gurwitsch's work is not generally known, it was basic in establishing some theoretical background for petroleum chromatography. The theory of adsorption, which allowed the classification of hydrocarbon groups according to their adsorbability on solids, was a major contribution. Gurwitsch also described the different types of adsorption fractionations, which we have come to know as frontal analysis, displacement development, and elution.

The second significant development was the advent of silica gel as an adsorbent. Silica gels had been known for many years, but new technology introduced in 1914 pointed the way to commercially feasible production of silica gels, thus making them readily available. A number of workers began to investigate silica gels as adsorbents during the period after the war. It also came into use in adsorption refining. Tarasov [19], in 1926, published the results of some percolation experiments with silica gel in which aromatics were quantitatively removed from the charge stock. The chromatographic isolation of a major hydrocarbon group thus seems to have been successfully accomplished.

IV. THE AMERICAN PETROLEUM INSTITUTE RESEARCH PROJECT 6

No discussion of chromatography in the petroleum industry could be complete without recognition of the work of the American Petroleum Institute (API) Research Project 6 and its contributors. The API Research Project 6 was started in 1927 to separate, identify, and determine the chemical constituents of commercial petroleum fractions. During the early years of the project, no use was made of chromatographic techniques. The first report by Mair and White [20] of fractionation by adsorption on silica gel appeared in 1935. Separations of a number of different hydrocarbon groups were reported. Aromatics and olefins were completely removed from the paraffins and naphthenes. A small but definite separation of naphthenes from paraffins occurred. A slight separation of normal paraffins of different molecular weights was also observed. Until the mid-1940s, adsorption continued to be used by the project as an adjunct to the overall separation scheme [21-32], mainly to simplify analytical distillation, etc.

The exact date when the fractionation work of the API project switched from simple adsorptive filtration to true chromatographic separation is somewhat hazy as a result of war-time restrictions on publication. Weil fixes the date as 1945 from a publication by Mair [33]. However, that particular work had been reported internally as early as 1943. This era does mark the first references to the work of Tswett and the use of the term chromatography. The method of differential adsorption fractionation described in this work is particularly interesting because it is undoubtedly the forerunner of the standard FIA method for determining paraffins, olefins, and aromatics which is still routinely used in the petroleum industry. The method was proposed to the American Society for Testing and Materials (ASTM) Committee D-2 in 1946 and eventually became Method D 936 [34].

In order to keep issues in their proper perspective, it is necessary to digress a bit at this point. The interests and research goals of the API Research Project 6 dictated that adsorption chromatography be utilized as a bulk separation process to aid in the compound-by-compound characterization of a representative crude petroleum. However, petroleum chemists in industrial laboratories recognized the analytical potential of the technique and developed practical methodology which is still in use today. In the late forties and early fifties Hirschler [35, 36] and Lipkin [37] and their co-workers at the Sun Oil Company developed several analytical applications of adsorption chromatography. About the same time Conrad [38] at Standard Oil of Ohio and Criddle and LeTourneau [39] at Standard Oil of California developed a more rapid refinement of ASTM D 936. They both incorporated the use of fluorescent dyes to mark the boundaries of the various hydrocarbon zones in a chromatographic column. This technique led to ASTM D 1319 [40], the FIA method for saturates, olefins, and aromatics.

Until the termination of all research work in 1967, the API Project 6 continued to make extensive use of adsorption and differential adsorption fractionation as methods for purification of pure compounds and as part of extensive separation schemes. In the last years of the project from 1956 on gas chromatography came into use both as an analytical tool and as a separation technique.

Several significant contributions from the API Project 6 in the period from 1945 to 1960 must be noted before moving on to other topics. In 1947, Mair, Gaboriault, and Rossini described the construction and testing of a 52-ft adsorption column consisting of five sections assembled in an elevator shaft [41]. These columns were later moved and reassembled at the Carnegie Institute of Technology. In this process, the length was increased to 63 ft. The columns probably were the longest ever built at that time.

Mair, Westhaver, and Rossini [42], in 1950, presented a paper which dealt with the theoretical analysis of the fractionation process of adsorption. This paper applied the "theoretical plate" concept to the adsorption process. In general, the concepts used were those common to distillation and provided a theoretical approach to practical applications. An excellent summary of this work was presented by Mair [43] in Chapter 34 of the "Treatise on Analytical Chemistry." In 1956, Mair, Montjar, and Rossini [44] reported on the "fractionation of hydrocarbons by adsorption with added components." This was in actuality an application of liquid-liquid partition chromatography as noted by the authors in a reference to Martin and Synge [45]. A report by Mair [46] on methods of separation of petroleum hydrocarbons, delivered at the Seventh World Petroleum Congress in 1967, gives extensive coverage to the techniques of liquid-liquid, liquid-solid, and gas-liquid chromatography, indicating that all were utilized in the work of the project.

V. MODERN CHROMATOGRAPHY

The era of modern chromatography can undoubtedly be dated from the work of Martin and Synge in 1941. The real impact on the petroleum industry, however, was not until nearly a decade later when Martin and James [47, 48] perfected gas chromatography.

Martin's interest in hydrocarbon mixtures resulted in his contacting S. F. Birch at the British Petroleum Company. Working for Birch was D. H. Desty, a man whose name was to become a by-word in connection with gas chromatography in petroleum analysis. Desty and Birch were inspired by the possibilities offered by Martin and James' work and immediately set out to build an apparatus to duplicate the equipment of Martin and James. Within a few months, Martin, James, and Desty were turning out retention data on over 150 hydrocarbons.

The fuse was lit, so to speak, and in the explosive development of gas chromatography over the past years, there has been a veritable multitude of contributors. We will attempt to cite some high spots; and to all of that legion who have made significant contributions and are not mentioned, we apologize.

The very first international symposium on vapor phase (gas) chromatography convened in 1956 and was sponsored by the Institute of Petroleum. Nearly all the application papers were of some significance to the petroleum industry. Some were quite specific: a paper by C. Rouit [49] described an analysis and control of refinery gas streams using chromatographic techniques; another by Freund, Benedek, and Szepesy [50] described a pilot plant unit for continuous gas chromatography. Others dealt with the distribution of aromatic compounds in coal tar naphthas, solvent analysis, and analysis of samples taken from internal combustion engines. An event of particular significance at that symposium was the presentation of a paper by H. H. Hausdorff [51] of the Perkin Elmer Corporation with a description of the first commercial instrumentation available. Almost all the other instruments described were home built. The availability of commercial instrumentation has since been a prime factor in the rapid growth of gas chromatography in the petroleum industry. Other significant contributors at this particular meeting were Boer [52], with a description of the first ionization detector, and Keulemans and Kwantes [53], working for Shell Oil in Amsterdam.

About this time, Eggertsen [54] and co-workers in the United States described the use of a 50-ft column packed with commercial carbon black containing 1.5% squalane to separate all ten of the commonly encountered C_5 and C_6 saturates in 2 hr. By 1958, the same authors had studied other packings and extended their saturate analysis through C_7. In the same year, Knight [55] reported the development of GC techniques for the analysis of C_5 and C_6 olefins. The year 1958 also marked the date of the Second International Symposium under the sponsorship of the Institute of Petroleum. At that symposium R. P. W. Scott [56] reported on the separation of C_7 and C_8 paraffin isomers using a highly efficient packed column capable of over 30,000 theoretical plates.

At this same symposium M. J. E. Golay [57] demonstrated the possibilities of open tubular capillary columns. The significance of these high-efficiency columns and their ability to separate close isomers such as meta- and paraxylene were soon recognized by the petroleum industry. Simmons, Richardson, and Dvoretzky [58] used capillary chromatography along with the methylene insertion reaction to analyze C_9 alkanes in an alkylate. In 1961, Desty, along with Goldup and Swanton [59], reported on the resolution of 122 peaks from C_3 to C_9 on a portion of the API Research Project 6 crude petroleum in a 20-hr run using a 900-ft glass open tubular column coated with squalane. More than half of these peaks were identified. By 1963, Schwartz and Brasseaux [60], utilizing a two-phase coating material made from hexadecane and a fluorocarbon oil, were able to produce a column which was quite successful in the resolution of naphtha-type materials, including paraffins, aromatics, and cycloparaffins through C_8.

A milestone in the development of capillary gas chromatographic methods was achieved in 1968 when Sanders and Maynard [61] published a method for determining C_3 through C_{12} hydrocarbons in a full-range gasoline. This analysis was accomplished on a 200-ft squalane column in less than 2 hr using both temperature and flow programming techniques; 240 peaks were observed in an average gasoline, of which 180 were identified. These accounted for over 230 individual hydrocarbons. At this point we should pause to recognize that while analytical gas chromatography does not allow the isolation of a pure individual hydrocarbon from a complex petroleum fraction, it does permit us to identify many individual components. By judicious use of preparative gas chromatography we may isolate some pure hydrocarbons, thus coming close to fulfilling the dream that Day expressed when he did his first experiments on filtration.

Gas chromatography has become a widely accepted technique in the petroleum industry. Many individual applications which are typical of those that are being carried out throughout the industry will be described in subsequent chapters.

VI. LIQUID CHROMATOGRAPHY

While liquid chromatography has not in the past enjoyed the extensive development of gas chromatography, it is currently on the threshold of such an evolution. We have shown how the early history of chromatography in petroleum analysis mainly involved the development of adsorption chromatography and how the API Project 6 in the mid-1950s began to utilize liquid-liquid chromatography. Certainly one of the most widely used chromatographic techniques in the petroleum industry is the FIA analysis or some modification thereof. This is truly a chromatographic procedure in the best sense, generally used to separate gasoline range mixtures into saturates, nonaromatic olefins, and aromatics.

A review on liquid chromatography would not be complete without taking note of the work of L. R. Snyder [62-72], who has published extensively on the application on linear elution adsorption chromatography in the petroleum industry. Snyder's work has covered the determination of nearly all the classes of materials that can be found in petroleum. He has made significant contributions to general compound class separations and to the determination of aromatics, polyaromatics, nonaromatic olefins, and sulfur and nitrogen types in petroleum. In a symposium on gel permeation chromatography at the Division of Petroleum Chemistry Meeting of the ACS in 1970, Altgelt contributed a paper [73] on the GPC separation and integrated structural analysis of petroleum heavy ends. The same symposium featured two papers [74, 75] contributed by members of the staff of the API Project 60 on the characterization of heavy ends of petroleum. API Project 60, in its quest to separate and characterize the heavy ends of petroleum, has made extensive use of liquid-solid, as well as gel permeation, chromatographic separation techniques. Two recent papers [76, 77] by the Project 60 staff describe the methods used.

In the last few years high-pressure automatic liquid chromatographic instruments have begun to appear in large numbers in petroleum laboratories.

These high performance units offer the promise of eventually making liquid chromatography as useful as gas chromatography is now, even though present applications of liquid chromatography are severely limited by the types of detectors available. Until now, the only two generally accepted detection systems have been the differential refractometer and the differential uv detector. The refractometer is limited by its sensitivity, and obviously the uv detector is limited to the detection of uv absorbers. However, some interesting separations have been performed on these automated liquid chromatographs. Schmitt, Henry, et al. [78] separated 12 aromatic compounds on a column of ODS Permaphase* using a reverse phase gradient of a water and methanol mixture as the mobile phase. This technique has been used to analyze the polycyclic materials and condensates from automobile exhaust gas. Descriptions of other applications are scattered throughout other chapters in this book.

There is no doubt that liquid chromatography has not, as yet, reached the degree of utilization that is enjoyed by gas chromatography. However, we are certain that better and more universal detectors will be developed. It is just a matter of time, and we fully expect that liquid chromatography in petroleum analysis in general will exceed the use of gas chromatography. This is a matter of simple mathematics. One can estimate that of all the known organic materials only 20% are volatile enough to be analyzed by gas chromatography, while high-speed liquid chromatography is applicable to the other 80%.

VII. PLANAR CHROMATOGRAPHY

Planar chromatography in petroleum analysis encompasses both thin layer and paper chromatography. Killer and Amos [79], in 1966, published an excellent paper on the uses of thin layer chromatography in the analysis of petroleum products. They pointed out that it can be used to separate polycyclic hydrocarbons and also is extremely useful in separating additive materials in petroleum products such as gasoline and lubricating oils. These are still the major uses for planar chromatography in the industry. Certainly, since thin layer and paper are more easily used as qualitative tools rather than quantitative, it would seem that high-speed liquid chromatography would soon supplant their use in these areas. However, planar chromatography is still very much in use and will probably continue to be so. As a rapid qualitative assay, it is easy to use and extremely inexpensive. Also, it is a valuable adjunct to automated liquid chromatography as a method of screening solvent systems, and the like.

---

*Trademark E. I. du Pont de Nemours and Company, Inc.

In conclusion, it is obvious that chromatography and the petroleum industry have become inseparable. The uses of chromatography in the industry will undoubtedly continue to multiply.

REFERENCES

1. M. Tswett, Ber deut. botan. Ges. 24:316 (1906).
2. H. Weil, Petroleum (London) 14:5 (1951).
3. H. Weil, Petroleum (London) 14:205 (1951).
4. H. Weil, Petroleum Processing 7:973 (1952).
5. D. T. Day, Proc. Am. Phil. Soc. 36:112 (1897).
6. D. T. Day, Paper submitted to the First International Petroleum Congress at Paris, August 16-28, 1900.
7. C. Engler and E. Albrecht, Z. Angew. Chemie 14:889 (1901).
8. L. Ubbelohde, Petroleum (Berlin) 4:1394 (1909).
9. D. T. Day, ed., "Handbook of the Petroleum Industry," Wiley, New York, 1922.
10. L. G. Gurwitsch, Petroleum (Berlin) 8(2):65 (1912).
11. L. G. Gurwitsch, Petroleum (Berlin) 20:903 (1924).
12. L. G. Gurwitsch, Kolloid Zeit. 11:17 (1912).
13. L. G. Gurwitsch, Kolloid Zeit. 22:80 (1923).
14. L. G. Gurwitsch, Kolloid Zeit. 38:247 (1926).
15. L. G. Gurwitsch, Z. fur Phys. Chem. 87:323 (1914).
16. L. G. Gurwitsch, J. Russ. Phys. Chem. Soc. 47:805 (1915).
17. L. G. Gurwitsch, J. Russ. Phys. Chem. Soc. 47:827 (1915).
18. L. G. Gurwitsch and N. L. Gurwitsch, Neft. Kohz. 8:636 (1925).
19. B. C. Tarasov, Azerb. Neft. Khoz. T1(10):47 (1926).
20. B. J. Mair and J. D. White, J. Research Natl. Bur. Standards 15:51 (1935).
21. B. J. Mair, S. T. Schicktanz, and F. W. Rose, J. Research Natl. Bur. Standards 15:557 (1935).
22. B. J. Mair and S. T. Schicktanz, J. Research Natl. Bur. Standards 17:146 (1936).
23. J. D. White and F. W. Rose, J. Research Natl. Bur. Standards 21:151 (1938).
24. C. B. Willingham, J. Research Natl. Bur. Standards 22:321 (1939).
25. B. J. Mair and A. J. Streiff, J. Research Natl. Bur. Standards 24:395 (1940).
26. B. J. Mair and A. J. Streiff, J. Research Natl. Bur. Standards 27:343 (1941).
27. F. D. Rossini and B. J. Mair, Proc. Am. Pet. Inst. 22(III):6 (1941).
28. F. D. Rossini et al., Proc. Am. Pet. Inst. 23(III):8 (1942).
29. F. D. Rossini, Pet. Eng. 14(5):41 (1943).

30. B. J. Mair and A. F. Forziati, J. Research Natl. Bur. Standards 32:151 (1944).
31. B. J. Mair and A. F. Forziati, J. Research Natl. Bur. Standards 32:165 (1944).
32. A. F. Forziati, J. Research Natl. Bur. Standards 32:11 (1944).
33. B. J. Mair, J. Research Natl. Bur. Standards 34:435 (1945).
34. ASTM D 936-55, "Annual Book of ASTM Standards, Petroleum Products," Am. Soc. Test. Mats., Philadelphia, 1971.
35. A. E. Hirschler and S. Amon, Ind. Eng. Chem. 39:1585 (1947).
36. A. E. Hirschler and R. L. James, Preprints, Div. of Petrol. Chem., ACS, 1(1):187 (1956).
37. M. R. Lipkin, W. A. Hoffecker, C. C. Martin, and R. E. Ledley, Anal. Chem. 20:130 (1948).
38. A. L. Conrad, Anal. Chem. 20:725 (1948).
39. D. W. Criddle and R. L. LeTourneau, Anal. Chem. 23:1620 (1951).
40. ASTM D 1319-69, "Annual Book of ASTM Standards, Petroleum Products," Am. Soc. Test. Mats., Philadelphia, 1971.
41. B. J. Mair, A. L. Gaboriault, and F. D. Rossini, Ind. Eng. Chem. 39:1072 (1947).
42. B. J. Mair, J. W. Westhaver, and F. D. Rossini, Ind. Eng. Chem. 42:1279 (1950).
43. B. J. Mair, Chromatography, in "Treatise on Analytical Chemistry," Vol. 3, Pt. 1 (Kolthoff, ed.), Interscience, New York, 1961, p. 1469.
44. B. J. Mair, M. J. Montjar, and F. D. Rossini, Anal. Chem. 28:56 (1956).
45. A. J. P. Martin and R. L. M. Synge, Biochem. J. (London) 35:1358 (1941).
46. B. J. Mair, Proc. Seventh World Pet. Cong. 9:39 (1967).
47. A. T. James and A. J. P. Martin, Biochem. J. (London) 50:679 (1952).
48. A. T. James and A. J. P. Martin, Analyst 77:915 (1952).
49. C. Rouit, in "Vapour Phase Chromatography" (D. H. Desty, ed.), Butterworths, London, 1957, p. 291.
50. M. Freund, P. Benedek, and L. Szepesy, in "Vapour Phase Chromatography" (D. H. Desty, ed.), Butterworths, London, 1957, p. 359.
51. H. H. Hausdorff, in "Vapour Phase Chromatography" (D. H. Desty, ed.), Butterworths, London, 1957, p. 377.
52. H. Boer, in "Vapour Phase Chromatography" (D. H. Desty, ed.), Butterworths, London, 1957, p. 169.
53. A. I. M. Keulemans and A. Kwantes, in "Vapour Phase Chromatography" (D. H. Desty, ed.), Butterworths, London, 1957, p. 15.
54. F. T. Eggertsen, H. S. Knight, and S. Groennings, Anal. Chem. 28:303 (1956).
55. H. S. Knight, Anal. Chem. 30:9 (1958).
56. R. P. W. Scott, in "Gas Chromatography" (D. H. Desty, ed.), Butterworths, London, 1958, p. 189.

57. M. J. E. Golay, in "Gas Chromatography" (D. H. Desty, ed.), Butterworths, London, 1958, p. 36.
58. M. C. Simmons, D. B. Richardson, and I. Dvoretzky, in "Gas Chromatography" (R. P. W. Scott, ed.), Butterworths, London, 1960, p. 360.
59. D. H. Desty, A. Goldup, and W. T. Swanton, ISA Proceedings, "1961 International Gas Chromatography Symposium of ISA," Michigan State University, East Lansing, Michigan, June 1961, p. 83.
60. R. D. Schwartz and D. J. Brasseaux, Anal. Chem. 35:1374 (1963).
61. W. N. Sanders and J. B. Maynard, Anal. Chem. 40:527 (1968).
62. L. R. Snyder, J. Chromatog. 5:430 (1961).
63. L. R. Snyder, J. Chromatog. 6:22 (1961).
64. L. R. Snyder, Anal. Chem. 33:1527 (1961).
65. L. R. Snyder, Anal. Chem. 33:1535 (1961).
66. L. R. Snyder, Anal. Chem. 34:771 (1962).
67. L. R. Snyder and W. F. Roth, Anal. Chem. 36:128 (1964).
68. L. R. Snyder and B. E. Buell, Anal. Chem. 36:767 (1964).
69. L. R. Snyder, Anal. Chem. 36:774 (1964).
70. L. R. Snyder, Anal. Chem. 37:149 (1965).
71. L. R. Snyder, Anal. Chem. 37:713 (1965).
72. L. R. Snyder, Preprints, Div. of Petrol. Chem., ACS, 15(2):C-44 (1970).
73. K. H. Altgelt and E. Hirsch, Preprints, Div. of Petrol. Chem., ACS, 15(2):A-262 (1970).
74. H. H. Oelert, D. R. Latham, and W. E. Haines, Preprints, Div. of Petrol. Chem., ACS, 15(2):A-204 (1970).
75. J. H. Weber and H. H. Oelert, Preprints, Div. of Petrol. Chem., ACS, 15(2):A-212 (1970).
76. D. E. Hirsch, R. L. Hopkins, H. S. Coleman, F. O. Cotton, and C. J. Thompson, Anal. Chem. 44:915 (1972).
77. D. M. Jewell, J. H. Weber, J. W. Bunger, H. Plancher, and D. R. Latham, Anal. Chem. 44:1391 (1972).
78. J. A. Schmit, R. A. Henry, R. C. Williams, and J. F. Dieckman, J. Chromatog. Sci. 9:645 (1971).
79. F. C. A. Killer and R. Amos, J. Inst. Petrol. 52:315 (1966).

Part II ANALYSIS OF DISTILLATES

Chapter 2

GAS ANALYSIS

Elizabeth A. Harvey
Research Services Department
Chevron Research Company
Richmond, California

I. INTRODUCTION

The ability to analyze the component in a gas mixture is vital to modern petroleum analysis. To ensure proper plant operation of a petrochemical process, the composition of feed, recycle, and product streams must be established. Gases feeding the furnaces must be analyzed for heat content. Air pollution regulations set limits on emissions of undesirable compounds to the atmosphere. This would include flue gases from refinery stacks and emissions from fuel delivery systems or combustion engines.

This chapter is written as a guide to the gas analysis methods in current use. It is not an exhaustive survey of the recent literature. An excellent review on this subject has appeared recently [1]. The basic principles of gas chromatography have been covered in a number of good reference books, including one by McNair and Bonelli [2].

II. SAMPLING AND STORAGE

Methods for sampling gases can be divided into three general categories. One involves portable containers; the second covers trapping on a solid support or in a liquid solvent; the third includes on-stream sampling into a process chromatograph.

A. Sample Containers

A glass bomb is recommended for samples near atmospheric pressure [3]. Usually the preevacuated vessel is connected to the stream outlet and then opened for filling. If the sample stream is at higher than atmospheric pressure, the gas is allowed to flow through the bomb and vented to the air. Even in this case it is helpful to preevacuate the bomb; otherwise, one may have to spend several minutes to purge residual air from a 200-cm sample container.

In some cases—generally with research samples—it may be advantageous to use water displacement to fill the container. This method is permissible only if the sample will not react with water and if water vapor will not interfere with the subsequent analysis. Alternate containers for samples near atmospheric pressure include rubber bladders or Teflon-coated plastic bags [4], which have been specially designed for sampling gases. Figure 1 shows some commonly used sample containers.

Samples at subatmospheric or at considerably above atmospheric pressure must be handled somewhat differently. Several manufacturers [5] make storage cylinders of 316 stainless steel which are appropriate for these occasions. When sampling low-pressure streams, the cylinder and sampling system are evacuated and the created vacuum is used to draw the

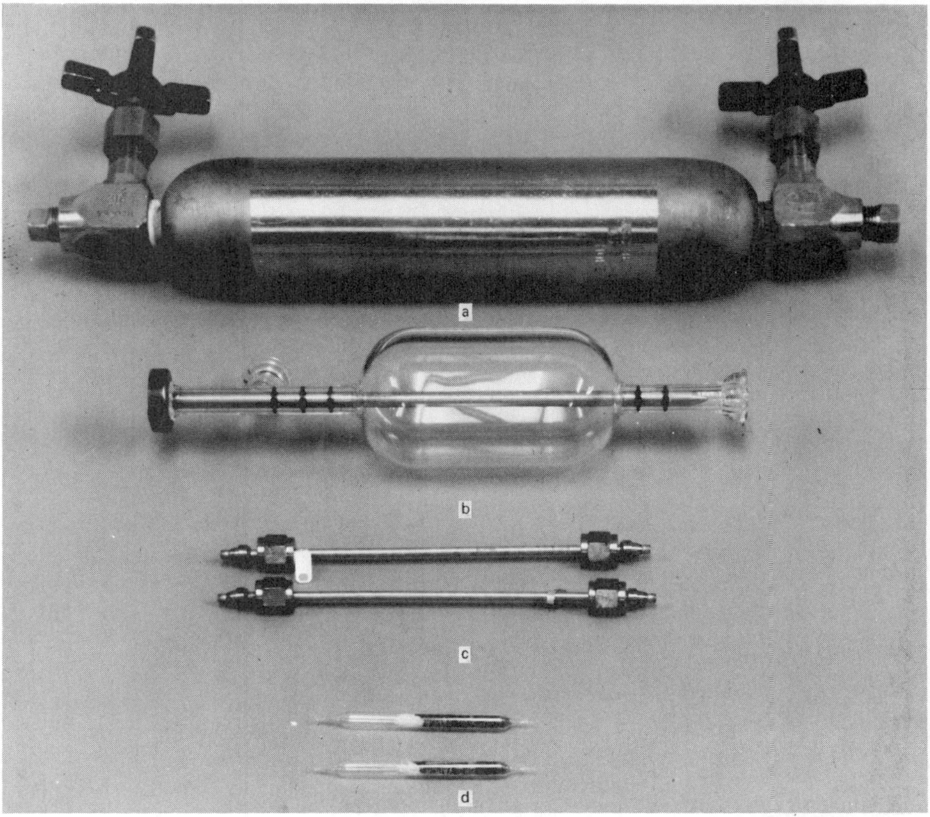

FIG. 1. Containers for gas samples. From top: (a) stainless steel bomb, (b) glass bomb, (c) adsorption tubes filled with TENAX-GC, (d) NIOSH-approved activated charcoal adsorption tubes.

sample into the bomb. Sampling of high-pressure streams is performed by attaching the container to the sampling point, opening the proper valves, and flushing the bomb with sample. When sampling high-pressure streams, one should take precautionary steps so as not to exceed the pressure rating stamped on the outside of the cylinder.

Some of these cylinders are fitted with a dip tube at one end. When the contents consist of a liquid and gas mixture, the dip tube should be considered to ensure getting a gas or a liquid sample, depending on what is desired.

Although it is a good practice to analyze samples soon after they are taken, mixtures stored in stainless steel bombs can be kept almost indefinitely unless they contain a reactive species. Samples stored in glass vessels should be analyzed within several days, since the valves or seals generally do not seal as well. This is especially the case with H_2-containing samples where there is greater danger of diffusion through the seals which, in turn, will allow air in.

Preevacuated aerosol sampling cans are commercially available to sample at atmospheric pressure such gases as stack gases, automobile exhaust, and atmospheric environment. The can is filled and sampled through a septum port and guide on the top.

The standard aerosol container, especially if it is under pressure, necessitates a different sampling approach. One possibility is to use a piercing unit [4] which attaches a septum to the side of the can allowing the use of a syringe to sample the contents.

B. Concentration of Trace Compounds

Some special samples may require a preconcentration step to permit trace compounds to be measured. This is often the case in pollutant analysis. Generally, the gases are passed through a trapping device to selectively retard the compounds of interest. Three types of traps have been used for this purpose: adsorption tubes, solvent traps, and cold traps.

1. Adsorption Tubes

A measured volume of the gas is pumped through a 4- to 12-in. long metal or glass tube filled with activated charcoal; Porapak Q, R, or T; Chromosorb 101, 2, 3, or 4 [6]; Tenax-GC [7]; or molecular sieves. These tubes must be baked out before use and must be kept sealed before and after sampling to prevent contamination. The volume of air to be sampled must be less than the retention volume of the least retained component. To desorb the sampled gases, the tubes are heated rapidly and carrier gas flowing through them drives the sample off either into the analytical column or into a cold trap for further concentration. Another approach is to pour the packing into a flask and extract it with solvent. This method is often employed to desorb activated charcoal tubes used for collecting air pollutants. One must keep in mind that the samples concentrated by any of these methods are liable to polymerization, pyrolysis, and isomerization reactions. Recovery is not always complete. The literature indicates, however, that 95% recovery rates are easily obtained if the correct packing is chosen [6-9].

A special application is reported by Mueller and Miller [10], who describe the industrial use of the NIOSH charcoal adsorption technique [11] for the analysis of potentially hazardous chemicals in air. They recommend Barneby's [12] method to estimate adsorption efficiencies of charcoal for organic vapors. Several laboratories provide the NIOSH-specified tubes and analyze them for a fee [13]. Also, several companies manufacture special concentrator/desorber units with portable sampling tubes [14].

2. Solvent Traps

In this approach, 10 to 100 ft^3 of the gas is pumped through several liters of an appropriate scrubbing solvent. The gas stream should be sparged through a fritted disk to increase mass transfer. This procedure may not be as efficient as using a sampling tube, but it can be of value if the compounds of interest are highly soluble in an easily handled solvent. If the compounds are not very volatile, the solvent can be evaporated readily for further concentration.

3. Cold Traps

This method is best for highly volatile materials [15, 16]. Dry ice-acetone is generally used as the coolant. As the gas is passed through the cold trap, the organic constituents in the gas stream condense in the well. When a reasonable amount of liquid has collected, the sample gas is shut off, the carrier gas is turned on, and the cold source is removed. The condensed liquid vaporizes and is swept into the head of a cold column where it is concentrated in a slug. This section is then heated rapidly to volatilize the material into the analytical column. In another approach, the cold liquid is taken up in a chilled syringe and injected into a hot injector.

The major problem with cold traps is that large quantities of water may accumulate along with the organics to be analyzed; thus, the method is of little use for wet gases. Also, aerosol formation may occur, lowering the trapping efficiency. On the other hand, this technique minimizes problems of sample oxidation or polymerization.

C. Direct On-stream Sampling

In some cases, one may desire to monitor a feed or product stream continuously. Process gas chromatographs have been developed to handle these analyses. The sample is continuously fed from the process stream to the sampling valve, which can be kept heated and free from any contamination. To ensure getting a representative mixture to the valve, the sample is taken through a slip tube which should run down the center of the process stream. If the sample is taken from the edge of the stream, the specimen may not be representative. With this method, it is easy to handle corrosive or high melting samples and to keep air and water out. Process gas chromatography is reviewed in Chapter 16 and in papers by Villalobos [17] and by Lipavsky and Berezkin [18].

III. INJECTION TECHNIQUES

A. Syringes

The traditional way to introduce samples into a gas chromatograph has been with a syringe. For gases, only specially designed syringes are acceptable [19] (Figure 2). These are larger than those used for liquids and can sample 0.1 to 10 cm^3 or more. Most have a locking device—either a plunger or a rotating tip—to prevent loss of sample through the needle. The plunger is often made of Teflon, with a small rubber O ring inserted in the plunger to create the additional pressure for a gas-tight seal. The needle has a large bore to allow large volumes of gas to be injected rapidly.

Despite their special design, gas syringes occasionally leak, especially past the plunger seal. To remedy, remove the plunger and slam the Teflon plug three or four times into a hard surface. When the plunger is replaced in the syringe, the fit should be noticeably better. However, this solution is only temporary. A leaking syringe is best returned to the manufacturer for repair.

Another problem occurs when some of the components of the gas sample are at their dew point. The compression of the sample on injection causes these components to condense on the walls of the syringe barrel. To counteract this problem, one should depress the syringe plunger twice during injection. When the plunger is retracted the first time, the condensed liquid will reevaporate; the gases are introduced into the chromatograph during the second stroke.

The large needles usually employed wear out the septa, which should be replaced frequently to prevent the occurrence of sample blowback or the

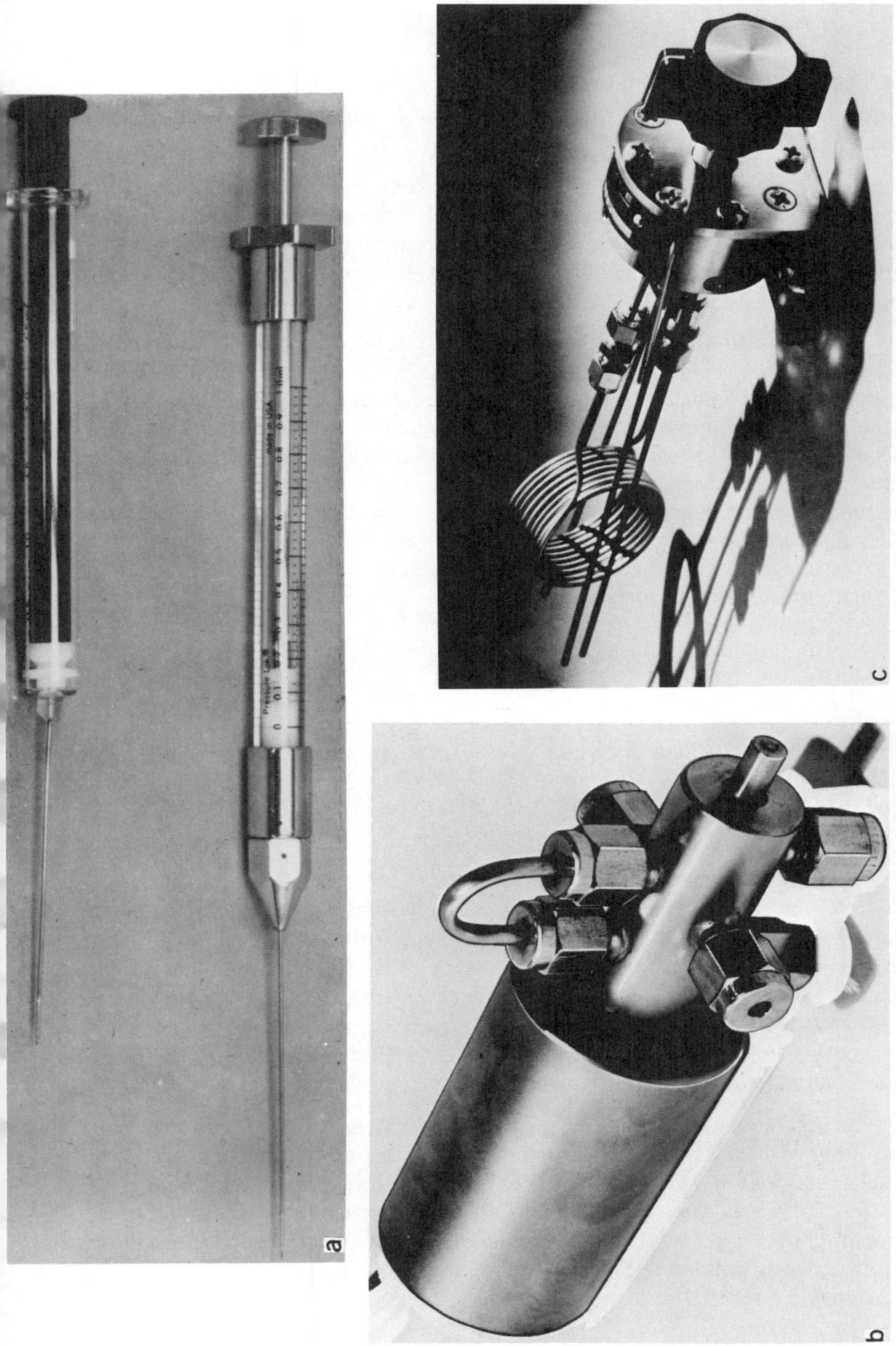

FIG. 2. Gas sampling devices: (a) two types of gas syringes, bottom syringe has a locking tip; (b) solenoid-operated gas sampling valve; and (c) manually operated gas sampling valve.

introduction of air along with the sample. Also, the needles often become plugged with small bits of rubber. They can be unplugged with fine needle-cleaning wires.

When improperly used, the gas syringe can be a safety hazard. Whenever an injection is made, the carrier gas exerts pressure against the syringe plunger. This force can become appreciable in large volume gas syringes. The best gas syringes have built-in stops to prevent plunger ejection.

In some situations, the pressure is sufficiently large to liquify some of the sample at room temperature, especially the C_4 and C_5 hydrocarbons. In this case, one has to choose between two alternatives. One can determine either the composition of the gas in equilibrium with condensed liquid at room temperature or the gas composition under plant conditions. In the latter case, the container is preheated before sampling. This is most conveniently done with heating mantles, heating tape, or heating sheets, which can be purchased to fit any size containers [20, 21]. Care must be taken when heating 316 stainless steel cylinders, since these containers lose strength at elevated temperatures.

B. Gas Sampling Valves

For repetitive samples of constant size, gas sampling valves simplify the injection procedure and eliminate some of the problems associated with syringes. Gas sampling valves, in their simplest form, usually have six ports [22]. Two types are pictured in Figure 2.

Problems of sample condensation, caused by wide fluctuations in room temperature, are overcome by small heating ovens, which maintain the injection valve at a constant elevated temperature. The sample loops may be exchanged to alter the sample size. Often the limiting factor on sample size is the valve itself, which may have as much as 0.2 to 0.3 cm^3 dead volume. Sample valves may be activated automatically, especially with on-stream process chromatographs or with computer-controlled chromatographs.

An important advantage of gas sampling valves is that they allow sampling without air contamination. Valves may of course leak and should be checked before use, both at the carrier gas line fittings and at the valve stem. Valve seals are either Buna-N O rings or washers made of Teflon, graphite, or Vespel. All of these materials may be damaged by excessive friction; when tightening the valve stem to stop leaking, one should proceed cautiously to avoid crushing the seal.

IV. DETECTORS

A. Thermal Conductivity Detector

The correct choice of a detector is vital to the success of a gas analysis. The most common detector used today is the hot wire or tungsten-rhenium filament thermal conductivity detector (TCD). The TC detector is not as sensitive to hydrocarbons and not linear over as large a concentration range as the flame ionization detector but it is the one commonly available detector which responds to any gaseous compound. With the exception of H_2, the response of all fixed gases in a helium carrier is linear over the sensitivity range of the detector. Outside the United States, hydrogen is often used as the carrier gas. Other gases, such as N_2, can also be used, but their lower thermal conductivity results in reduced sensitivities.

B. Flame Ionization Detector

The second most often used detector is the flame ionization detector (FID), which is very sensitive to any hydrocarbon gas. Unfortunately, the FID will not respond to the fully oxidized inorganic gases, to the fixed gases, or to H_2O. This is perfectly acceptable if an airfree analysis is desired, or if hydrocarbons are the only components of interest. If the hydrocarbons are to be determined quantitatively in a sample containing air, an external standard is required.

The FID will cover a linear range of 10^7 concentration units and is approximately 1000 times more sensitive to hydrocarbons than the TC detector. It requires less sample than the TC detector, but it is easily overloaded. The FID is often used in the trace analysis of low molecular weight hydrocarbon gases. In the analysis of CO and CO_2, the two compounds are separated and then reduced over a Rh or Ni catalyst to CH_4, which the FID easily detects to 5 ppm.

C. Other Detectors

Various other detectors are used to detect special classes of compounds. The electron capture detector operates on the principle that some of the eluting molecules will absorb electrons, thus lowering a fixed electron current in the detector cell. This detector is most sensitive to molecules containing atoms with high electron affinity, such as halogens.

The gas density balance is particularly useful for the analysis of corrosive gases since these gases never pass over the flow-sensitive elements.

For fixed gas analysis, argon is the carrier of choice. Since it has a high molecular weight, it does not exhibit the diffusion problems of H_2 or He, and it will give most fixed gas peaks a positive polarity.

For (the heavier) hydrocarbon gas samples, H_2 and He are preferred as carrier gases because of their low molecular weight (mol. wt.). With these gases all peaks on the chromatogram have the same polarity. Response factors (rf) for gas density balance detectors are particularly easy to obtain. They are calculated by the formula [23]:

$$rf_i = \frac{\text{mol. wt.}_i}{\text{mol. wt.}_i - \text{mol. wt.}_{\text{carrier}}} \qquad (1)$$

A commercially available chromatograph featuring dual gas-density balance detectors has recently been introduced [14]. This instrument allows the user to determine the molecular weight of most unknown compounds in a sample.

The helium ionization detector is little used at present. It was originally developed to detect trace amounts of gases. Ionizing radiation raises ultrapure helium to a metastable state. All compounds with a lower ionization potential are ionized by the excited helium. The ion current rises, giving a positive signal. This detector can measure fixed gases at low ppb levels.

Ion specific detectors are often employed in the analysis of special gas mixtures. Trace amounts of nitrogen compounds, e.g., can readily be detected with an alkali flame detector. This detector operates much like the flame ionization detector, except that an alkali metal salt tip is placed above the flame jet. The actual flame burns on the salt surface. The result is a highly enhanced ionization of nitrogen and phosphorous compounds relative to hydrocarbons. Signal enhancement by a factor of more than 30,000 has been reported.

Sulfur compounds can be analyzed by a flame photometric detector. This instrument measures the intensity of the 394 nm chemiluminescent emission line. This sensitive method is capable of detecting low ppb levels of sulfur [24].

A newly marketed detector, the microwave plasma detector [25], appears to demonstrate great versatility as an elemental analyzer for hetero-organic compounds. In this system, the helium carrier gas is excited by microwaves to form a plasma. The eluting components are excited by this plasma, and the resultant uv spectra are detected by a bank of up to 14 photomultiplier tubes. More than one element can be detected at the same time.

One last detector, not often used for routine GC analyses, but invaluable for qualitative identification of unknown compounds in the sample, is the mass spectrometer. A GC-MS system can be extremely useful for identifying unusual or unknown species or for quantifying very low-level trace impurities in a gas mixture. These systems are currently quite expensive and are not readily available in most laboratories.

V. CALIBRATION AND CALCULATION

A. Calibration Mixtures

The detector response is generally assumed to be proportional to the weight of the eluting compound. This is not always the case, especially for air-hydrocarbon mixtures. The TC response factors for some compounds in these systems may vary by as much as 30% from the expected values.

Response factors, the correction factors used to normalize the response of the components in a mixture, have been published in the literature. However, because each detector is operated in a slightly different way, it is wise to calibrate and determine response factors for each specific system. The simplest approach is to order specially prepared gas mixes of known composition from a reputable gas supply company to calibrate the unit. Once response factors have been established, the accuracy of the results from a TC detector will be approximately ±3 to 4%. Reproducible TC results with a standard deviation of ±2% can be obtained if correlations between detector response and occasional variations in instrument parameters are known and monitored [26]. For this performance, instrument parameter changes are limited to:

Bridge current, %	±0.1
Carrier gas flow rate, %	±0.3
Cell temperature, °C	±0.3
Oven temperature, °C	±0.3

The brave and the unwary may be tempted to prepare their own gas blends. This is fairly difficult to do well. If relatively large volumes of the gases are available, one can use a gas blender [17]. These are surge tanks equipped with pressure valves which allow regulated flows of two or three gases to mix evenly. An even more accurate procedure is to weigh the amounts of gas added to a blending tank. The tank must then be rolled for some time to mix the gases thoroughly.

Low-level standards, containing trace components at the ppb level or lower, may be prepared by the use of an exponential dilution flask (Figure 3).

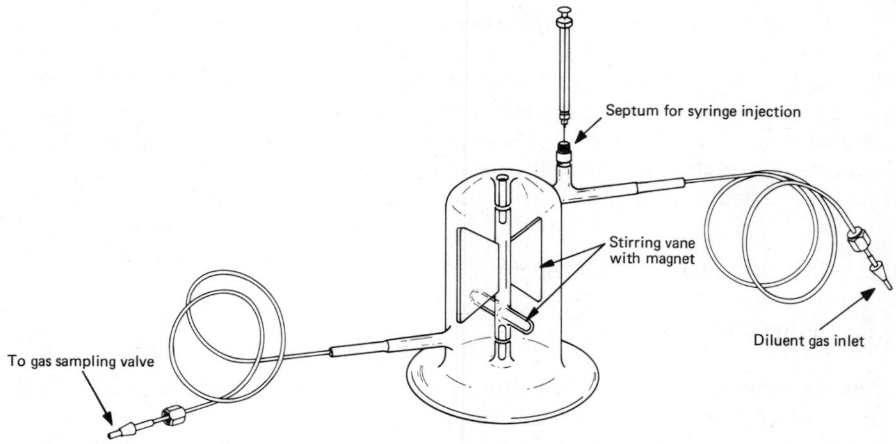

FIG. 3. Exponential dilution flask.

These devices are available commercially [28]. The flask is swept by a constant flow of carrier gas or anything which can be separated from the trace compound and does not react with it. The magnetic stirrer is rotated rapidly to ensure thorough mixing. The trace material is injected into the carrier stream as it enters the mixing chamber. The concentration of this compound at any one time will be:

$$C = C_0 e^{-ft/V} \tag{2}$$

where C_0 is the initial concentration of the trace compound injected, f is the carrier gas flow rate, t is the time elapsed since the introduction of the trace compound, and V is the volume of the flask. By taking a sample of the gas in the flask at the appropriate time, one can obtain almost any trace concentration desired.

B. Quantitation

Unless all components in the gas sample give measurable peaks and unless the response factors of these components are known, the method most often used to quantitate gas analyses is by an external standard. A standard is run each time a set of samples is to be analyzed. The composition of the standard should be close to that of the samples. Results are calculated from:

$$\text{weight }\%_{i,\text{sample}} = \text{weight }\%_{i,\text{std.}}\left(\frac{\text{area}_{i,\text{sample}}}{\text{area}_{i,\text{std.}}}\right) \tag{3}$$

or, if a component, j, is present in the sample but not in the standard:

$$\text{weight }\%_{j,\text{sample}} = \text{weight }\%_{i,\text{std.}}\left(\frac{\text{area}_{j,\text{sample}} \times \text{rf}_j}{\text{area}_{i,\text{std.}} \times \text{rf}_i}\right) \tag{4}$$

The accuracy of the analysis depends on the accuracy with which the composition of the standard has been established.

C. Hydrogen

The most difficult gas to quantitate is hydrogen. It cannot be detected by an FID, and it exhibits an anomalous behavior in a TCD because its thermal conductivity is higher than that of any other compound. H_2 will, therefore, produce negative peaks with any carrier gas.

The response of a TCD to hydrogen has been discussed in detail by Villalobos and Nuss [29]. Figure 4 shows its behavior in helium gas streams. As the concentration of H_2 increases, the output signal increases, fairly linearly, up to 8.5% hydrogen (a, b) at which point the thermal conductivity of the He/H_2 mix reaches a minimum (c). With further increasing H_2 concentration, the thermal conductivity of the mix increases; and the TCD signal becomes negative (d-g). This negative response is, unfortunately, not linear with respect to concentration.

A solution to this problem is to use a carrier gas which is 8.5% H_2 in He. Any H_2 in the sample will cause the thermal conductivity of the H_2/He mix to increase. By switching the signal polarity for the H_2 peak and calibrating the H_2 detector response, one can measure the H_2 concentration accurately.

An alternative approach [30] utilizes a sintered palladium alloy tubing which, when heated, will allow H_2 to diffuse into a larger concentric tube through which N_2 flows as the carrier gas. The mixture is subsequently introduced into the reference side of a TCD eliminating the necessity for switching polarity. Provided the He flow is kept low (25 to 55 $cm^3\ min^{-1}$) and the temperature of the sintered tube is maintained between 500°C and 625°C, the tube will exhibit maximum transfer efficiency for H_2 and essentially none for He or any other fixed gases. Although H_2 has a nearly linear detector response in N_2 carrier gas, it is advisable to calibrate the response for samples containing less than 25 μl of H_2.

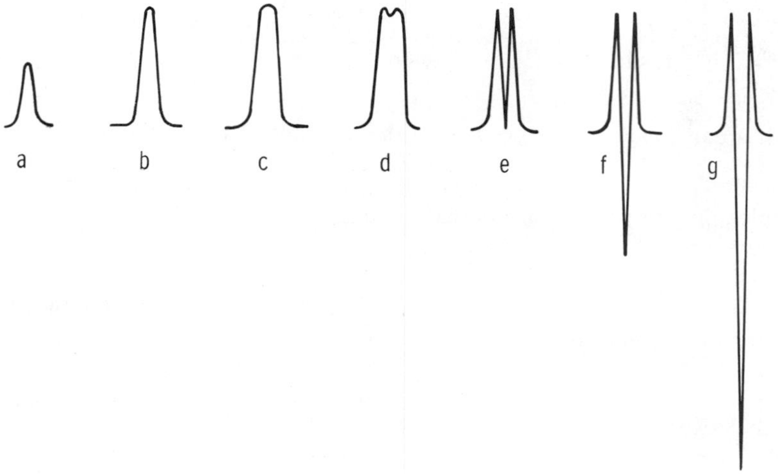

FIG. 4. Thermal conductivity detector response to H_2 in He carrier gas. The H_2 concentration increases from a to g; c represents 8.5% H_2 in He [29].

VI. ANALYSIS OF SPECIFIC GASES AND GAS MIXTURES

No one has been able to devise any single column that will separate all of the gases in a refinery or petrochemical plant. Hundreds of methods have been reported on how to analyze gas mixtures. We only cite Leibrand's [31] paper, which is particularly useful in this regard. His atlas of gas analyses shows the separation of gases which have been achieved on 60 different columns.

A. Fixed Gases; Low Molecular Weight Hydrocarbon Gases; Flue Gases

Many of the recognized methods for the analysis of fixed gases and low molecular weight hydrocarbons make use of porous polymer or molecular sieve columns. Most commonly mentioned are columns containing molecular sieve 5A. These are either operated at isothermal ambient conditions (25 to 30°C) or are temperature programmed at slow rates from 35°C up. Temperature-programmed runs are not recommended for wet gases. H_2O will elute in a broad, irregularly shaped peak, interfering with proper integration.

Column lengths normally vary from 2 to 10 ft in length, but the conditions of analysis and the length of the column can be changed according to the gases one wishes to separate. These may include N_2, O_2, Ar, Ne, CO, CO_2, NO, and NO_2. Oxygen and Ar are normally eluted together unless one uses either a long column and/or low temperatures [32]. Davidson [33] reports that a 15-cm molecular sieve 13X column can separate O_2, N_2, CH_4, and CO in five seconds (Figure 5). Molecular sieve 13X will separate the same gases as molecular sieve 5A but in half the time. Molecular sieve 13X gives a more rapid analysis; 5A a more efficient one. The major advantage of molecular sieve columns is that one can separate the fixed gases at room temperature. A disadvantage is that molecular sieve is deactivated by water and large amounts of CO_2. Type 5A is deactivated more rapidly than Type 13X. The recommended reactivation procedure is to heat the column under flow in an oven at 350°C for 8 to 10 hr to remove H_2O or at 400 to 425°C for 8 to 10 hr to activate the column sufficiently for an O_2-Ar split.

Another column packing which is often used for analyzing fixed gases and low molecular weight hydrocarbons is Porapak Q. This porous polymer will also separate the gases mentioned above. Water does not interfere with the analysis. In fact, one can even measure the water content of a gas on this column [34]. Hydrocarbons through propane and propene can be

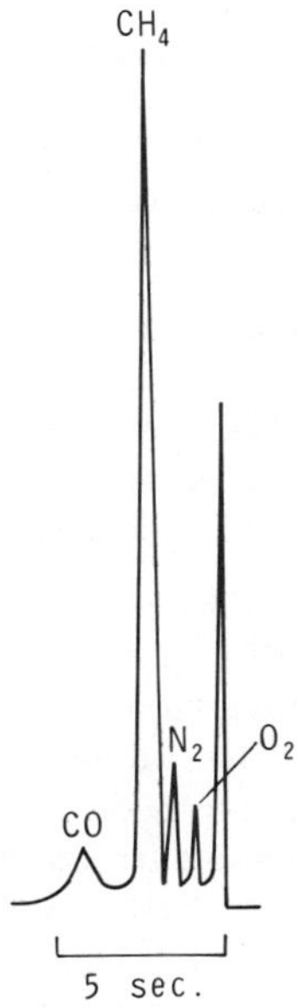

FIG. 5. Analysis of fixed gases plus light hydrocarbons: 15 cm by 1 cm ID molecular sieve 13X, 80-100 mesh, H_2 carrier gas, 10-m min^{-1} flow, 21°C isothermal temperature, 2-μl sample size [33].

separated satisfactorily. If one needs to separate the components of air, however, one must use cryogenic cooling. The analysis may be temperature programmed, but it should start out near or below -60°C. If a N_2, O_2, Ar, CO split is not needed, the Porapak column can separate air from NO, CO_2, and N_2O at ambient temperatures. A table of relative retention times of 68 low-molecular-weight compounds on various Porapak polymers may be found in a book on packed columns [35] or in a product bulletin on Porapaks [34].

The third type of packing in this series is carbon molecular sieves [36, 37]. A 9-ft long carbon molecular sieve column can separate H_2, O_2, N_2, CO, CH_4, and CO_2 in 16 min [36]. Oxygen is eluted before N_2. Ar is not separated from N_2. It is recommended that the carrier gas be oxygen-free to prevent oxidation of the packing and dry to prevent the elution of a broad H_2O peak close to the CO_2 peak.

B. Hydrocarbon Gases

Hydrocarbon gases may be processed by the same methods as hydrocarbon liquids. Nonpolar columns have proved to be very successful in this respect. On a column such as squalane, hydrocarbons can be eluted in nearly boiling point order. With the exception of 1- and isobutene, all paraffins and monoolefins can generally be separated quite easily. Jacobs [38] reports the separation of hydrocarbons in automobile exhaust using a 150-ft by 0.01-in. DC-200 silicone-coated capillary column at -55°C, which yields an almost complete separation of the C_1-C_4 hydrocarbons in less than three minutes.

Unusual orders of elution can be achieved by using polar rather than nonpolar columns. Depending on the material to be analyzed, these columns can be of great use. Mallik and Khurama report that a highly polar 700-cm β-methoxypropionitrile column in series with a 200-cm β-pentoxypropionitrile column will separate 1- and isobutene [39].

Diolefins and triolefins are also readily analyzed on porous polymer, packed, or capillary columns. The presence of C_4 and C_5 paraffins and monoolefins may, however, interfere with their detection. On most analytical columns methylacetylene and allene are separated from other C_3 hydrocarbons. To resolve them from each other, however, is more difficult. This separation has been achieved on carbon molecular sieves [40]. Vorachek and Koob [41] report separating a mix of alkanes, olefins, and acetylene using a 35% squalane column, 25 ft long by 1/4 in. OD. Similar gas mixes have been separated on Porapak Q [40] and Porapak N [41]. Note that acetylenic hydrocarbons, especially diacetylene, C_4H_2, may react to form explosive acetylides with certain metals. Avoid using Cu or brass metals with this type of sample.

Because of its high sensitivity, a flame ionization detector is usually employed for the detection of hydrocarbon gases alone provided that the linear range of the detector is not exceeded. Mixtures which contain fixed gases as well as hydrocarbon gases are often monitored with thermal conductivity detectors or gas-density meters.

C. Mixtures of Hydrocarbons and Fixed Gases

Blends of hydrocarbon and fixed gases are exceptionally difficult to analyze. Most analytical methods involve multiple columns which are switched in and out of the system at the appropriate times. Some analyses even involve several detectors. Although generally a single column is not adequate to separate all of the components in the mixture, many single-column analyses have been published. Stufkens and Bogaard [42] reported the separation of N_2, O_2, CO_2, CH_4, C_2H_6, C_3H_8, butanes, pentanes, and hexanes on a Porapak R column, temperature programmed from -10°C to 230°C. The fixed gases were directed to a TCD, the balance went to an FID. Sarkar and Haselden [43] reported the separation of H_2, N_2, O_2, CO, CO_2, and hydrocarbons through C_4 on a temperature-programmed 6-ft Porapak Q column. Ottenstein and Supina [40] reported that carbon molecular sieve at 175°C will separate H_2, N_2, O_2, CO, CH_4, CO_2, SO_2, NO and the hydrocarbons through C_3. None of these systems were able to separate C_4 and C_5 olefins.

A separation of the fixed gases, C_5 olefins, and all of the C_4 olefins except 1- and isobutene is shown in Figure 6. In this procedure, which we have developed for the analysis of refinery gas streams, all of the compounds are eluted in boiling point order. The analysis conditions are shown in Table 1; the flow diagram is shown in Figure 7. Column A is temperature programmed from -60°C; Column B remains outside the oven at ambient temperature. This analysis can also be used to analyze fixed gases and low molecular weight hydrocarbons (flue gases) by bypassing Column B.

One very popular method for analyzing the more complex full-range refinery gas mixtures has been developed by Drews and Pfeiffer [44]. Three columns, diethylene glycol adipate plus diethylene glycol sebacate, Porapak Q, and 13X molecular sieve, are set up as shown in Figure 8. The chromatogram of a full-range refinery gas, plus the timing sequence for valve switching, is shown in Figure 9. It has become evident that if methods such as this one are to be used efficiently, switching valves must be automated. Several manufacturers have developed chromatographic systems incorporating similar automatic valve switching arrangements. These are discussed in Sec. VII. Thompson has written a book useful for chromatographers in developing their own gas analysis systems [45].

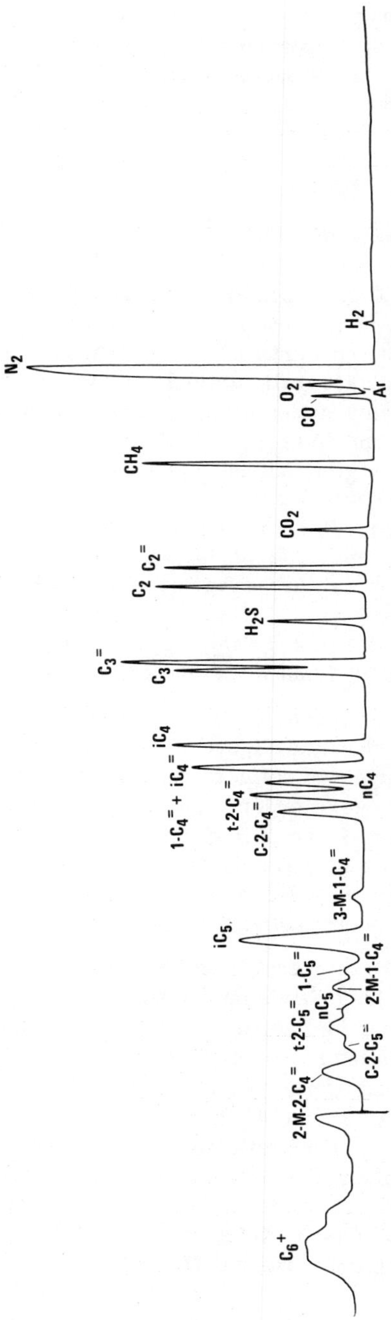

FIG. 6. Chromatogram of a refinery gas stream.

TABLE 1. Refinery Gas Stream Analysis Conditions

Detector	TCD
Carrier gas	7.5 % H_2; 91.5% He
Column A	10-ft × 1/8-in. OD SS, Porapak Q, 80/100 mesh in series with 1-ft × 1/8-in. OD SS, Chromosorb 104, 80/100 mesh
Column B	10-ft × 1/8-in. OD SS, 15% Squalane on Chromosorb W-HP, 80/100 mesh in series with 1-ft × 1/8-in. OD SS, 15% Ucon 550X on Chromosorb W-HP, 80/100 mesh
Flow rate	40 cm^3 min^{-1}
For hydrocarbon gases: (Column B at ambient temperature)	
Sample size, ml	~1.0
Initial temp., °C	-70
Initial time, min	2
Programming rate, °C/min	16
Final temp., °C	210
Final time, min	30
Length of run, min	~50
For flue gases: (Column B switched out of operation)	
Sample size, ml	~0.5
Initial temp., °C	-70
Initial time, min	2
Programming rate, °C/min	32
Final temp., °C	162
Final time, min	4
Length of run, min	15

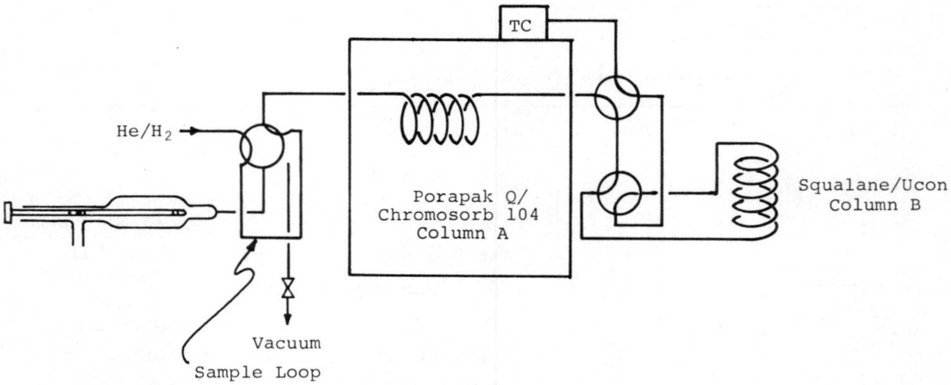

FIG. 7. Flow diagram of gas analysis system used at Chevron Research Company.

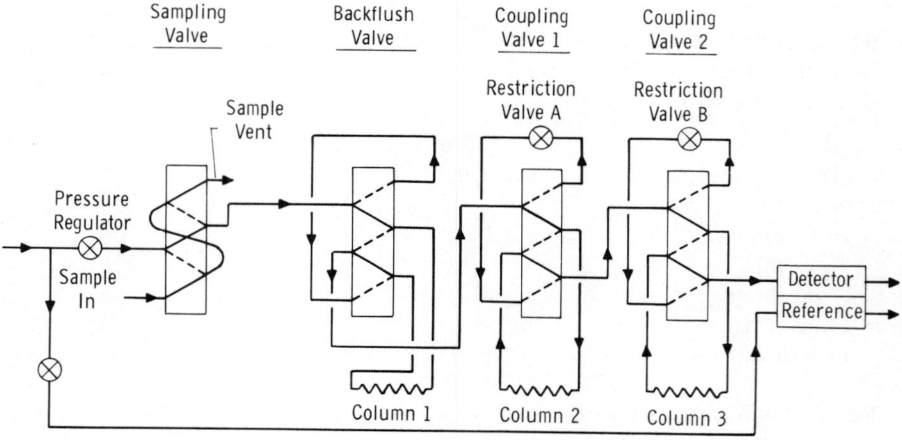

FIG. 8. Flow diagram of UOP gas analyzer [44].

D. Nitrogen Compounds: NO, NO_2, N_2O, NH_3

With the exception of nitrous oxide, the nitrogen compounds listed are somewhat reactive and, thus, will need special care if quantitative analyses are desired. In most cases, it will be necessary to condition the column before use. This involves the injection of several samples to allow the reactive gas to cover all active sites in the column by reversible chemisorption on the walls of the column and on the packing. Once the column is conditioned, reproducible and quantitative results can be obtained. Many columns will

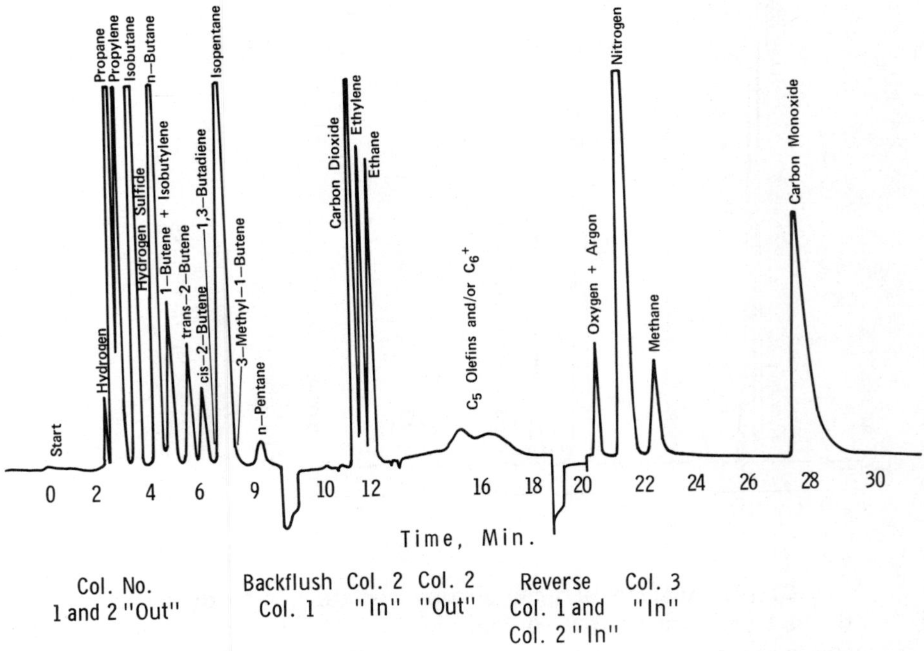

FIG. 9. Typical chromatogram from UOP gas analyzer [44].

require reconditioning each day; some will need it after every sample or two, especially if the sample is highly reactive. One approach to continuously condition a column for basic nitrogen compounds is to use a carrier gas doped with 2 to 3 % NH_3.

Nitrous oxide, N_2O, is rather inert. It can be separated from fixed gases and other oxides of nitrogen by using molecular sieve [43], QF-1 on Chromosorb T [44], Polypak 1 [46], or Porapak Q [34]. Figure 10 shows a chromatogram obtained on a combination Porapak Q-Porapak R column [47].

Nitric oxide, NO, may be difficult to measure because of the equilibrium between nitric oxide and nitrogen dioxide:

$$2NO + O_2 \rightleftharpoons 2NO_2 \tag{5}$$

$$2NO_2 \rightleftharpoons N_2O_4 \tag{6}$$

Jeffery and Kipping [48] describe an analytical method based on the dilution of the NO-containing samples to minimize the NO-NO_2 oxidation problem.

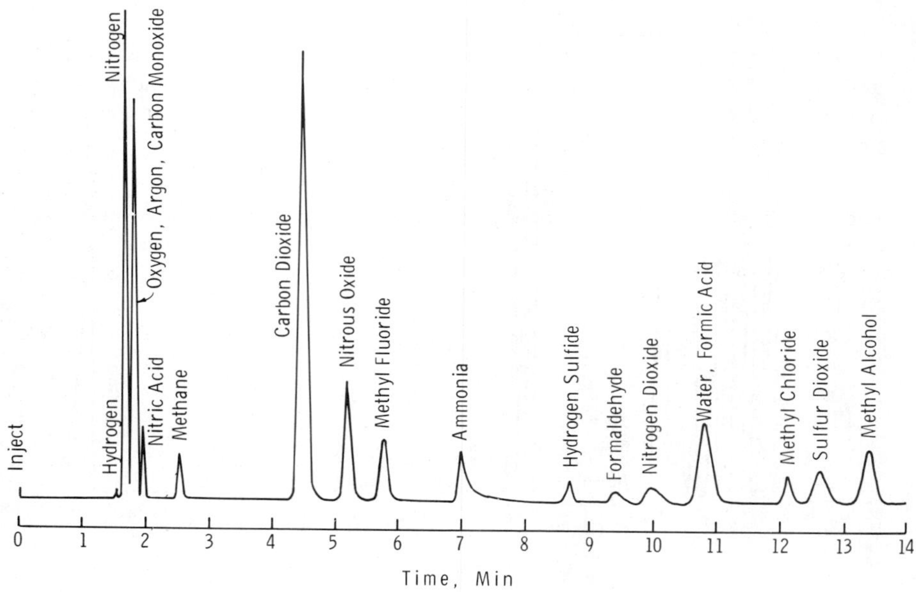

FIG. 10. Analysis of a gas mixture containing nitrogen compounds. 8 ft × 0.1 in. Porapak Q 60-80 mesh plus 8 ft × 0.1 in. Porapak R 60-80 mesh in series, He carrier gas, 40-ml min^{-1} flow, 25°C initial temperature, 150°C final temperature, 12°C min^{-1} program rate, 100-μl sample size [47].

It was also reported that small amounts of NO can be determined with an argon ionization detector. A column conditioning method to eliminate NO peak tailing is described by Clay and Lynn [49]. Perhaps the most widely used packing for the determination of NO is Porapak Q [34], which easily separates NO from CO_2.

Nitrogen dioxide, NO_2, is difficult to analyze because of its reactivity. As early as 1958, Greene and Pust [50] warned that NO_2 is not eluted intact from molecular sieve columns but that it reacts with water in these columns, forming one-third its volume of NO:

$$2NO_2 + H_2O \longrightarrow HNO_2 + HNO_3 \qquad (7)$$

$$3HNO_2 \longrightarrow HNO_3 + 2NO + H_2O \qquad (8)$$

One can, of course, measure the NO level, from which the volume of NO_2 originally present can be calculated. This approach is obviously fraught with uncertainties. Since then, several investigators have reported successful analyses for NO_2 using Porapak Q columns [47, 51]. However, Trowell [52] has found that NO_2 also reacts with Porapak Q, releasing NO and water. NO_2 also attacks Chromosorb 102. Bethea and Meador [46] report that Kel-F 90 Grease or Arochlor 1232 work well for NO_2 as long as no acid gases are present in the sample.

Ammonia, NH_3, is difficult to analyze because of its reactive nature and its tendency to tail. In their survey of the performance of 20 columns for gas analysis, Bethea and Meador [46] report that 10% triacetin on Chromosorb T or 10% N,N-dodecylphthalate on Chromosorb T are the best choices for separating NH_3 from air, CO_2, oxides of nitrogen, oxides of sulfur, and Cl_2. Jones [53] reports that Porapak Q can separate NH_3 from air, CO_2, H_2S, H_2O, and the C_2-C_5 hydrocarbons. Wilhite and Hollis [47] separated NH_3 easily on their Porapak Q column in series with a Porapak R column (Figure 10). NH_3 can be separated from methylamine and H_2O on Chromosorb 103, a porous polymer packing specifically developed for amines [55]. Jeffery and Kipping suggest using an argon ionization detector for ammonia.*

E. Sulfur Compounds: SO_2, H_2S, COS, CS_2

Sulfur compounds resemble the nitrogen compounds in their reactive properties. The use of copper tubing or brass fittings should be avoided in favor of stainless steel or glass. Teflon tubing is considered best for trace sulfur analysis. Column conditioning will probably be necessary for reproducible results. It is fairly simple to separate sulfur compounds from each other and from a mixture of hydrocarbons. Horton and Guerin [54] have used silica gel to separate a mixture of sulfur compounds—COS, H_2S, CS_2, and SO_2—from cigarette smoke. Bitterfield [55] has separated SO from SO_2 on a 6-ft Chromosorb 104 column. Bailey et al. [56] have used Porapak R to separate SO_2 and CS_2 from furnace gases. This author routinely separates H_2S and SO_2 from fixed gases and light hydrocarbon gases using a Porapak Q column in series with a short Chromosorb 104 column.

Thermal conductivity detectors are sensitive to these sulfur compounds down to around 50 ppm. If lower concentrations must be measured, one might consider using a flame photometric detector.

VII. COMMERCIAL SYSTEMS

A number of manufacturers have developed chromatographic systems for the analysis of various types of gas mixtures. Perhaps the simplest system is the ubiquitous Fisher gas partitioner [57], the flow diagram of which is shown in Figure 11. Column 1, 20 ft by 1/8 in. aluminum, is packed with

---

*Reference 1, p. 124.

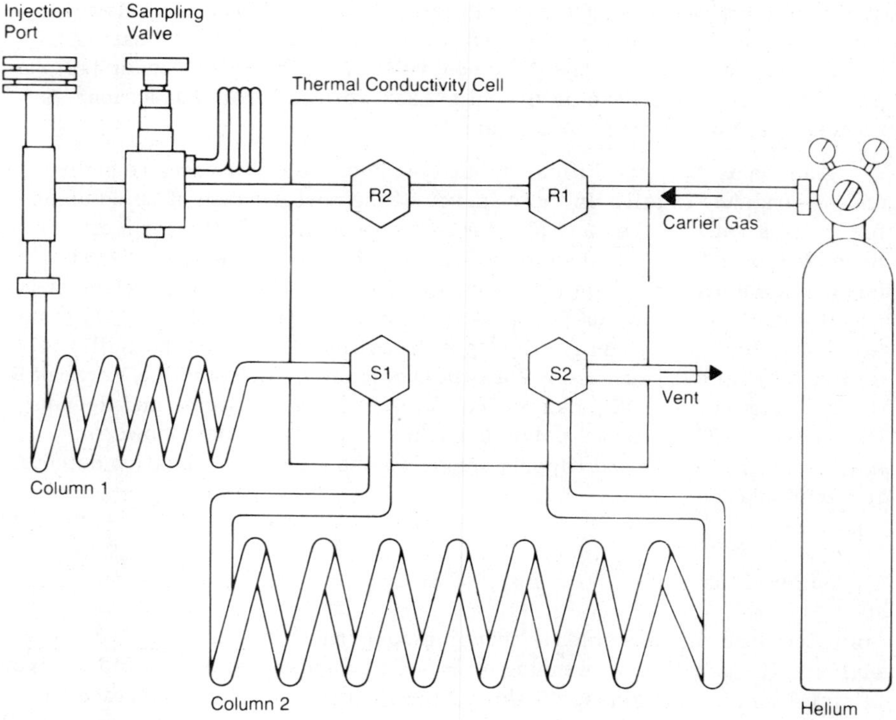

FIG. 11. Fisher gas partitioner flow diagram [59].

37.5% DC-200/500 on 80/100 mesh Chromosorb P-AW. Column 2, 6 ft by 3/16 in., is packed with 60/80 mesh molecular sieve 13X. Samples can be introduced by syringe or by sample loop. This instrument can analyze natural gas, refinery gas mixes through C_5, and dissolved gases in liquids. A backflush valve is available for samples containing C_{6+} material. Gases containing C_4 olefins will produce a number of unresolved peaks.

A more complex and more versatile system is offered by Carle [58]. This instrument can be set up to analyze sulfur-rich flue gas, natural gas or a full-range refinery gas by either FID or TCD. A chromatogram of a refinery gas is shown in Figure 12. This analytical system can handle the C_4 and C_5 olefins and will separate the 2-pentanes if the backflush is delayed. The three columns required for this analysis are a bis-2-ethoxy ethyl adipate plus squalane column in series with a modified Porasil C column, a Porapak Q plus Porapak N column, and a 13X molecular sieve column.

Gas analysis systems are also available from Hewlett-Packard [59]. which also furnishes an analytical system equivalent to the UOP method.

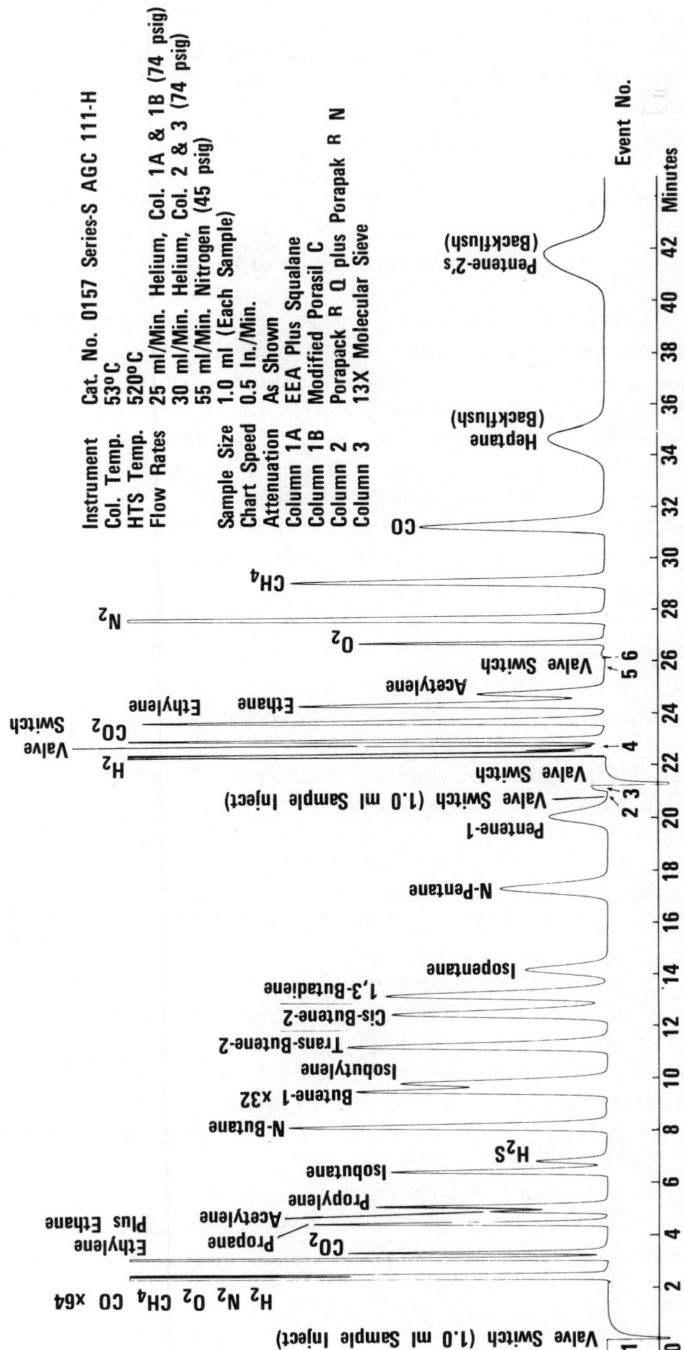

FIG. 12. Typical chromatogram of refinery gas obtained with a Carle system [58].

Their instruments require three columns, packed with the following: bis(2-methoxyethyl)adipate plus DC200, Porapak Q, and molecular sieve 5A. Their current systems make extensive use of microprocessor circuits and minicomputers to automate the analyses.

ACKNOWLEDGMENTS

My thanks go to the following for permission to reproduce figures from their publications: Figures 2b, 2c, and 3 are reproduced by permission of the Instrument Society of America [29]. Figure 4 is reproduced from ISA Transactions by permission of the Instrument Society of America [29]. Figure 5 is reproduced from Chromatographia by permission of Friedr. Vieweg und Sohn [33]. Figures 8 and 9 are reproduced from the Division of Petroleum Chemistry Preprints by permission of the American Chemical Society [44]. Figure 10 is reproduced from the Journal of Gas Chromatography by permission of Preston Publications, Inc. [47]. Figure 11 is reproduced by permission of Fisher Scientific Co. [57]. Figure 12 is reproduced by permission of Carle Instruments, Inc. [58].

REFERENCES

1. P. G. Jeffery and P. J. Kipping, "Gas Analysis by Gas Chromatography," 2nd ed., Pergamon, New York, 1972.
2. H. M. McNair and E. J. Bonelli, "Basic Gas Chromatography," Consolidated Printers, Berkeley, 1968.
3. Available from: Alltech Associates, 202 Campus Drive, Arlington Heights, Illinois 60004, (312) 392-2670; Supelco, Inc., Supelco Park, Bellefonte, Pennsylvania 16823, (814) 359-2784.
4. Available from: Alltech Associates, 202 Campus Drive, Arlington Heights, Illinois 60004, (312) 392-2670.
5. Available from: Air Products and Chemicals, Inc., Specialty Gas Department, P.O. Box 351, Tamaqua, Pennsylvania 18252, (717) 467-2981; Matheson Gas Products, P.O. Box 85, East Rutherford, New Jersey 07073, (201) 933-2400; and Union Carbide Corporation, Linde Division, P.O. Box 372, South Plainfield, New Jersey 07080, (201) 753-5800.
6. L. D. Butler and M. F. Burke, J. Chromatog. Sci. 14:117 (1976).
7. J. P. Mieure and M. W. Dietrich, J. Chromatog. Sci. 11:559 (1973).
8. E. D. Pellizzari, J. E. Bunch, B. H. Carpenter, and E. Sawicki, Env. Sci. and Tech. 9:552 (1975).
9. E. D. Pellizzari, B. H. Carpenter, J. E. Bunch, and E. Sawicki, Env. Sci. and Tech. 9:556 (1975).

10. F. X. Mueller and J. A. Miller, American Laboratory 6(5):49 (1974).
11. NIOSH Sampling Data Sheet 6, June 16, 1972.
12. H. L. Barnaby, ASHRAE Trans. 64(1946):481 (1958).
13. One such laboratory is Dow Chemical U.S.A. Interpretive Analytical Services, 2040 Dow Center, Midland, Michigan 48640.
14. Available from: Chromalytics Corporation, Box 798, Metuchen, New Jersey 08840, (201) 549-7144.
15. R. A. Rasmussen, American Laboratory 4:19 (1972).
16. L. Rohrschneider, A. Jaeschke, and W. Kubik, Chem. Ing. Tech. 43: 1010 (1971).
17. R. Villalobos, Anal. Chem. 47:983A (1975).
18. V. N. Lipavsky and V. G. Berezkin, J. Chromatog. 91:583 (1974).
19. Available from: Hamilton Company, P. O. Box 10030, Reno, Nevada 89410, (702) 786-7077; Precision Sampling Corporation, P. O. Box 15119, Baton Rouge, Louisiana 70815 (504) 927-1128.
20. Available from: Glas-col Apparatus Company, 711 Hulman Street, Terre Haute, Indiana 47802, (813) 235-6167.
21. Available from: Watlow Electric Manufacturing Company, 12001 Lackland Road, St. Louis, Missouri 63141, (314) 878-4600.
22. Available from: Carle Instruments, Inc., 532 South Rose Street, Anaheim, California 92805, (714) 533-4600; Valco Instruments, P.O. Box 19032, Houston, Texas 77024, (713) 465-0574; and Varian Instrument Division, 611 Hansen Way, Palo Alto, California 94303, (415) 493-8100.
23. P. G. Jeffery and P. J. Kipping, "Gas Analysis by Gas Chromatography," 2nd ed., Pergamon, New York, 1972, p. 44.
24. P. G. Jeffery and P. J. Kipping, "Gas Analysis by Gas Chromatography," 2nd ed., Pergamon, New York, 1972, p. 51.
25. Available from: Applied Research Laboratories, Ltd., 9545 Wentworth Street, Sunland, California 91040.
26. H. W. Durbeck and B. Telin, Chromatographia 8:180 (1975).
27. Available from: Air Products and Chemicals, Inc., Specialty Gas Department, P.O. Box 351, RD No. 2, Tamaqua, Pennsylvania 18252, (717) 467-2981.
28. Available from: Varian Instrument Division, 611 Hansen Way, Palo Alto, California 94303, (415) 493-8100.
29. R. Villalobos and G. R. Nuss, ISA Transactions 4(3):281 (1965).
30. R. A. McCormack, ed., Carle Current Peaks 7(2):1 (1974).
31. R. J. Leibrand, J. Gas Chromatog. 5(10):518 (1967).
32. G. W. Heylum, J. Gas Chromatog. 3:82 (1965).
33. E. Davidson, Chromatographia 1:43 (1970).
34. Brochure PB 205, Waters Associates, Inc., Milford, Mass., July 1971.
35. W. R. Supina, "The Packed Column in Gas Chromatography," Supelco Bellefonte, Pennsylvania, 1974.
36. "Analysis of Permanent Gases," Supelco Bulletin 760, Supelco, 1976.

37. Spherocarb Packing Brochure, Analabs, Inc., Northhaven, Conn.
38. E. S. Jacobs, Anal. Chem. 38:43 (1966).
39. K. L. Mallik and M. L. Khurama, ISI Bulletin 25(8):327 (1973).
40. D. M. Ottenstein and W. R. Supina, Paper presented at the 162nd ACS National Meeting, Washington, D.C., September 13-17, 1971, Abstract No. 102, Division of Analytical Chemistry.
41. J. H. Vorachek and R. D. Koob, Can. J. Chem. 51(3):344 (1973).
42. J. S. Stufkens and H. J. Bogaard, Anal. Chem. 47(3):383 (1975).
43. M. K. Sarkar and G. G. Haselden, J. Chromatog. 104(2):425 (1975).
44. A. W. Drews and C. H. Pfeiffer, Preprints, Div. of Petrol. Chem., ACS, 18(3):522 (1973); and UOP Method 539-73, "Laboratory Test Methods for Petroleum and Its Products," UOP Process Division, 30 Algonquin Road, Des Plaines, Illinois 60016.
45. B. Thompson, "Fundamentals of Gas Analysis by Gas Chromatography," Varian Associates, Inc., 1977.
46. R. M. Bethea and M. C. Meador, J. Chromatog. Sci. 7:655 (1969).
47. W. F. Wilhite and O. L. Hollis, J. Gas Chromatog. 6(2):84 (1968).
48. P. G. Jeffery and P. J. Kipping, Nature 200:1314 (1963).
49. D. T. Clay and S. Lynn, Anal. Chem. 47(7):1207 (1975).
50. S. A. Greene and H. Pust, Anal. Chem. 30:1039 (1958).
51. O. van Cleemput, J. Chromatog. 45:315 (1969).
52. J. M. Trowell, J. Chromatog. Sci. 9:253 (1971).
53. C. N. Jones, Anal. Chem. 39:1858 (1967).
54. A. D. Horton and M. R. Guerin, J. Chromatog. 90(1):63 (1974).
55. Z. Bitterfield, cited in Johns-Manville Chromosorb Century Series Bulletin, Englewood Cliffs, New Jersey, p. 10.
56. J. B. W. Bailey, N. E. Brown, and C. V. Phillips, Analyst (London) 96:477 (1971).
57. Bulletin No. 335, Fisher Scientific Company, Pittsburgh.
58. R. A. McCormack, ed., Carle Current Peaks 8:3 (1975).
59. Technical Data, Hewlett-Packard Company, Palo Alto, November 1974.

Chapter 3

HIGH RESOLUTION GAS CHROMATOGRAPHY OF THE GASOLINES AND NAPHTHAS

Irville M. Whittemore
Research Services Department
Chevron Research Company
Richmond, California

I. INTRODUCTION

The volatile liquid products and byproducts formed in the production and refining of petroleum present a myriad of challenges to the analytical chromatographer. Not the least of these is the selection of the proper method of analysis. By this, we mean finding the method which gives the required answers in a speedy, economical (but not necessarily elegant) way. This, in turn, implies the necessity of the analyst's understanding of his customer's needs as well as knowledge of the analytical tools available. In fact, it is our experience that more analytical failures occur because of lack of communication between the analyst and his customer than because of lack of skill or tools on the part of the analyst.

The nature of this challenge has shaped the form of this chapter. The techniques and the areas of applicability are considered as much as possible together. The multiplicity of techniques and objectives make their careful matching the essence of the analytical process.

II. THE MATERIALS

We are concerned here with liquid hydrocarbon mixtures which have true boiling point (TBP) endpoints up to about 450°F (230°C). The lightest materials found may include methane and the heaviest components may, of course, boil much lower than 450°F.

This range of materials is dominated in the petroleum industry by two classes. One is motor gasoline and gasoline blending components. The other is naphtha, a term which we will use here to describe hydrocarbon fractions within our boiling range which are not intended for direct use in motor fuel.

In terms of amount of product, only water exceeds gasoline in importance as a consumer item, at least in the United States. In composition, gasoline must meet many (and increasing) requirements. To product quality standards, such as volatility, octane number, storage stability, there are being added regulatory standards, such as lead level, olefin, and sulfur content. Manufacturing acceptable gasoline products within mandated requirements requires more and more sophisticated knowledge of the blending components for gasoline.

Most naphthas are used as feeds to reactors where chemical conversions take place. The most important of these conversions is reforming. Worldwide, about half a trillion pounds of naphtha are processed by reforming each year, making it the largest synthetic chemical process.

The properties of reformate, the product of reforming, are as much a function of the composition of the naphtha feed as they are of the reforming process. Some of the important decisions in refining are based on how to operate reformers: what feeds to use; at what severity to operate; what use to make of the reformate. Knowledge of the composition of the naphthas, insofar as it can lead to predictions of reformate compositions, becomes an important economic tool.

Chromatography can provide much of the information and knowledge required to convert crude petroleum into profitable products. Nowhere is this more often the case than in the areas of naphtha utilization and gasoline production. Much has been accomplished in these areas, but yet more remains to be done.

III. THE TECHNIQUES AND THEIR APPLICATIONS

The hydrocarbon mixtures considered here are amenable, in varying degrees, to two different concepts of chromatographic analysis. These are high-resolution and low-resolution methods. The former nearly always involve capillary column gas chromatographic techniques, and the latter involve packed column methods.

The major objective of a high resolution chromatographic analysis is to obtain a quantitative breakdown of the individual components in the mixture. If this could be done both economically and perfectly, i.e., with no important errors, then no other kind of analysis would be necessary. But, in fact, important errors can beset these high-resolution methods. And it also makes no sense to employ a method which costs in excess of $100 and hours to carry out when the <u>necessary</u> information can be obtained for $10 to $20 in minutes.

The low resolution methods can be used effectively where speed and reliability are of particular importance. Advantage can be taken of the wide variety of equipment and packings available to provide surprisingly sophisticated analyses which can ultimately provide nearly as much information as the "perfect" high-resolution chromatogram.

This arbitrary grouping into high- and low-resolution (or capillary and packed column) method is more a convenience for discussion than otherwise. The analytical aim, as stated in the introduction, is for the analyst to provide his or her customer with the needed information. The benzene and toluene content of a gasoline can be ascertained by high resolution chromatography or by packed column methods. Use of the former may discard 99% of the information generated but might be the best choice if a special setup for a packed column analysis were required.

IV. CHROMATOGRAPHIC ANALYSIS OF GASOLINES AND NAPHTHAS

The modern motor gasoline is almost entirely a synthetic product—synthetic in the sense that nearly all of it has undergone one or more chemical transformations between distillation from crude oil and final blending into a salable product. The product is complex, containing up to 400 individual compounds in measurable amounts (arbitrarily here, >100 ppm). Up to 15 or so refinery streams (each one complex) may be blended to make the finished gasoline.

Chromatography can be used to analyze the whole gasoline or the component streams. Table 1 gives a partial listing of some of the reasons for analyzing gasoline and an indication of one or more usable techniques to achieve the desired purpose. (One should keep in mind that nonchromatographic methods may frequently be better than chromatography.)

As can be seen from this table, high-resolution capillary gas chromatography (HRCGC) is the most generally useful method. The ultimate (but not always necessary) aim of HRCGC is complete component analysis.

TABLE 1. Analysis of Gasoline and Components

Purpose	Method	Comments
Octane number	High resolution capillary gas chromatography (HRCGC)	Actual engine determination most accurate. Calculation algorithm important.
Reid vapor pressure	HRCGC, simulated distillation (SD)	SD works best if lightest components are individually determined.
Boiling range	HRCGC, simulated distillation, D 86	SD method of choice
Olefin content	FIAM, olefin absorption methods, HRCGC	
Benzene, toluene, and total aromatics	HRCGC, FIAM, packed-column techniques with polar columns and backflushing	ASTM D 3606 method of choice for benzene and toluene.
Competitive surveys	HRCGC, lead alkyls by GC-AA	
Lead alkyl analysis	GC-AA	
Leak source identification	HRCGC, GC-AA	
Forensic studies	HRCGC, GC-AA, simulated distillation	

Use of capillary columns of up to 300,000 theoretical plates is routine in many laboratories. But even the best columns are not up to task of completely resolving the complex mixture of gasoline. We try, then, to achieve a useful analysis, one that resolves the most important components. As noted above, a gasoline may contain more than 400 detectable components. Fortunately, we need not resolve or even identify all of them. In a typical gasoline, 100 to 150 compounds account for 90% of the volume,* 150 to 200 compounds for 95%, and 200 to 250 compounds for 98% of the total. If we can then resolve, identify, and quantify only a minority of compounds in a gasoline, many of its properties can then be estimated by assigning typical properties to the relatively unimportant majority.

Compared to gasoline, naphthas are both simpler and more complex. In the boiling range up to about 200°F (100°C), naphthas contain a third or fewer components than a typical modern gasoline. This means that there are much less stringent requirements on column resolution. However, beyond this point gasolines tend to become simpler and naphthas more complex. The reason for this paradox is the reforming process which converts naphthas into gasoline components. In this process, most of the naphthenes and the heavier paraffins in the naphtha are converted to aromatic hydrocarbons. There are substantially fewer aromatics than naphthenes and paraffins in this particular boiling range with a consequent simplification of the chromatogram of the gasoline.

The increased complexity of the 300°F+ (150°C+) portion of a naphtha severely restricts the applicability of HRCGC as an analytical tool. The upper tolerable endpoint for naphtha analysis will depend somewhat on their sources. Useful analyses can be made on naphthas from paraffinic crudes with endpoints as high as 350°F (175°C), while highly naphthenic crudes may be limited to as low as 250°F (120°C).

For analysis of naphthas with endpoints too high for HRCGC, mass spectrometry or the chromatographic PNA (paraffin, naphthene, aromatic) analyzer may be used.

A. Column and Column Packings

The conditions under which the HRCGC is carried out can influence which peaks are resolved. Obviously, the selection of the liquid phase has the greatest influence on the elution order of components from a column, but even one column can be manipulated to alter elution order. Most practical

*All percentages are given in liquid volume percent, unless otherwise stated.

analytical conditions represent compromises between resolution, speed of analysis, and the ability to elute all the components of interest.

Squalane capillary columns were among the first to be used. As early as 1961, Desty et al. [1] reported the use of a 273-m by 152μ ID glass capillary column coated with this substrate to analyze the gasoline fraction of the Ponca crude which had been extensively studied by Rossini and co-workers [2]. As a liquid phase, squalane has many virtues. It remains liquid to below 0°C and can make highly efficient columns. It is the least polar of all liquid phases as measured by both Rohrschneider's and McReynold's methods. It is, therefore, the closest we can come to making a column that elutes hydrocarbons in boiling point order.

Other liquid phases also useful in capillary analysis are polyethylene glycols and polypropylene glycols [3] because they make stable, efficient columns; and their high polarity make a complementary contrast to the non-polar squalane. Silicone oils have also been widely used [4]. Because they have much higher upper temperature limits than squalane, they are especially useful in shortening analysis times by programming to higher temperature. A somewhat unusual phase which gained transitory prominence because of good work by Schwartz and Brasseaux [5], consisted of a mixture of hexadecene, hexadecane, and Kel-F_3. In our experience, however, no other HRCGC column has proved as generally useful as squalane columns for analyzing gasoline-type samples.

Squalane columns are not, however, without faults. Principal among these is the maximum temperature to which they may be programmed. Sanders and Maynard [6], in their classic paper on the chromatographic analysis of whole gasolines, had to limit the upper operational temperature to 100°C. This severely limits the analysis of the heaviest components in gasoline. In our laboratory, we have developed post-column techniques which have enabled us to program squalane columns to 140°C and to extend the analytical method to the methyl naphthalenes and n-tridecane. This has not been without problems, however, as column life is shortened. In addition, the resultant change in column characteristics can shift elution orders making computer identification of the peaks more difficult.

Squalane is a very useful liquid phase for another reason. More published data on the retention properties of hydrocarbons in the gasoline range have been published for squalane than for any other liquid phase [7, 8]. Some, like Sanders and Maynard's study, were carried out for immensely practical reasons [6]. Others, like Matukuma's, were carried out for systematic data collection [9]. The result is a body of data of great use in the creation of a HRCGC method based on squalane column.

A word of caution is in order. Most published retention data have been obtained under isothermal conditions, while most practical analyses are temperature programmed. Most hydrocarbons exhibit a change in retention relative to, say, the normal hydrocarbons as elution temperature is changed. The practical result is that relative elution times and even elution order can change with different chromatographic conditions.

B. Packing Capillary Columns

We standardized on a 1000-ft by 0.02-in. ID stainless steel column coated with squalane for gasoline more than 10 years ago. Since then we have processed perhaps 5000 gasoline and naphtha samples. We have found that our gasoline analysis columns can process at least 500 samples before loss of resolution creates serious problems. They are then recoated after washing out the old coating. Our ability to coat these columns with squalane has evolved over the years. The columns which we now make have perhaps twice the number of theoretical plates as observed in our original efforts.

The literature on the preparation of metal capillary columns is extensive. Some involve very complicated and precise steps. The simplest, rinsing a tube with a solution of the liquid phase, works well (at least occasionally) for others. After nearly 20 years of application, the manufacture of these columns is still a mixture of science, art, and luck. With our method, with which we attain a good rate of success with squalane columns, we can usually obtain columns having about 600,000 theoretical plates computed for n-octane under isothermal conditions at 50°C. Corrected for the dead volume of the column, the efficiency for this peak is around 400,000 plates. If upon test a column has poorer-than-typical efficiency, it can seldom be improved. However, a good column can be recoated after its useful life again and again with almost invariable success.

Columns are prepared from 347 stainless steel capillary tubing, 0.02-in. ID "Chromat. ID" grade (Handy and Harman Corporation, Norristown, Pennsylvania). The tubing is first cleaned by flushing with the following solvents at 140 psi inlet pressure:

1. CH_2Cl_2
2. Acetone
3. Water
4. N-methylpyrrolidone-acetone 1:1
5. CH_2Cl_2

Sufficient amounts of each solvent are used until a clear effluent is observed. To reclean used columns, it is sufficient to use a solvent wash with CH_2Cl_2 and with acetone only.

Packing is carried out dynamically with a single 10-ml charge of a 10% solution of squalane in CH_2Cl_2 at 40 psig inlet pressure. The calculated average initial film thickness, based on material balance of the liquid phase, should be about 0.5 to 0.8μ. The columns are baked out overnight at 100°C and then heated to 140°C for a few hours before use.

When starting with new tubing, this procedure does not always yield good columns. The quality of the tubing is the most important parameter in this respect; careful and extensive cleaning procedures can only improve the final quality of the column to a modest degree. Tubing from the same manufacturer can show distinct variations in ultimate column qualities. We have split a number of our columns to examine their interior surface by optical and by scanning electron microscopy. The results, so far, are not very conclusive, although it appears that the smoother the interior surface the better the performance of the column.

C. Glass Capillary Columns

The last two years have seen a veritable explosion in the use and availability of glass capillary columns. Many companies now supply these columns in an ever-increasing number of liquid phases, including squalane. Their manufacture, once frustrating and chancy, and their stability, once tenuous and uncertain, have succumbed to determined research. For many, if not most, applications glass is the column material of choice. For gasoline and naphtha, metal columns of high efficiency are still useful and competitive with glass columns.

Glass capillary columns, however, are almost sure to become the standard analytical tools for HRCGC in the future. They are so much less costly than metal capillary columns that this fact alone will ensure their widespread adoption. It is worth noting that the ASTM Committee D-2 is conducting tests to develop a standard analysis for gasoline based on glass capillary columns.

D. Identification Techniques

Up to now, we have assumed that the peaks in our lovely, high-resolution chromatograms represent known compounds. But chromatograms do not come with labels on each peak, and we must determine the identity of the peaks by some means. The classical approach to this problem is to prepare synthetic blends of increasing complexity and to analyze these mixtures under standardized conditions. Many pure hydrocarbons are available from various sources. This technique can be tedious and costly, however.

The combination of the gas chromatograph with a mass spectrometer (GC-MS) is by all odds the premier technique for identifying or at least classifying unknown peaks in a chromatogram. GC-MS, coupled with a computer to gather and manipulate the data, comes as close to an ideal analytical tool for volatile compounds as we are likely to have in the near future. The use of GC-MS can range from single runs to identify components in an analysis where the range of composition can be expected to vary only a little (e.g., gasolines) to routine or semiroutine analyses for naphthas from varying sources.

One limitation on GC-MS in the identification of hydrocarbons is the great similarity of the mass spectra of isomeric hydrocarbons. A second limitation is that there are no reference mass spectra of many prominent hydrocarbons found in the heavier fractions of gasolines and naphthas. In the analysis of naphthas, these limitations are not as important as might be expected. The practical goal here is usually a group-type analysis. For this it is adequate that each peak in the chromatogram be classified as either an aromatic, paraffin, or naphthene and that its carbon number is determined. Most mass spectra allow this type of determination to be made.

As noted earlier, however, identification of high endpoint naphthas by HRCGC can be difficult because no column in the end will have sufficient resolution to resolve all the compounds into separate peaks.

Perhaps the biggest drawback to the application of GC-MS to HRCGC is the difficulty in exactly duplicating the analytical chromatogram in the GC-MS combination. Chromatographs differ notoriously in their nominal temperatures and abilities to follow programmed temperatures. Added to the degradation in signal usual in interfacing devices, the chromatogram analyzed by GC-MS may differ significantly from the analytical chromatogram. Where the analytical chromatogram is "crowded" with peaks, the intercomparison of the chromatograms may be difficult.

Not everyone who must chromatograph hydrocarbons is fortunate enough to have a GC-MS. However, more and more consulting firms have GC-MS combinations available. Contracting work out to resolve peaks of questioned identity may be an economical alternative if the analyst does not have a GC-MS at his disposal.

Several other methods of establishing identities of peaks in a HRCGC are available. Operating similar columns under similar conditions to those found in the literature is very helpful. Tabulated lists of retention times, often with schemes for predicting retention times, are available, especially for the popular liquid phases such as squalane [7, 8]. Much can be accomplished without a GC-MS and without having to resort to endless matching of chromatograms spiked with pure compounds. Frequently, simplifications can be made in the chromatogram to aid in the identification process. Olefins and aromatics can be removed from gasolines or naphthas by liquid chromatography on a column and then recovered as separate fractions for subsequent analysis. The simpler chromatograms of the saturate, aromatic, and olefinic parts may be much easier to interpret than the chromatogram of the whole sample.

Similarly, samples from different sources can be used to help identify peaks at least by type. Straight run naphthas will have saturates and aromatics. Cracked gasoline stocks will be highly olefinic, especially the lighter fractions. Reformates will be highly aromatic and especially deficient in the easily dehydrogenated naphthenes. Alkylate stocks will consist almost entirely of highly branched paraffins. Their presence or absence in an unknown or unidentified gasoline may provide a useful clue to its origin.

As one builds up a repertory of identified stocks on a useful columns, patterns are obvious and become the major method of identification. Sole reliance upon pattern, however, can lead to difficulty when an unfamiliar stock must be analyzed.

V. HIGH RESOLUTION CAPILLARY GAS CHROMATOGRAM OF A GASOLINE

Figure 1 shows a chromatogram of a gasoline which we have obtained on a squalane column. Table 2 shows the identification of the peaks. Identification is based on a combination of all the identification methods listed above. Our principal tool has been GC-MS, but the other approaches have all been used to some extent. We believe that the table contains no important errors. We cannot of course claim total correctness. The names attached to the peaks represent our best current knowledge.

The experimental details are as follows: The carrier gas flow control system is shown schematically in Figure 2. Helium from cylinders is used as the carrier gas. After the pressure is reduced to about 160 psi, the gas enters a tee. A pressure regulator set at 50 psig outlet pressure is mounted on one line, and another pressure regulator set at about 30-psig outlet pressure is mounted on the other line.

These two lines enter a three-way solenoid, the outlet of which goes into the injector-splitter system of the gas chromatograph. In the initial period of the program, the gas flow is as shown in Figure 2. At the end of the temperature-programmed portion of the run, the solenoid is activated to the alternate position where the 50-psi pressure regulator controls the carrier gas flow. A surge vessel and small bleed have been incorporated in this section. In the absence of this relief system, the pressure regulator would be subjected to a shock wave as the solenoid is actuated, often resulting in a change in the pressure setting.

The split outlet system is a 1-ft by 1/4-in. piece of tubing packed with 5A molecular sieves. A micrometering valve is mounted at the end of this tube. The column retains the split portion of the sample, and the micrometering valve remains clean under these conditions. Very reproducible split ratios over long periods of time can be maintained with this approach.

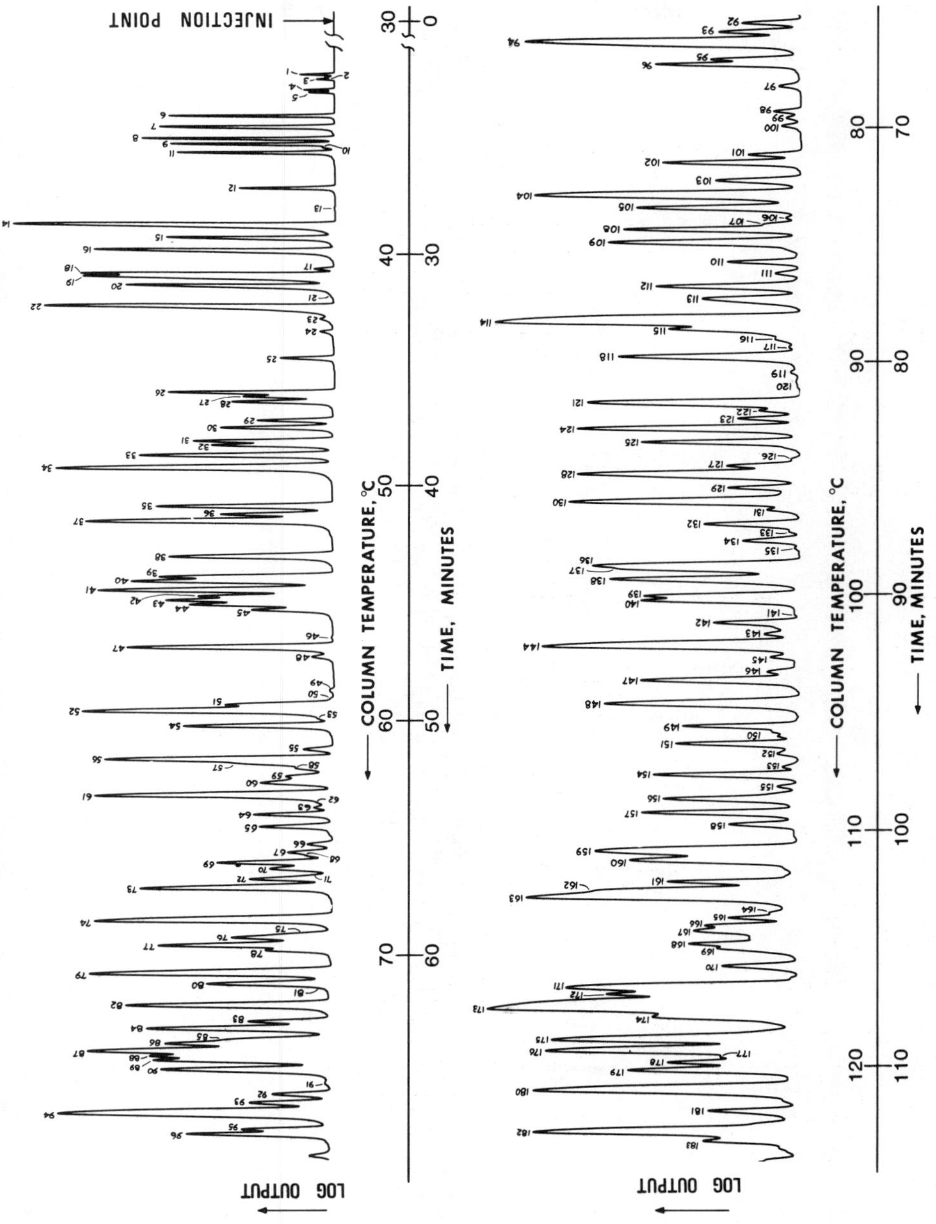

FIG. 1. Chromatogram of whole gasoline.

(continued)

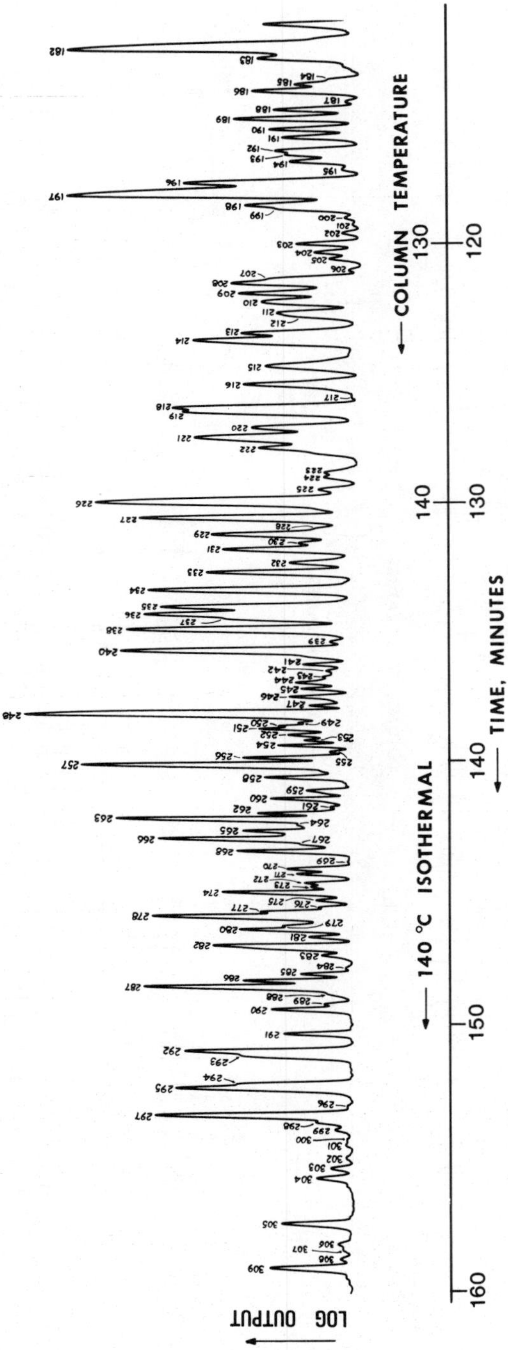

FIG. 1 (continued)

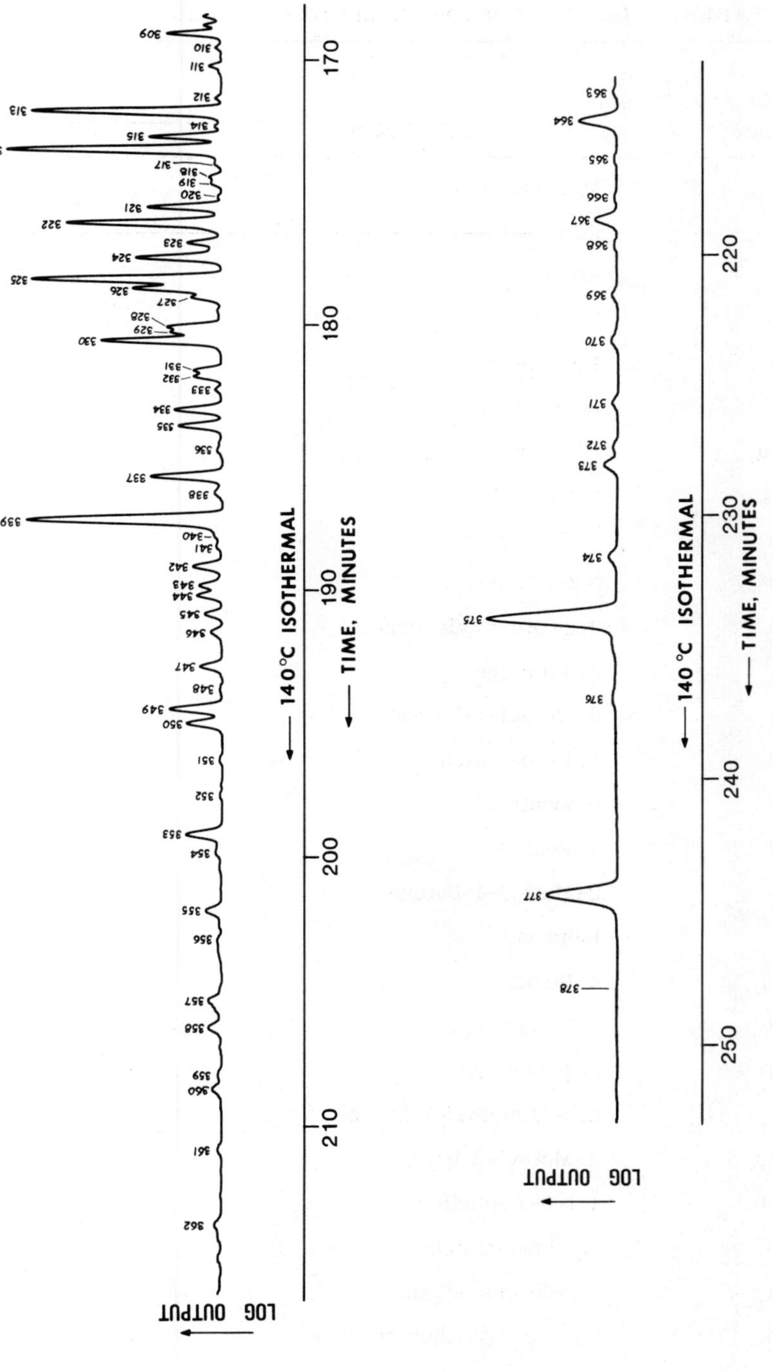

FIG. 1 (continued)

TABLE 2. Gasoline Components in Order of Elution

Code number	Description	Boiling point (°C)
1	Methane	-161.50
2	Ethene	-103.71
3	Ethane	-88.60
4	Propene	-47.70
5	Propane	-42.07
6	Isobutane	-11.73
7a	1-Butene	-6.26
7b	Isobutene	-6.90
8	n-Butane	-0.50
9	t-2-Butene	0.88
10	2,2-Dimethylpropane	9.50
11	c-2-Butene	3.72
12	3-Methyl-1-butene	20.06
13	1,4-Pentadiene	25.96
14	Isopentane	27.85
15	1-Pentene	29.97
16	2-Methyl-1-butene	31.16
17	Isoprene	34.07
18	n-Pentane	36.04
19	t-2-Pentene	36.35
20	c-2-Pentene	36.94
21	3,3-Dimethyl-1-butene	41.25
22a	2-Methyl-2-butene	38.57
22b	1-t-3-Pentadiene	42.02
23	Cyclopentadiene	41.5
24	1-c-3-Pentadiene	44.07
25	2,2-Dimethylbutane	49.74

TABLE 2 (continued)

Code number	Description	Boiling point (°C)
26	Cyclopentene	44.24
27	4-Methyl-1-pentene	53.87
28	3-Methyl-1-pentene	54.18
29	c-4-Methyl-2-pentene	56.39
30	2,3-Dimethyl-1-butene	55.62
31	t-4-Methyl-2-pentene	58.61
32	Cyclopentane	49.26
33	2,3-Dimethylbutane	57.99
34	2-Methylpentane	60.27
35	2-Methyl-1-pentene	62.12
36	1-Hexene	63.48
37	3-Methylpentane	63.28
38a	2-Ethyl-1-butene	64.68
38b	c-3-Hexene	66.45
38c	t-3-Hexene	67.08
39	t-2-Hexene	67.88
40	2-Methyl-2-pentene	67.29
41	n-Hexane	68.74
42	3-Methylcyclopentene	64.91
43a	c-3-Methyl-2-pentene	67.70
43b	4,4-Dimethyl-1-pentene	72.49
44	c-2-Hexene	68.89
45	4-Methylcyclopentene	75.22
46	C_6 Diolefin	
47	t-3-Methyl-2-pentene	70.44
48	t-4,4-Dimethyl-2-pentene	76.73
49	Methylcyclopentadiene I	

TABLE 2 (continued)

Code number	Description	Boiling point (°C)
50	Methylcyclopentadiene II	
51a	2,2-Dimethylpentane	79.20
51b	2,3-Dimethyl-2-butene	73.21
52	Methylcyclopentane	71.81
53	2,3-3-Trimethyl-1-butene	77.89
54	2,4-Dimethylpentane	80.50
55	4,4-Dimethyl-c-2-pentene	80.43
56a	Benzene	80.10
56b	3,4-Dimethyl-1-pentene	80.80
57a	2,4-Dimethyl-1-pentene	81.61
57b	3,3-Dimethyl-1-pentene	77.48
58	2,2,3-Trimethylbutane	80.87
59	C_7 Diolefin	
60a	2,4-Dimethyl-2-pentene	83.26
60b	c-2-Methyl-3-hexene	83.6
61	1-Methylcyclopentene	75.49
62	3-Methyl-1-hexene	83.94
63	3-Ethyl-1-pentene	84.11
64	2-Methyl-t-3-hexene	85.90
65a	2,3-Dimethyl-1-pentene	84.28
65b	5-Methyl-1-hexene	85.31
66	C_7 Diolefin	
67	c-4-Methyl-2-hexene	86.31
68	C_7 Diolefin	
69a	4-Methyl-1-hexene	86.73
69b	4-Methyl-t-2-hexene	87.56
70	3,3-Dimethylpentane	86.07

TABLE 2 (continued)

Code number	Description	Boiling point (°C)
71	t-5-Methyl-2-hexene	88.11
72	3-Methyl-2-ethyl-1-butene	86.37
73	Cyclohexane	80.74
74	2-Methylhexane	90.05
75	c-5-Methyl-2-hexene	89.9
76	Cyclohexene	82.98
77a	3,4-Dimethyl-c-2-pentene	89.25
77b	2,3-Dimethylpentane	89.78
78	1,1-Dimethylcyclopentane	87.85
79	3-Methylhexane	91.85
80	2-Methyl-1-hexene	91.95
81	3,4-Dimethyl-t-2-pentene	91.58
82a	1-c-3-Dimethylcyclopentane	90.77
82b	1-Heptene	93.64
83	t-3-Methyl-3-hexene	93.53
84a	1-t-3-Dimethylcyclopentane	91.72
84b	3-Ethylpentane	93.48
85	t-3-Heptene	95.67
86	1-t-2-Dimethylcyclopentane	91.87
87	2,2-4-Trimethylpentane	99.24
88	t-3-Methyl-2-hexene	95.22
89a	c-3-Heptene	95.75
89b	2,2-Dimethyl-t-3-hexene	100.85
90a	2-Methyl-2-hexene	95.44
90b	c-3-Methyl-3-hexene	95.33
91	3,5-Dimethyl-1-hexene	105.4
92	3-Ethyl-2-pentene	96.01

TABLE 2 (continued)

Code number	Description	Boiling point (°C)
93	t-2-Heptene	97.95
94a	c-3-Methyl-2-hexene	97.26
94b	Heptane	98.43
95	c-2-Heptene	98.41
96a	2,3-Dimethyl-2-pentene	97.40
96b	2,4,4-Trimethyl-1-pentene	101.46
97	Cyclopentene C_7	
98	Cyclopentene C_7	
99	Cyclopentene C_7	
100	Cyclopentene C_7	
101	2,2-Dimethylhexane	106.84
102	1-c-2-Dimethylcyclopentane	99.53
103a	1,1,3-Trimethylcyclopentane	104.89
103b	2,2,3,3-Tetramethylbutane	106.47
104	Methylcyclohexane	100.93
105	2,5-Dimethylhexane	109.10
106	Octene	
107	3-Methylcyclohexene	102.5
108	2,4-Dimethylcyclohexane	109.43
109	Ethylcyclopentane	103.47
110	2,2,3-Trimethylpentane	109.84
111	Octene	
112	1-t-2-c-4-Trimethylcyclopentane	109.29
113	3,3-Dimethylhexane	111.97
114	Toluene	110.63
115	1-t-2-c-3-Trimethylcyclopentane	110.2
116	Octene	

TABLE 2 (continued)

Code number	Description	Boiling point (°C)
117	Octene	
118	2,3,4-Trimethylpentane	113.47
119	Octene	
120	Octene	
121a	2,3-Dimethylhexane	115.61
121b	2,3,3-Trimethylpentane	114.78
122	2-Methyl-3-ethylpentane	115.65
123	1,2,2-Trimethylcyclopentane	113.73
124	2-Methylheptane	117.65
125	4-Methylheptane	117.71
126	Octene	
127	3,4-Dimethylhexane	117.72
128a	3-Methylheptane	118.92
128b	3-Ethylhexane	118.53
128c	3-Methyl-3-ethylpentane	118.26
129	1-c-2-t-4-Trimethylcyclopentane	116.73
130a	2,2,5-Trimethylhexane	124.08
130b	1-c-2-c-4-Trimethylcyclopentane	117.78
131	1,1-c-3-t-4-Tetramethylcyclopentane	121.22
132	1-c-2-t-3-Trimethylcyclopentane	117.5
133	Octene	
134	Isoparaffin C_9 (alkylate)	
135	Octene	
136	1-c-3-Dimethylcyclohexane	120.09
137	1-t-4-Dimethylcyclohexane	119.35
138a	1,1-Dimethylcyclohexane	119.55
138b	1-Methyl-c-3-ethylcyclopentane	121.1

TABLE 2 (continued)

Code number	Description	Boiling point (°C)
138c	2,2,4-Trimethylhexane	126.54
139	1-Methyl-t-3-ethylcyclopentane	120.78
140	1-Methyl-t-2-ethylcyclopentane	121.22
141	c-4-Octene	122.53
142	1-Methyl-ethylcyclopentane	121.52
143	1-Octene	121.27
144	n-Octane	125.67
145	c-2-Octene	125.64
146	Isoparaffin C_9 (alkylate)	
147	1-t-2-Dimethylcyclohexane	123.43
148a	1-t-3-Dimethylcyclohexane	124.46
148b	1-c-4-Dimethylcyclohexane	124.32
149	2,3,5-Trimethylhexane	131.33
150	Isoparaffin C_9 (alkylate)	
151	Isopropylcyclopentane	126.41
152	Cyclopentane C_9	
153	Isoparaffin C_9 (alkylate)	
154	2,4-Dimethylheptane	132.89
155	2,2-Dimethylheptane	132.69
156	1-Methyl-c-2-ethylcyclopentane	128.05
157	2,6-Dimethylheptane	135.21
158	Cyclopentane C_9	
159a	2,5-Dimethylheptane	136.0
159b	n-Propylcyclopentane	130.94
160a	1-c-2-Dimethylcyclohexane	129.74
160b	3,5-Dimethylheptane	136.0
161	3,3-Dimethylheptane	137.01

TABLE 2 (continued)

Code number	Description	Boiling point (°C)
162	Ethylcyclohexane	131.79
163	Ethylbenzene	136.18
164	Isoparaffin C_{10} (alkylate)	
165	Isoparaffin C_{10} (alkylate)	
166	Isoparaffin C_{10}	
167	1,1,3-Trimethylcyclohexane	136.63
168	1-c-3-c-5-Trimethylcyclohexane	138.41
169a	Isoparaffin C_9 (alkylate)	
169b	1,1,4-Trimethylcyclohexane	135
170	Cyclopentane C_9	
171	p-Xylene	138.35
172	2,3-Dimethylheptane	140.34
173a	m-Xylene	139.10
173b	4-Ethylheptane	141.22
174	3,4-Dimethylheptane	140.61
175	4-Methyloctane	142.42
176	2-Methyloctane	143.26
177	1-t-2-t-4-Trimethylcyclohexane	141.22
178	1-c-3-t-5-Trimethylcyclohexane	141.22
179	3-Ethylheptane	143.0
180	3-Methyloctane	144.21
181	Isoparaffin C_{10} (alkylate)	
182	o-Xylene	144.41
183	Isoparaffin C_{10} (alkylate)	
184	Unclassified	
185	Cyclopentane C_9	

TABLE 2 (continued)

Code number	Description	Boiling point (°C)
186	1-t-2-c-4-Trimethylcyclohexane	145
187	Cyclopentane C_9	
188	Cyclopentane C_9	
189	1-t-2-c-3-Trimethylcyclohexane	143.9
190	Cyclopentane C_9	
191	1-c-2-c-4-Trimethylcyclohexane	148
192	1-c-2-t-4-Trimethylcyclohexane	148
193	1,1,2-Trimethylcyclohexane	145.2
194	Cyclopentane C_9	
195	Cyclopentane C_9	
196	1-Methyl-c-3-ethylcyclohexane	148
197	n-Nonane	150.82
198	1-Methyl-t-4-ethylcyclohexane	148
199	Cumene	152.39
200	Paraffin C_{10}	
201	Paraffin C_{10}	
202	Naphthene C_{10}	
203	Paraffin C_{10}	
204	Paraffin C_{10}	
205	Dinaphthene C_9	
206	Paraffin C_{10}	
207	Isoparaffin C_{10}	
208	1-Methyl-t-3-ethylcyclohexane	151
209	Isoparaffin C_{10}	
210a	1-c-2-t-3-Trimethylcyclohexane	151
210b	1-Methyl-t-2-ethylcyclohexane	151.7
211	1-Methyl-c-4-ethylcyclohexane	151

TABLE 2 (continued)

Code number	Description	Boiling point (°C)
212	1-c-2-c-3-Trimethylcyclohexane	153
213	1-Ethyl-1-methylcyclohexane	152.16
214	Isoparaffin C_{10}	
215	Isoparaffin C_{10}	
216	Isoparaffin C_{10}	
217	Isopropylcyclohexane	154.76
218	2,6-Dimethyloctane	160.38
219	n-Propylbenzene	159.22
220	Isoparaffin C_{10}	
221	n-Propylcyclohexane	156.72
222	1-Methyl-c-2-ethylcyclohexane	156.09
223	Paraffin C_{10}	
224	Naphthene C_{10}	
225	Naphthene C_{10}	
226	1-Methyl-3-ethylbenzene	161.30
227	1-Methyl-4-ethylbenzene	161.99
228	Unclassified C_{10}	
229	Isoparaffin C_{10}	
230	Isoparaffin C_{10}	
231	Isoparaffin C_{10}	
232	Cyclohexane C_{10}	
233	5-Methylnonane	165.11
234	4-Methylnonane	165.72
235	1-Methyl-2-ethylbenzene	165.15
236	2-Methylnonane	167.00
237	Unclassified	
238	1,3,5-Trimethylbenzene	164.72

TABLE 2 (continued)

Code number	Description	Boiling point (°C)
239	Naphthene C_{10}	
240	3-Methylnonane	167.8
	Isoparaffin C_{11}	
241	tert-Butylbenzene	169.15
242	Naphthene C_{10}	
243	Naphthene C_{10}	
244	Naphthene C_{10}	
245	Naphthene C_{10}	
246	Isoparaffin C_{11}	
247	Unclassified C_{11}	
248	1,2,4-Trimethylbenzene	169.35
249	Naphthene C_{10}	
250	sec-Butylbenzene	173.33
251	Isobutylbenzene	172.76
252	Naphthene C_{11}	
253	Paraffin C_{11}	
254	Naphthene C_{10}	
255	Unclassified C_{11}	
256	Unclassified C_{11}	
257	n-Decane	174.15
258	m-Cymene	175.18
259	Cyclopentane C_{11}	
260	Cyclopentane C_{11}	
261	Cyclopentane C_{11}	
262	p-Cymene	177.13
263	1,2,3-Trimethylbenzene	176.08
264	Cyclopentane C_{11}	

TABLE 2 (continued)

Code number	Description	Boiling point (°C)
265	Cyclopentane C_{11}	
266	Indane	177.85
267	Indene	182.43
268a	o-Cymene	178.18
268b	Dimethylnonane	
269	Cyclopentane C_{11}	
270	Unclassified C_{11}	
271	Unclassified C_{11}	
272	Unclassified C_{11}	
273	Unclassified C_{11}	
274	1,3-Diethylbenzene	181.14
275	Unclassified C_{11}	
276	Unclassified C_{11}	
277	Unclassified C_{11}	
278	1-Methyl-3-n-propylbenzene	182.0
279	Unclassified C_{11}	
280	n-Butylbenzene	183.31
281	Unclassified C_{11}	
282a	1-Methyl-4-n-propylbenzene	183.34
282b	1,4-Diethylbenzene	183.79
282c	1,2-Diethylbenzene	183.46
283	Unclassified C_{11}	
284	Unclassified C_{11}	
285	Cyclohexane C_{10}	
286	1-Methyl-2-n-propylbenzene	184.83
287	1,3-Dimethyl-5-ethylbenzene	183.6
288	Unclassified	

TABLE 2 (continued)

Code number	Description	Boiling point (°C)
289	Unclassified C_{11}	
290	5-Methyldecane	186.11
291	4-Methyldecane	187.89
292	1,4-Dimethyl-2-ethylbenzene	186.94
293	2-Methyldecane	189.19
294	3-Methyldecane	187.78
295	1,3-Dimethyl-4-ethylbenzene	188.44
296	Unclassified	
297	1,2-Dimethyl-4-ethylbenzene	189.78
298	1,3-Dimethyl-2-ethylbenzene	190.04
299	Aromatic C_{11}	
300	Unclassified	
301	1-Ethyl-3-isopropylbenzene	192.10
302	2-Phenyl-2-methylbutane	192.38
303	1-Methyl-3-sec-butylbenzene	194
304	1-Methyl-3-isobutylbenzene	194
305	1,2-Dimethyl-3-ethylbenzene	193.96
306	2-Phenylpentane	192.78
307	1-Methyl-4-sec-butylbenzene	197.22
308	1-Ethyl-4-isopropylbenzene	196.61
309a	n-Undecane	195.89
309b	1-Phenyl-2-methylbutane	197
310	Unclassified	
311	1-Phenyl-3-methylbutane	197.78
312	Unclassified	
313	1,2,4,5-Tetramethylbenzene	196.84
314	Aromatic C_{11}	

TABLE 2 (continued)

Code number	Description	Boiling point (°C)
315	1-Ethyl-3-n-propylbenzene	201
316	1,2,3,5-Tetramethylbenzene	198.05
317	Unclassified	
318	Aromatic C_{11}	
319	1-Ethyl-2-n-propylbenzene	203
320	Unclassified C_{12}	
321	1-Methyl-3,5-diethylbenzene	200.14
322	Methylindane	
323	1,3-Dimethyl-5-n-propylbenzene	202
324	1,2-Dimethyl-3-isopropylbenzene	202.6
325	Methylindane	
326	Aromatic C_{11}	
327	Unclassified	
328	Aromatic C_{11}	
329	Aromatic C_{11}	
330	1,2,3,4-Tetramethylbenzene	205.09
331	1-Methyl-3-n-butylbenzene	205.0
332	Tetralin	207.65
333	Unclassified	
334	1,3-Dimethyl-4-n-propylbenzene	208.9
335	1-Methyl-3,5-diethylbenzene	200.7
336	Aromatic C_{11}	
337	1,4-Dimethyl-2-n-propylbenzene	204.3
338	Aromatic C_{11}	
339	Naphthalene	217.94
340	Unclassified	
341	Unclassified	

TABLE 2 (continued)

Code number	Description	Boiling point (°C)
342	Dimethylindane	
343	1,2-Dimethyl-3-n-propylbenzene	210.7
344	1,2-Dimethyl-4-n-propylbenzene	207.6
345	Dimethylindane	
346	1,3,5-Trimethyl-2-ethylbenzene	212.33
347	C_2 Indane	
348	Unclassified C_{12}	
349	1,2,4-Trimethyl-5-ethylbenzene	213.0
350	1,2,5-Trimethyl-3-ethylbenzene	212.4
351	Unclassified C_{12}	
352	Unclassified C_{12}	
353a	n-Dodecane	216.3
353b	1,2,3-Trimethyl-5-ethylbenzene	215.8
354	1,2,4-Trimethyl-3-ethylbenzene	216.6
355	C_{12} Alkylbenzene	
356	Unclassified	
357	Ethylindane	
358	Ethylindane	
359	C_{12} Alkylbenzene	
360	1,2,3-Trimethyl-4-ethylbenzene	220.4
361	Aromatic C_{12}	
362	Aromatic C_{12}	
363	Unclassified	
364	Dimethylindane	
365	Unclassified C_{13}	
366	Unclassified C_{13}	
367	Dimethylindane	

TABLE 2 (continued)

Code number	Description	Boiling point (°C)
368	Unclassified C_{13}	
369	Unclassified C_{13}	
370	Indane C_{11}	
371	Aromatic C_{12}	
372	Aromatic C_{12}	
373	Indane C_{11}	
374	Pentamethylbenzene	231.8
375	2-Methylnaphthalene	241.10
376	Aromatic	
377	1-Methylnaphthalene	244.68
378	n-Tridecane	235.47

Split ratios may be changed and reset as desired without changing any plumbing.

A 1000-ft by 0.02-in. column coated with squalane is used to carry out the separation. A 15-ft by 0.02-in. column coated with OV-101 is connected to the back of the main column. This post column removes the column bleed from the main column during the final portion of the run.

The post column is changed after about every 10 analyses. It is cleaned, repacked with a single 5-ml chanrge of a 10% solution of OV-101 in CH_2Cl_2, and finally baked out at 225°C. The direction of carrier gas flow through the 1000-ft column is reversed each time the post-column is changed.

Nitrogen is used as the make-up gas and is added at the end of the analytical column. A back-reference controller maintains the flow of make-up gas at 25 ml min^{-1}. Hydrogen is introduced into this stream at a rate of 35 ml min^{-1}. Air, which has been thoroughly purified [10], is metered into the flame ionization detectors at a rate of 400 ml min^{-1}.

The size of the injected sample is generally 3 μl. The splitter operates at a 50:1 split ratio. The column temperature is maintained isothermal

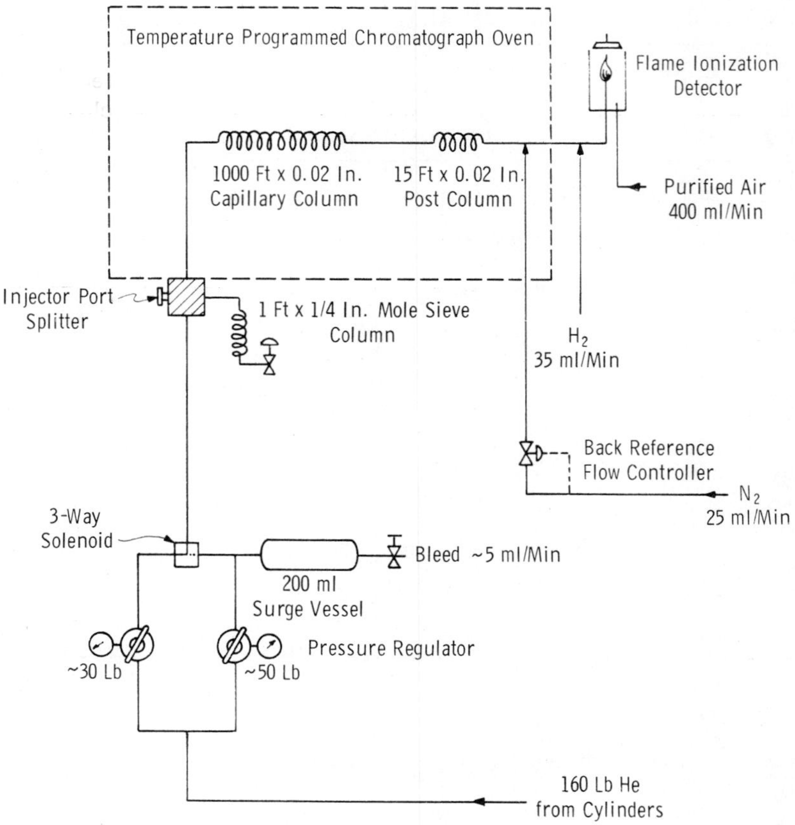

FIG. 2. Carrier gas flow diagram.

at 30°C for 20 min, after which it is programmed to 140°C at a rate of 1°C min^{-1}.

At the end of the temperature-programmed section of the run, the solenoid in the flow system is activated to increase the column inlet pressure to 50 psi. The duration of this final isothermal portion of the program at the higher flow rates depends on the material to be analyzed. It can range from 0 to 120 min. The increase in column inlet pressure results in a significant reduction in elution times of the heavier components; it also sharpens the peak shapes considerably with a resultant increase in integration accuracy. The sudden increase in the carrier gas flow rate does not affect the baseline stability because of the distance between the point of pressure change and the detectors. The increase in column bleed is hardly noticeable because of the presence of the post column and because the increment in total inert gas to the detector per unit time is small.

VI. LOW-RESOLUTION (PACKED-COLUMN) METHODS

Depending on the purpose of the analysis, many packed-column techniques are more practical and useful than the ultimate high-resolution chromatographic approach. Frequently, only a limited amount of data is needed—a vapor pressure, a boiling range, yield of C_5+—which can be obtained quickly and economically without determining each component individually.

Trace analyses are often best carried out using packed columns. Here, as in many cases where it is desired to isolate specific components, highly selective columns or combinations of columns are usually needed. An example of a multiple-column method is the ASTM D 3606 benzene and toluene in gasoline procedure. A sample of the gasoline is chromatographed on a nonpolar (boiling point) column until about one-half of it, including toluene, has eluted. The other half is backflushed from this first column to waste as it contains nothing of interest to the analysis. The first half containing essentially all the gasoline components through the C_8 saturates is passed to a second, highly polar column [e.g., tris(cyanoethoxy)propane in the D 3606 method]. Here the benzene and toluene are completely separated from the less polar light gasoline components and are readily measured. Quantitation is carried out by using an internal standard (methyl ethyl ketone). The method can be readily adapted to the measurement of benzene and toluene in other matrices, such as in vapor analysis and in trace atmospheric studies.

The principle of using one column to isolate the analyte(s) of interest from most interferences and a second column to complete the simplified analysis probably represents one of the most important areas of analytical development in gas chromatography.

A second area of increasing use of packed columns in the analysis of gasolines and naphthas is that of simulated distillation (SD). Simulated distillation has been formulated into a standard ASTM method as D 2887. As currently written up, the method only determines a boiling range by a standard method. However, much additional data can be derived, including vapor pressure measurements and conversion of the SD curve to physical distillation equivalents (ASTM D 86 or D 1160). Although D 2887 specifies columns of very limited resolution, the resulting chromatogram represents greater resolution than can be obtained by any but the most efficient, time-consuming physical distillations. Because the SD data can be obtained automatically, in short times and reduced by computers, it is expected that SD will largely replace routine analytical distillations where physical collection of the cuts is not necessary.

One point to be remembered about SD: It is not meant to be a substitute for distillation, but rather a more informative replacement. Although the SD data can be converted to, say, D 86 equivalents this, in fact, represents a degradation of the SD data which is not only not necessary but unproductive. A more detailed description of this technique is given in Chap. 4 of this volume.

A. The PNA (Paraffin, Naphthene, Aromatic) Analyzer

This chromatographic analysis depends upon the special properties of 13X molecular sieve which enables it to separate paraffins and naphthenes. In a commercial version of this instrument, aromatic hydrocarbons are separated from the saturates on a highly polar column and finally analyzed on a nonpolar column to give an aromatics analysis by carbon number. The saturates are trapped on a cold 13X Molecular Sieve column. After the aromatics are eluted, the 13X column is programmed to over 500°C. The saturates elute during the program in carbon number order but with a unique feature. Within each carbon number, the naphthenes and paraffins are split into peaks or bands permitting separate quantitation of the two types. This, coupled with the aromatic separation, permits the construction of an analytical matrix giving the total breakdown of each type: paraffins, naphthenes, and aromatics by carbon number. The information in the matrix is of inestimable value in predicting performance of a naphtha in reforming.

There are several limitations on the method. The endpoint of the naphtha must be less than 400°F (210°C) on a true boiling point basis. Stocks containing olefins cannot be analyzed and, in the commercial instrument, reformates can cause detector linearity problems if sample sizes are too large.

The 500°C+ upper temperature of the 13X Molecular Sieve column poses some problems. In order to avoid cracking at these temperatures, the carrier gas must be ultrahigh purity hydrogen, and the sieve must be carefully treated to remove any acidic sites.

The PNA analysis is another example of the great power of the use of multiple columns in the analysis of complex mixtures. Although this analysis is complex with multiple column switching, trapping, and reinjecting separated fractions and precise timing, it is carried out in the commercial version completely automatically. A chromatogram obtained by this technique is shown in Figure 15 of Chapter 5.

VII. SPECIALIZED DETECTORS

Over two dozen gas chromatography detectors have been described. Only two of these are customarily used in the general analysis of gasolines and naphthas—the flame ionization detector and the thermal conductivity detector.

Other detectors, however, have proved useful in specialized areas. Most petroleum products are extremely complex, and a selective detector can be combined with chromatography to provide the analytical discrimination that cannot readily be obtained by the separation alone.

The most ubiquitous nonhydrocarbons in gasolines are the lead alkyls, i.e., tetraethyl- and tetramethyl lead and their cross-substitution compounds. Total lead in gasoline, of course, may be determined by excellent x-ray fluorescence or atomic absorption methods. Determination of the isomer distribution, however, is a gas chromatography problem. The lead alkyls are readily chromatographed and separated. They can be detected by selective detectors. The first to be used for this was the electron capture detector (ECD). Its use is complicated by differing sensitivities to the various lead alkyl isomers. It is also sensitive to the alkyl halides present in gasoline as lead scavengers.

Both these difficulties are obviated by the use of an atomic absorption spectrophotometer as the detector. It is quantitative, selective, and readily amenable to flow programming to speed the analysis [11]. We use a flow-programmed GC-AA combination to routinely analyze the lead alkyl isomer distribution in gasolines in about 2 min total analysis time per sample.

Sulfur and nitrogen compounds in gasolines and naphthas can be determined with selective detectors. The flame photometric detector (FPD) and the nitrogen/phosphorus detectors are most often used. Both have excellent sensitivities and selectivities. However, trace analyses for nitrogen or sulfur are hampered by the fact that the FPD and the NPD do respond in a limited way to hydrocarbons. For this reason, high-resolution columns are favored to separate as nearly as possible the hetero compounds from the hydrocarbon matrix.

REFERENCES

1. D. H. Desty, A Goldup, and W. T. Swanton, Proc. 3rd Intern. Symp. Gas Chromatog. Instr. Soc. Am., Michigan State University, 3:105 (1961).
2. F. D. Rossini, B. J. Mair, and A. J. Streiff, Hydrocarbons from Petroleum, API Project 6, Reinhold, New York, 1953.
3. R. L. Martin and J. C. Winters, Anal. Chem. 35:1930 (1963).
4. A. G. Polgar, J. J. Holst, and Sigurd Groennings, Anal. Chem. 34:1227 (1962).
5. R. D. Schwartz and D. J. Brasseaux, Anal. Chem. 35:1374 (1963).
6. W. N. Sanders and J. B. Maynard, Anal. Chem. 40:527 (1968).
7. N. Dimov, J. Chromatog. 119:109 (1976); 119:307, 321.

8. (A) O. E. Shupp, III and J. S. Lewis, eds., Gas Chromatographic Data Compilation; (B) ASTM D S25A (1967); AMD25AS-1 (1971).
9. A. Matukuma, in "Gas Chromatography 1968" (C. L. A. Harbourn and R. Stock, eds.), Institute of Petroleum, London, 1969.
10. W. H. King, Jr., Anal. Chem. 43:984 (1971).
11. P. R. Ballinger and I. M. Whittemore, Preprints, Div. of Petrol., ACS, 13(3):133 (1968).

Chapter 4

SIMULATED DISTILLATION BY GAS CHROMATOGRAPHY

Ronald D. Butler
Fuels Branch
Air Force Aero Propulsion Laboratory
Wright-Patterson Air Force Base
Ohio

I. INTRODUCTION

Distillation is by far the most widely used separation process in the petroleum industry. Knowledge of the boiling range of crude oils and finished products is essential for quality assurance purposes, for control of plant and refinery processes, and for predicting certain other physical properties.

Routine laboratory-scale distillation tests are widely used for determining the boiling ranges of crude oils and their products. The most familiar quality control test for boiling range distribution, the so-called Engler distillation, was adopted by the American Society for Testing and Materials* as standard test method D86 in 1926. Essentially a one-plate distillation, the D86 test method is generally applicable to most commercial products. A reduced-pressure version, ASTM D1160, is available for higher boiling and/or thermally labile materials. Samples having an extremely wide boiling range, such as crude oils, can be characterized by the "true boiling point" method D2892 which involves a lengthy (100-hr) separation in a

*Details regarding all of the cited ASTM test methods may be found by consulting the most recent annual book of ASTM Standards, Parts 23 and 24.

high-efficiency still. Part of the D2892 procedure involves vacuum distillation for materials having very high endpoints.

Most of the routine distillation tests are intentionally performed under conditions of limited efficiency; detailed fractionation of the sample is usually of secondary importance. This compromise is made largely for economic reasons. Highly refined data would require almost endless testing, whereas the general shape of a one-plate distillation curve is often adequate for making engineering calculations, correlating with other physical properties, and predicting the product slate [1].

In 1960, Eggerston et al. [2] reported that a low-resolution, temperature-programmed gas chromatographic analysis could be used to simulate the time-consuming true boiling point distillation. The GC method relies on the observation that hydrocarbons generally are eluted from a nonpolar column in order of their boiling points. In essence, the chromatograph is operated simply as a very efficient microdistillation apparatus. Because of the regularity of the elution order of the hydrocarbon components, retention times can be converted with confidence to equivalent distillation temperatures; the accumulated detector response is related to the amount of sample distilled.

This distillation model deviates from conventional gas chromatographic practice in that the conditions are selected to give only limited column efficiency and resolution. This approach admittedly violates the fundamental motivation of most practicing chromatographers, namely, a compulsive urge to ever greater resolution. Nonetheless, Green et al. [3] in 1964 confirmed that low-resolution gas chromatographic analysis does indeed provide distillation data that are in excellent agreement with the precise D2892 analytical distillation. These authors coined the term "simulated distillation by gas chromatography," which is still universally used in referring to the technique.

The full development of simulated distillation as a routine procedure has been made possible by the veritable explosion in GC instrumentation during the last decade. Beginning with the introduction of automatic temperature programming, there have been remarkable advances in such features as sensitive and stable detectors, automatic sampling, electronic integration, and, in particular, microprocessors with capabilities for online acquisition, postrun calculations, and remote control of instrument parameters. Today, a fully automated simulated distillation system, under computer control, can operate continuously to provide finished reports in a choice of formats that agree well with true boiling point data. With the aid of established correlations, the computer can also provide the corresponding Engler (D86) profile [4] and predict other properties such as vapor pressure and flash point [5-7]. In what is nearly the ultimate application of simulated distillation, the computer proceeds to adjust processing conditions, particularly blending operations, to yield products meeting tighter specifications [8, 9]. A complete analysis can be performed in well under one hour—less than one percent of the time required for a conventional true boiling point distillation—and requires almost negligible operator time.

Simulated distillation achieved formal status as as ASTM standard method in 1973 under the designation D2887, "Boiling Range Distribution of Petroleum Fractions by Gas Chromatography."

II. THEORY AND PRINCIPLES

The technique of simulated distillation is based upon the assumption that individual components of a sample are eluted from a GC column in order of their boiling points. Chromatographic experience indicates that this assumption is valid under two important conditions: the sample consists of "nonpolar" hydrocarbons, and the column contains a nonpolar stationary phase. Fortunately, these conditions are inherent in the routine chromatographic analysis of most crude oils and their commercial products. Thus it is valid to develop an empirical correlation between observed retention times and the normal boiling points of the sample components.

However, experience also demonstrates that even the so-called nonpolar stationary phases do discriminate to some extent among different classes of hydrocarbons. Retention time is a complex function of the affinity of a sample component for the stationary phase, and of the vapor pressure of the component, which in turn is dependent upon its heat of vaporization. These factors vary appreciably among hydrocarbons. Thus an empirical correlation between retention time and boiling point, usually established by running a series of n-alkanes, is not strictly valid for other hydrocarbon classes. Aromatic hydrocarbons and, to some degree, cycloalkane derivatives (naphthenes) show the effect of eluting earlier than n-alkanes having the same boiling points.

Fortunately, in simulated distillation these effects are smoothed out by the requirement of low resolution. The relationship between retention time and distillation temperature is usually valid, and the method can be used with confidence. Adjustments may be necessary only when dealing with samples having high concentrations of aromatic or heterocyclic compounds, or when more highly polar stationary phases are used.

Just as retention time is used as an analog of distillation temperature, the accumulated area under the chromatogram is related to the amount of sample recovered in a bulk distillation. This analogy implies that the detector response is identical for all eluting species. It is true that the flame ionization detector responds fairly consistently on a mass basis to most hydrocarbons, even those of different classes. The thermal conductivity detector, on the other hand, shows an approximately linear molar response within a homologous series [10]. The difference between the thermal conductivity of the pure carrier gas, and that of the carrier gas and component, is often assumed to be proportional to the mole fraction of the component. Most gas mixtures, however, do not show this proportionality, and the thermal conductivity-concentration isotherm is curved.

In simulated distillation, the eluting "species" at any instant is actually a mixture of several, indeed many, components—again, because of the requirement of low resolution. This mixture effectively represents an average component that typifies the entire sample, so that in actual practice the detector response does remain reasonably uniform throughout the analysis.

It might be expected, on intuitive grounds, that the accumulated area under the chromatogram corresponds to the accumulated weight of sample recovered. McTaggart et al. [11] report that this expectation is verified in the case of crude oils, with the simulated distillation technique giving essentially the same results, in terms of weight percent recovered, as a high-efficiency true boiling point distillation. Green [4], on the other hand, claims that simulated distillation most closely reflects the volume percent recovered in the true boiling point distillation. The difference, of course, lies in the fact that density increases regularly with molecular complexity and boiling point. This discrepancy has not been resolved, but it does not appear to affect the applicability of the method.

III. DESCRIPTION OF THE METHOD

Simulated distillation is performed routinely in accordance with the procedure described in ASTM standard method D2887. It is applicable to all petroleum products boiling below 1000°F and having distillation ranges at least 100 degrees Fahrenheit wide. These permissive limits allow the characterization of most commercial petroleum products.

A calibration sample, containing a series of known n-alkanes, serves both for establishing the correlation between retention time and distillation temperature and for determining the resolution of the column. The calibration mixture must cover the complete boiling range of the sample to be analyzed and must contain at least one component boiling below, and one boiling above, the range of the sample. A satisfactory calibration plot, relating retention time and boiling point, is shown in Figure 1. If the carrier flow and temperature programming rate have been properly selected in relation to the column, the plot will be essentially linear, with some negative curvature at the front end.

Resolution of the column is an important parameter and must be carefully controlled. Column resolution is determined by measuring the retention times (rt) and widths (w) of the n-hexadecane and n-octadecane peaks, and is defined in the conventional way by

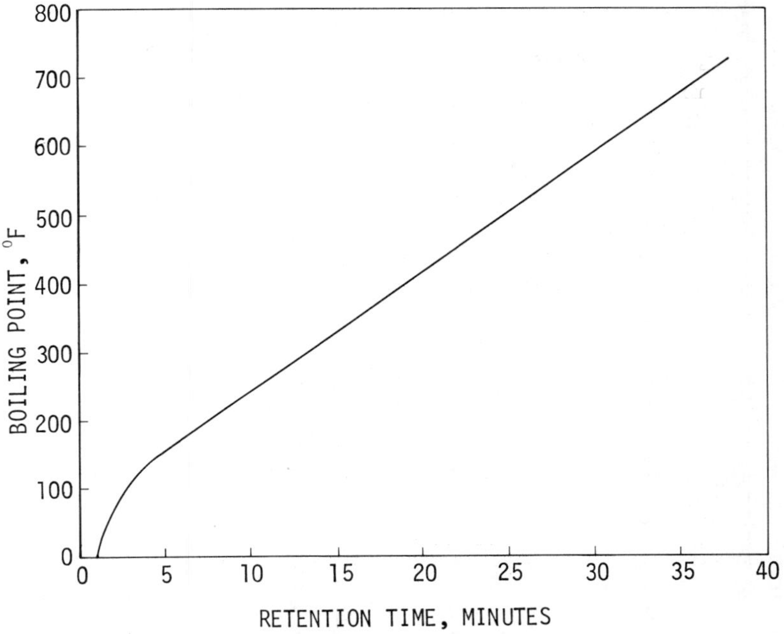

FIG. 1. Calibration plot for correlating retention time and distillation temperature from the boiling points of n-alkanes.

$$r = \frac{2(rt_{18} - rt_{16})}{w_{18} + w_{16}} \tag{1}$$

Resolution is approximately the number of peaks, of intermediate width, that can be fitted between the C_{16} and C_{18} peaks. In simulated distillation, this value must be between 3.0 and 8.0.

The column and operating conditions are selected to meet this criterion. The method permits great latitude in the choice of stationary phase, column geometry, carrier gas, detector, and temperature program. Consequently, different laboratories report a wide variety of operating conditions, as shown in Table 1.

The stationary phase may be any of the familiar nonpolar silicones, but should be selected so as to function satisfactorily throughout the entire temperature program. The majority of operators appear to prefer the methyl vinyl silicone UCW-98 as a stationary phase. This material is attractive because it exhibits superior performance at both low and high temperatures. Furthermore, columns prepared with UCW-98 can be remarkably

TABLE 1. Examples of Reported Instrument Parameters for Simulated Distillation

Laboratory	Detector	Carrier gas	Flow rate (ml min^{-1})	Temperature program	Stationary phase	Column dimensions
A	FID	N_2	30	-50 to 350°C, 10°C min^{-1}, 8 min hold at 350°C	UCW-98	1/8 in. × 20 in.
B	TCD	He	14	0 to 325°C, 8°C min^{-1}	OV-101	1/8 in. × 10 ft
C	TCD	Not reported	Not reported	-20 to 250°C, 10°C min^{-1}	UCW-98	1/2 in. × 4 ft
D	TCD	Not reported	Not reported	-30 to 250°C, 10°C min^{-1}	UCW-98	1/4 in. × 18 in.
E	TCD	He	60	1 min isothermal at 50°C, then 50 to 200°, 10°C min^{-1}	UCW-98	1/4 in. × 2 ft
F	FID	N_2	50	-30 to 380°C, 11°C min^{-1}	UCW-98	1/4 in. × 18 in.
G	FID	He	38	-30 to 350°C, 8°C min^{-1}	OV-1	1/8 in. × 2.5 ft
H	TCD	He	110	-40 to 365°C, 10°C min^{-1}	UCW-98	1/4 in. × 18 in.
I	FID	N_2	30	-40 to 360°C, 10°C min^{-1}	UCW-98	1/8 in. × 1.7 ft
J	FID	He	Not reported	-20 to 300°C, 8°C min^{-1}	SE-30	1/8 in. X 25 ft
K	FID	N_2	50	-30 to 230°C, 10°C min^{-1}	UCW-98	1/4 in. × 18 in.
L	TCD	He	35	1 min isothermal at 50°C, then 50 to 350°C, 20°C min^{-1}	UCW-98	1/4 in. × 2 ft
M	TCD	He	30	0 to 250°C, 15°C min^{-1}	OV-1	1/8 in. × 7 ft
N	TCD	He	16	2 min isothermal at 25°C, then 25 to 250°C, 16°C min^{-1}	SE-30	1/8 in. × 8 ft

durable: one such column has been in use in the author's laboratory for more than 4 years with essentially no change in efficiency or retention properties, despite regular and repeated temperature cycling between -20° and 300°C.

The physical dimensions of the column, and the choice of carrier gas and flow rate, seem to be intensely personal, judging from the information in Table 1. There is nearly universal agreement, however, that subambient temperature programming is essential in order to ensure that at least 1 min is required for elution of the first 0.5% of the sample, as required by the method. A low starting temperature also has the effect of minimizing the curvature at the low-temperature end of the calibration plot, thereby increasing the precision of the time-temperature correlation. Dual-column operation is essential, because for highest accuracy the signal must return to baseline at the end of the run, and column bleed is nearly impossible to eliminate, even with exhaustive conditioning.

With one exception, which detector to use is a matter of preference; the method permits the use of either the thermal conductivity detector (TCD) or flame ionization detector (FID). Careful testing has shown that both detectors give nearly equivalent results for most sample types. A serious discrepancy was found, however, when the FID was used in the analysis of gasoline samples. It appears that the high concentrations of very volatile gasoline components (it is not uncommon for gasolines to contain as much as 10% butane, for example) overload the detector, thereby distorting the results. The method, therefore, recommends the use of the TCD for gasoline samples.

The flame ionization detector has numerous advantages, notably its high sensitivity (it is several orders of magnitude more sensitive than the TCD). However, in view of the ease with which the FID can be overloaded, it is important to observe a number of precautions to ensure that the linear range of the detector is not exceeded. Gas flow rates, for example, must be carefully monitored and maintained well within the ranges specified by the manufacturer. Using nitrogen instead of helium as a carrier gas is not only economical, but also extends the linear range of the FID. Much smaller samples must also be used with the FID. Experiments should be performed periodically to confirm the linearity of the flame detector. The linearity check can be performed simply by injecting varying volumes of a known hydrocarbon mixture; the normalized peak areas should be easily repeatable to within 1 percent and should show no systematic variation with sample size.

Generally, for complex samples, particularly those having very wide boiling ranges, sample volumes less than 1 μl are satisfactory with FID instruments. However, if tests indicate that the sample volume cannot be conveniently reduced to ensure detector linearity, then dilution with a noninterfering solvent is beneficial. Carbon disulfide is good for this purpose because it gives a very low response with the FID. If the initial boiling point of the sample is high enough, dilution with a low-boiling solvent (pentane or hexane) is satisfactory.

Similar precautions should be taken with the TCD. Thermal conductivity detectors are usually designed for a specific carrier gas flow rate, and this rate usually determines the time constant and flow sensitivity of the detector. With proper attention to gas flow rates and sample volumes, good results should be obtained with either the FID or TCD.

Once the instrument parameters have been optimized and the boiling point-retention time calibration has been established, samples are run under identical conditions. Figure 2 illustrates the types of chromatograms that are suitable for simulated distillation. Under the conditions used for generating these chromatograms, resolution was near the lower limit of 3.0.

The area under the chromatogram is monitored continuously and accumulated as a function of time. Any convenient method of integration is satisfactory, but the most accurate—and certainly the most convenient—is the electronic integrator. It is important, however, that the integrator use the true baseline as a reference: some integrators have logic functions that construct artificial baselines below unresolved peaks, thereby discarding significant areas, and for this reason are unsuited to simulated distillation.

The time at which 0.5% of the total area has been accumulated is converted to an equivalent temperature from the calibration plot, and this temperature is reported as the initial boiling point of the sample. Similarly, the time for 99.5% of the total accumulated area is converted, and this temperature is reported as the final boiling point. Intermediate distillation temperatures are determined in the same way. Depending upon the amount of information desired, any number of intermediate temperature-recovery readings may be taken.

With currently available instrumentation and data processing equipment, a great deal of sophistication is possible. Figure 3 illustrates a final report that was generated for quality control purposes by a GC system with electronic integration and simulated distillation software. For reasons that will be discussed presently, it is sometimes desirable to know the equivalent distillation temperatures for a laboratory (D86) distillation of the sample. In general, simulated distillation data do not correspond directly to values obtained from a D 86 distillation. Extensive correlation studies have been performed for the two methods, and at least one manufacturer [4] offers a software package providing computation of the equivalent D 86 temperatures. The printout shown in Figure 3 includes this kind of additional information.

The entire simulated distillation can be performed fully automatically by very stable instrumentation. It is therefore not surprising that the precision of the method is quite good. Both the repeatability and reproducibility

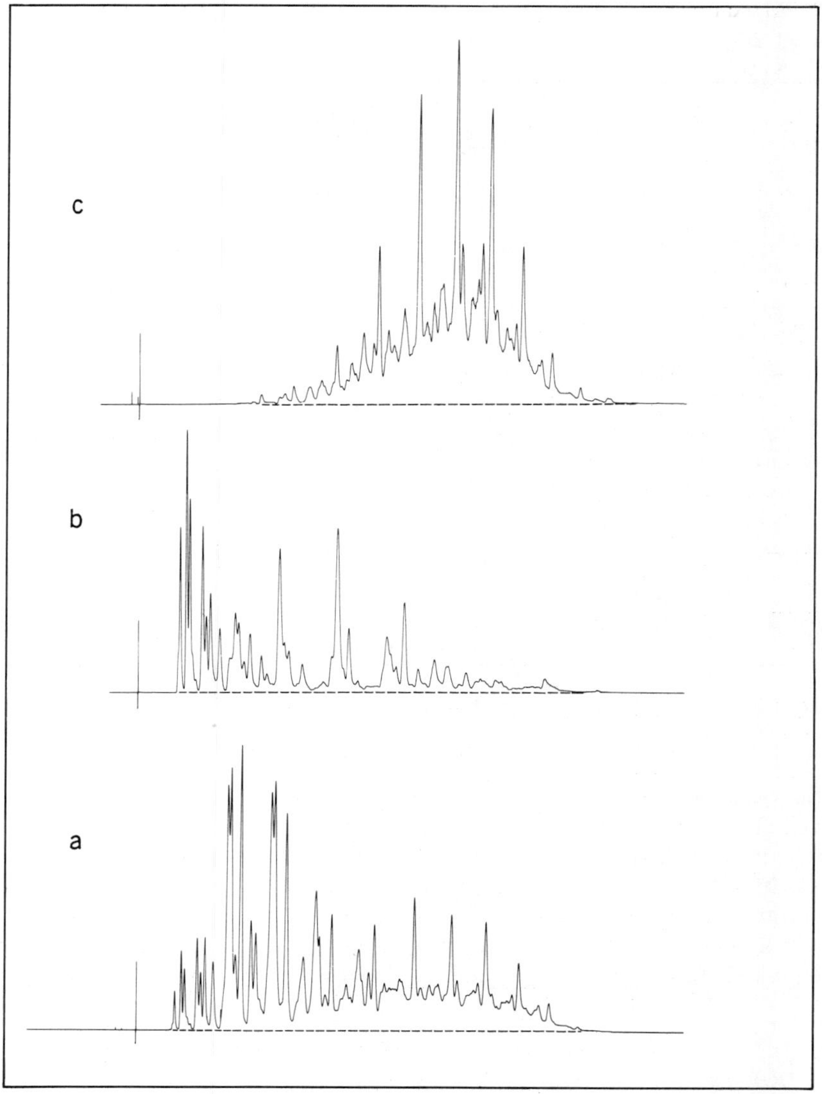

FIG. 2. Chromatograms illustrating satisfactory resolution for simulated distillation. (a) Widecut aircraft fuel (military grade JP-4). (b) Regular grade gasoline. (c) Kerosine-range aircraft fuel (military grade JP-5).

Boiling Point Distribution by ASTM D2887-73

Sample: JP-4- AFFB-16-73 Date: 7-12-75

Run #1 of 3 Vial #14

% Rec	Temp (°F)	% Rec	Temp (°F)
IBP	46	50	321
1	58	65	350
2	73	75	368
3	86	85	391
5	122	90	405
7	157	93	417
10	183	95	428
15	213	97	442
20	235	98	456
25	254	99	482
35	285	FBP	496

ASTM D86 Correlation

% Rec	Temp (°F)
IBP	126
10	228
20	257
50	326
90	388
FBP	431

FIG. 3. Example of simulated distillation report with D86 correlation.

of the technique are comparable to those of a carefully performed Engler distillation. For reasons that will be discussed presently, the greatest uncertainties lie in establishing the initial and final boiling points. Not only is this difficulty inherent in any distillation test, but of course any uncertainty in defining the endpoints also affects all of the intermediate points. The fact that duplicate determinations give close agreement indicates that simulated distillation is a very good method indeed.

IV. CORRELATIONS AND APPLICATIONS

A simulated distillation plot for a wide-cut petroleum product is shown in Figure 4, together with the corresponding D86 distillation curve. It will be noted that the two curves superficially have the same general shape; however, there are significant differences between them, particularly at the extremes. This relationship is typical for D86-D2887 correlations, with

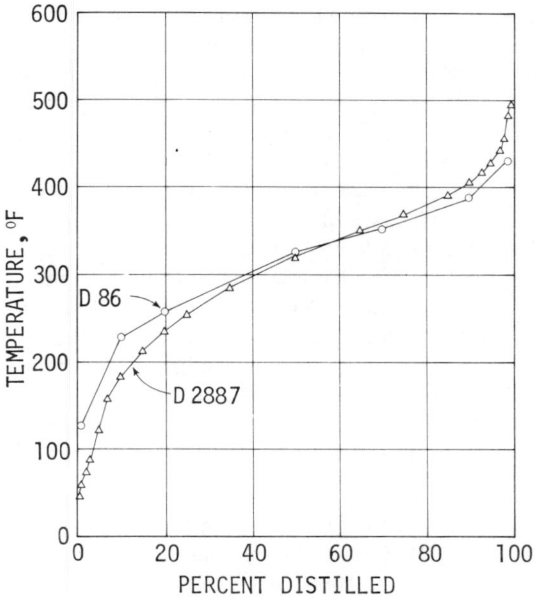

FIG. 4. Simulated (D2887) and Engler (D86) distillation curves for wide-cut aircraft fuel.

simulated distillation consistently predicting lower initial, and higher final, boiling points. The curves usually cross at about the midpoint.

The observed differences lie in uncertainties that are inherent in both methods. Precision of the D86 method, for example, is limited by small but finite loss of sample through column holdup and incomplete condensation, whereas the chromatographic technique necessarily accounts for 100% of the sample. Another reason for the discrepancy is that simulated distillation provides a definition of the initial boiling point that is not strictly accurate. While the GC method does accurately report the boiling points of the most volatile components, these temperatures do not correspond to the lowest boiling temperature of the bulk sample, which is, of course, the temperature at which the vapor pressure of the bulk material becomes equal to the barometric pressure. The "true" boiling point of a mixture can be measured under conditions of total reflux, but this value is also a function of the ratio of the vapor space to the volume of the liquid phase. Because the boiling point is defined differently in each type of test, the reported values must also differ.

Much attention has been paid to correlating the results of the two techniques. Bradley and Kennard [12] question the wisdom, or even the desirability, of attempts to reconcile differences in data produced by Engler and simulated distillations; they base their opinion on the observations that the D86 procedure itself is less precise, less sensitive, less repeatable, and less reproducible, and, therefore, not usable as an absolute standard by which to evaluate the new technique. This position admittedly has much to justify it. However, there are important reasons for attempting to correlate the two data sets. In the first place, it is often necessary to make changes in refinery processes on the basis of little more information than a distillation curve; even with experienced hands on the valve, changing the process conditions on the basis of an unfamiliar D2887 curve could easily spell the difference between profit and loss. Second, product specifications usually include distillation limits as determined by D86; the direct substitution of D2887 temperatures would quite obviously define a difference product. Finally, over the years a number of correlations have been developed that allow the calculation of other physical properties from D86 data; these correlations would not be valid for simulated distillation data.

A number of efforts have been made to correlate not only the two distillation methods, but also simulated distillation data with other physical properties. Ford et al. [13] describe two successful approaches for a wide range of product types. In the first, a straightforward multiple linear regression analysis is used to compute D86 distillation temperatures from a series of known D2887 points. The predicted temperature for 20% recovery in a D86 distillation, for example, would be given by an equation of the form

$$20\%(\mathrm{D86}) = A + B[30\%(\mathrm{D2887})] + C[20\%(\mathrm{D2887})] + D[10\%(\mathrm{D2887})] \qquad (2)$$

where 30%(D2887), 20%(D2887), and 10%(D2887) are temperatures for 30, 20, and 10% recovered, respectively, as found by simulated distillation.

The second correlation reported by Ford et al. is based upon a procedure first proposed by Geddes [14], in which each D86 point is computed from the slope of the D2887 plot at several points. Both of the correlations are reported to work equally well.

Martel [7] also used a multiple linear regression analysis technique to determine the D2887 temperatures that correspond to the current D86 distillation limits imposed by U.S. Air Force turbine fuel specifications. As a result of that study, the Air Force now permits the use of simulated distillation as an acceptable alternative to D86 in specification testing. As a bonus, Martel also found that, for both commercial and military kerosine-range aircraft fuels (Jet A, Jet A-1, JP-5), flash point can be calculated from D2887 data with a standard error of estimate of only 2.1°F; viscosity at -30°F can be predicted with a standard error of only 0.3 cSt.

DeBruine and Ellison [5, 6] have developed models, based upon fundamental physical chemical principles, to calculate both the D86 distillation curves and Reid vapor pressures (approximately the vapor pressure of the sample at 100°F) for gasolines from simulated distillation data. By inferring compositional data from the GC analysis, DeBruine and Ellison are able to compute these volatility characteristics with good reliability. The information can then be used to define specification criteria for gasolines, which will improve engine performance aspects such as cold starting, evaporation loss, vapor lock, hot starting, hot idling, carburetor icing, warm-up, acceleration, and fuel economy.

According to Bird and Kimball [8], at least one large refinery has been able effectively to discontinue vapor pressure and vapor lock inspections of gasoline blends because of successful correlations with simulated distillation data. Moreover, the computer that reduces the raw GC distillation data also automatically controls the refinery's gasoline blending operations to deliver products meeting preprogrammed volatility specifications. Kimball and McCracken [9] report that the single, automated simulated distillation test now replaces Engler distillation, flash point, and API gravity measurements of heating oil and diesel fuel.

Despite the rapid development of simulated distillation as a routine technique and its widespread acceptance, workers in the field continue to report modifications and improvements to the technique. Jackson et al. [15], for example, describe the use of short capillary columns for the simulated distillation of crude oils. The obvious advantage in this variation is that the higher-boiling fractions can be eluted within a reasonable time, thereby reducing the analysis time. Excellent agreement again is found with true boiling point data.

Historically, one area of concern has been the problem of crude oils that contain fractions boiling so high that they are simply not eluted from the column. A number of techniques have been proposed for dealing with these materials. Philyaw et al. [16] determined high-boiling materials by backflushing the column and oxidizing the recovered fractions, which is then measured as CO_2. Trusell and Mikulas [17] have evaluated a number of methods, and recommend that an auxiliary analysis be used, from which the fraction of sample eluted below the 1000°F limit can be calculated.

McTaggart et al. [11, 18] also find it useful to perform a preliminary distillation of the crude oil and follow the semimicro distillation by a GC analysis. However, a more accurate method is based upon the recommendation of Worman and Green [19], who analyzed the crude oil neat and then with a series of four normal alkanes as internal standards; from the difference in the two sets of data, it is possible to determine the fraction of the crude oil boiling below the 1000°F cutoff point of the GC method. The internal standard approach has been gaining wide acceptance, and a simulated distillation method along this line is under development as a proposed ASTM standard procedure.

The greatest success of simulated distillation has been in its application to the control of refinery processes (including blending operations) and quality control for gasolines. Gasoline is, after all, the major item of commerce and is therefore of highest interest in product analysis and testing. Yet, as discussed previously, it is the one material that requires extreme care in simulated distillation. In order to avoid distortions from nonlinear detector response and appreciable variations in response factors of the very volatile gasoline components, Hickerson [20] has developed and proposed a modified simulated distillation method designed specifically for gasolines. In this "GC distillation" technique, isopentane and lighter components are resolved discretely, and response factors are applied so that their concentrations can be accurately determined. The results can then be used in making better calculations of the volatility characteristics of the gasolines.

REFERENCES

1. W. L. Nelson, "Petroleum Refinery Engineering," 4th ed., McGraw-Hill, New York, 1969, Chapter 4.
2. F. T. Eggerston, S. Groennings, and J. J. Holst, Anal. Chem. 32: 904-909 (1960).
3. L. E. Green, L. J. Schmauch, and J. C. Worman, Anal. Chem. 36: 1512-1516 (1964).
4. L. E. Green, Fully automatic boiling range distribution of petroleum fractions by gas chromatography, Preprint, Pittsburgh Conference on Analytical Chemistry and Applied Spectroscopy, Cleveland, March 1-5, 1976.
5. W. DeBruine and R. J. Ellison, J. Petr. Inst. 59:146-153 (1973).
6. W. DeBruine and R. J. Ellison, Calculation of ASTM Method D86-67 distillation and Reid vapor pressure of a gasoline from the gas-liquid chromatographic true boiling point, "Calculation of Physical Properties of Petroleum Products from Gas Chromatographic Analysis," ASTM Special Technical Publication 577, Am. Soc. Test. Mats., Philadelphia, 1975, pp. 9-19.
7. Charles R. Martel, "Investigation of ASTM D2887 Test Method for Use with Aircraft Turbine Engine Fuels," Technical Report AFAPL-TR-74-122, Air Force Aeropropulsion Laboratory, Wright-Patterson Air Force Base, Ohio, March, 1975.
8. W. L. Bird and J. L. Kimball, Application of gas chromatographic distillation to motor gasoline blending, "Calculation of Physical Properties of Petroleum Products from Gas Chromatographic Analyses," ASTM Special Technical Publication 577, Am. Soc. Test. Mats., Philadelphia, 1975, pp. 51-64.

9. J. L. Kimball and E. A. McCracken, Quality control of petroleum products with GC distillation, Report to ASTM Committee D-2, R&D Division IV, Section L, Montreal, June 1975.
10. C. M. Drew, Detectors, in "Principles and Practice of Gas Chromatography" (R. L. Pecsok, ed.), Wiley, New York, 1959, p. 126.
11. N. G. McTaggart, P. Glaysher, and A. F. Harding, Correlation of simulated true boiling point curves by gas liquid chromatography and 15-plate distillation data (ASTM Method D2892-73) on crude oils, "Calculation of Physical Properties of Petroleum Products from Gas Chromatographic Analysis," ASTM Special Technical Publication 577, Am. Soc. Test. Mats., Philadelphia, 1975, pp. 81-94.
12. M. P. T. Bradley and C. E. Kennard, Volatility control by ASTM Method D2887-73, a feasibility study, ASTM Special Technical Publication 577, Am. Soc. Test. Mats., Philadelphia, 1975, pp. 95-107.
13. D. C. Ford, W. H. Miller, R. C. Thren, and R. Wertzler, Correlation of ASTM Method D2887-73 boiling range distribution data with ASTM Method D86-67 distillation data, ASTM Special Technical Publication 577, Am. Soc. Test. Mats., Philadelphia, 1975, pp. 20-30.
14. R. L. Geddes, Ind. Eng. Chem. 33:795-801 (1941).
15. B. W. Jackson, R. W. Judges, and J. L. Powell, J. Chrom. Sci. 14: 49-51 (1976).
16. L. E. Philyaw, A. E. Krc, and M. J. O'Neal, Anal. Chem. 43:787-789 (1971).
17. F. Trusell and J. L. Mikulas, Another approach to boiling range distribution of crude oils, Preprint, Division of Petroleum Chemistry, American Chemical Society 18:517-521 (1973).
18. N. G. McTaggart and P. Glaysher, Simulated true boiling points of total crude oils by gas chromatography, "Recent Analytical Developments in the Petroleum Industry" (D. R. Hodges, ed.), Wiley, New York, 1974, pp. 79-95.
19. J. C. Worman and L. E. Green, Anal. Chem. 37:1620-1621 (1965).
20. J. F. Hickerson, Proposed method for boiling range distribution of gasoline and gasoline components by gas chromatography, Report to ASTM Committee D-2, R&D Division IV, Section L, Montreal, June 1975.

Chapter 5

GAS CHROMATOGRAPHY OF MIDDLE AND HEAVY DISTILLATES

Fred C. Trusell
Marathon Oil Company
Denver Research Center
Littleton, Colorado

I. INTRODUCTION

The complexity of the middle and heavy distillate fractions of crude oil is at once challenging and frustrating to the petroleum chemist. Considerable information concerning its origin and the potential difficulties in refining the oil can be obtained through a knowledge of the composition of these fractions. High-resolution gas chromatographic analyses, such as those discussed in Chap. 3 are, at the current state-of-the-art, not available for these boiling ranges. Nonetheless, it is possible to glean a wealth of information from low-resolution gas chromatograms of these fractions. The approximate boiling range, the presence or absence of specific compounds or classes of compounds, or an indication of a possible impurity type or adulteration can be observed from the chromatogram. These traces are often aptly called fingerprint chromatograms. Chap. 7 describes a special and particularly important application of this form of chromatography for the identification of the sources of oil spills.

One approach is to employ appropriate preseparation techniques to simplify the mixture to be analyzed or to concentrate the components of interest. Chromatography of these concentrates will ordinarily result in a more complete resolution of these components. The informational gain is obviously offset by the additional complexity of the procedure. In the examples discussed in this chapter, there may be some emphasis on the direct analyses for the components of interest in the original matrix; but preseparation steps are described if they appear to offer the most valid approach to the desired solution.

II. FINGERPRINT CHROMATOGRAMS

Because of the high endpoint of some of the materials, it is necessary to use short columns to ensure complete or fairly complete elution in a reasonable period of time. One particularly important application in this field is simulated distillation by gas chromatography, as described in Chap. 4. Fingerprint chromatograms are also obtained to determine other properties of the sample. The presence of the homologous series of n-paraffin peaks can be observed quite readily even in a low-resolution chromatogram. The appearance of a binodal distribution may indicate that the sample is a mixture of two or more products. The technique is very effective for determining small amounts of lower molecular weight contaminants, such as residual solvents, in the bulk material.

In this technique, one would generally use an 18-in. to 3-ft-long column packed with a nonpolar substrate, such as OV-101 or SE-30. These short columns are adequate for a rapid and surveylike evaluation of the general characteristics of a hydrocarbon fraction. An extremely versatile column for these applications is the short capillary column. Good results have been reported using a 32-ft by 0.01-in. column [14]. Figure 1 shows a chromatogram of a synthetic mixture of predominantly n-hydrocarbons used to calibrate the retention time of the column. Figure 2 shows a lubricating oil fraction which has been contaminated by a few percent of a low boiling material. Even though this is a simple chromatographic analysis, the chromatogram shows sufficient detail to differentiate between two different lube oil fractions of similar boiling range.

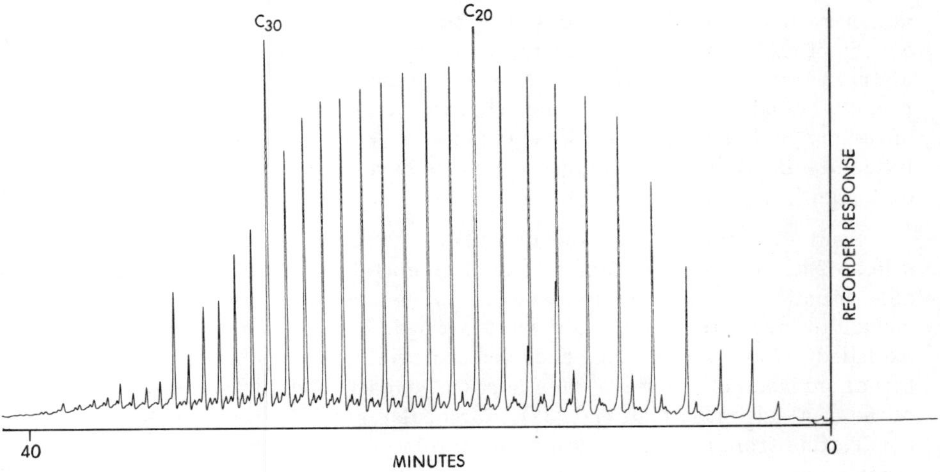

FIG. 1. Mixture of n-paraffins to calibrate retention times; 0.01 in. × 32 ft capillary column.

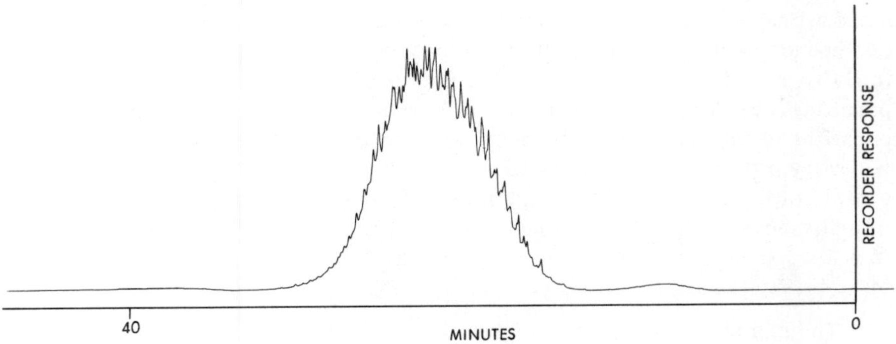

FIG. 2. Fingerprint chromatogram of a contaminated lubricating oil.

III. n-PARAFFINS

The n-paraffins occur in relatively high levels in the distillate cuts of most crude oils, especially those oils from geological formations of fairly recent age. The concentration level of these compounds affects a number of important properties of distillate fuels, such as pour point and viscosity. To obtain a good indication of the level of these compounds in a matrix of hydrocarbons, a preseparation step is generally necessary.

The two common approaches for isolating n-paraffins from other hydrocarbons are urea adduction and adsorption in molecular sieves. The former is generally the fastest; it also offers the most convenient method for recovery of the separated paraffins. Urea crystallizes from solutions in a matrix where the n-paraffins, if they are present, are included within roughly cylindrical pores. Branched paraffins will be included if one end of the molecule has an unbranched chain of at least 14 carbons. The analytical applications of inclusion compounds have been very lucidly reviewed by Baron [1].

Synthetic zeolites, or molecular sieves, are crystalline metal alumina silicates and can be manufactured so as to contain regularly spaced channels of uniform molecular dimensions. Molecules with diameters small enough to invade these channels are adsorbed, but larger molecules are excluded. While some of the excluded material will be adsorbed on the external surfaces of the sieve, this portion typically constitutes less than 1% of the total adsorbed material [32]. Molecular sieve 5A selectively adsorbs n-paraffins from hydrocarbon mixtures; this property constitutes the most common use of molecular sieves in petroleum analysis.

A. Separation of n-Paraffins by Urea Adduction

Marquart et al. have evaluated a method for determining the distribution of n-paraffins in heavy petroleum distillates based on the isolation of these compounds as urea adducts [20]. The hydrocarbon-urea mixture is shaken initially at 55 to 60°C to increase the rate of adduction of the heavier n-paraffins. Shaking is continued at a temperature which depends on the lowest paraffins of interest. At 25°C n-hexadecane is completely recovered, but recovery of lighter n-paraffins is not quantitative. At 0°C n-dodecane is completely recovered, but enough isoparaffins may be adducted to cause interferences. The adduct is removed by filtration, washed, and dissolved in water, from which the liberated paraffins are extracted into n-decane or some other suitable solvent.

The gas chromatographic separation is carried out on a 9-ft by 0.125-in. stainless steel column packed with 2% Carbowax 1000 on Chromosorb W. The column temperature is programmed from 75°C to 250°C at 6°C min^{-1}.

Marquart et al. recommend squalane as the internal standard. If its peak is not resolved from that of n-C_{27}, the contribution of the latter is found by linear interpolation of the n-C_{26} and n-C_{28} peak areas and subtracted from the squalane peak. Recoveries of 8 n-paraffins between C_{16} and C_{30} averaged 100.3%, with the repeatability averaging ±6% of the amount present.

B. Vapor Phase Separation of n-Paraffins with Molecular Sieves

Early methods on the vapor phase separation of n-paraffins with molecular sieves generally involved the use of a silicone rubber column and a molecular sieve column in series with a detector after each column. The reduction in area caused by the passage through the second column is ascribed to the adsorption of the n-paraffins. Results were often erratic. Barrall and Baumann demonstrated that these were caused by trace amounts of water in the gas used to purge the sieves during activation [2]. Reproducible results were achieved by packing a 0.25-in. by 15-ft aluminum column with molecular sieve 5A, 42-60 mesh, activating it at 475°C followed by rapid cooling to room temperature. The helium purge gas is dried by passing through a molecular sieve trap at liquid nitrogen temperature. Sieve columns prepared in this manner retain n-paraffins from C_{10} to C_{20} and normal olefins from C_{11} to C_{18} under the conditions of the analysis.

Subtractive methods frequently lead to the inherently inaccurate process of finding a small difference between two large numbers. Barrall and Baumann found that while samples containing at least 35% n-paraffins could be analyzed without problems, samples containing less require special treatment to achieve acceptable results.

A more recent paper by Mortimer and Luke [22] describes a technique whereby the molecular sieves used to remove the n-paraffins are destroyed with hydrofluoric acid; the paraffins are recovered for direct analysis. The adsorber unit is shown in Figure 3. The tube of borosilicate glass is 15 cm long, 3 mm ID, and 6 mm OD. It is packed, as shown, with 7 cm of 40 to 60 mesh 5A molecular sieves (approximately 400 mg) and 2 cm of 60-80 mesh Celite. Stainless steel gauze and glass wool separators are used as indicated. A brass union fitted with a gas inlet tube and an O ring seal holds the septum against the slotted end of the tube. The adsorber tube is fitted into a metal block drilled to accept a cartridge heater and a thermocouple.

With nitrogen carrier gas flowing at 10 ml min^{-1}, a sample containing approximately 10 mg of n-paraffins is injected far enough into the Celite to be in the heated zone. The block temperature is raised to 300°C over a 20-min interval and held until no more smoke issues from the tube, indicating complete elution of the nonlinear hydrocarbons. After cooling, the sieve is removed and destroyed in a small amount of hydrofluoric acid.

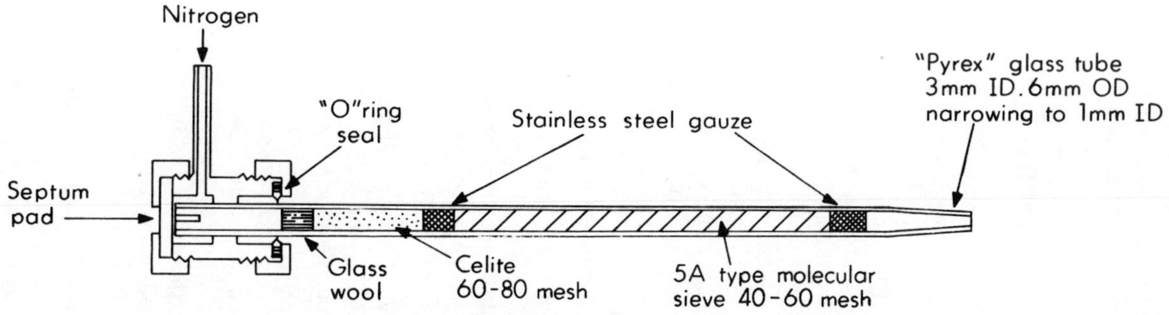

FIG. 2. Adsorber tube and heater for the vapor phase separation of n-paraffins with molecular sieves. (Reprinted with permission from Anal. Chim. Acta.)

The liberated paraffins are extracted into iso-octane and a gas chromatogram of the mixture is obtained using a 0.125-in. by 3-ft silicone gum column starting at 50°C and programmed up at 6.5°C min^{-1}. The method described has been applied to distillates in the C_9 to C_{40} range.

C. Liquid Phase Separation of n-Paraffins with Molecular Sieves

Liquid phase separations of n-paraffins tend to be much more time consuming and involved than the vapor phase procedures described above. The presence of aromatic compounds, particularly highly substituted monoaromatics, inhibits adsorption of the n-paraffin by the sieve. Prior removal of aromatics is, obviously, one solution to the problem. Most workers opt for long reaction times at elevated temperatures. Benzene and iso-octane are the solvents commonly employed, and refluxing times up to 48 hr are frequently recommended.

Washall et al. [38] describe the isolation of n-paraffins from microcrystalline waxes using the solvent decalin. By taking advantage of its high boiling point, the adsorption time is shortened to overnight. The sieve, with its adsorbed paraffins, is separated from the reaction mixture, exhaustively washed, and destroyed with hydrofluoric acid. The liberated paraffins are extracted into benzene. After removal of the solvent, the residue is chromatographed on a 0.125-in. by 2-ft stainless steel column packed with 2% microcrystalline wax bottoms on 100-200 mesh Chromosorb W temperature programmed from 225 to 330°C at 6.5°C min^{-1}.

D. Determination of n-Paraffins Without Preliminary Separation

A single-step analysis for n-paraffins without the necessity of a preliminary separation from the matrix would offer a great saving of time. In waxes and in distillate fractions from highly paraffinic crudes, a fairly good indication of the n-hydrocarbon composition can be obtained by analyzing the sample straight on, e.g., a 50-ft SCOT column, a 200-ft by 0.01-in. capillary column, or a 6-ft to 10-ft-long packed column. The analyses should, in general, be carried out by temperature programming the column in a temperature range depending on the boiling range of the product. Samples with an endpoint below 1000°F should be completely eluted in this analysis; for higher boiling samples, an internal standard can be used to determine the fraction of material eluted. To estimate the n-hydrocarbon composition of the sample, an auxiliary baseline is drawn below each n-hydrocarbon peak by connecting the lowest points of the adjacent valleys to each peak. The area above this line is then assumed to be proportional to the amount of n-hydrocarbon present. This approach is obviously inapplicable to samples with more than a few percent of non-n-paraffin constituents. A potentially more precise but more complex method has been proposed by Folmer [7] and extended by Stuckey [31]. The instrument consists of essentially two parallel gas chromatographs adjusted to give identical operation. The detectors are connected so that the output of one is subtracted from that of the other, yielding a differential signal.

Analyses are carried out on a 0.187-in. by 6-ft aluminum column packed with 20% SE-52 on Anakrom ABS operated isothermally at 170°C. Both the sorber, packed with 30-60 mesh 5A molecular sieve, and the dummy, packed with 30-60 mesh untreated Celite, are maintained at 320°C.

The retention times, peak amplitudes, and peak shapes of the two parts of the sample emerging from the sorber and the dummy must be exactly matched to obtain a null signal from a nonsorbed substance. Adjusting flow rates and amplifier response compensates adequately for the first two. Matching of peak shapes, however, is very difficult to achieve in practice; and the analyst must be prepared to settle for some minimum, rather than a true null signal. With patience, a chromatogram, such as that shown in Figure 4 showing the n-paraffins of a typical kerosene, can be obtained.

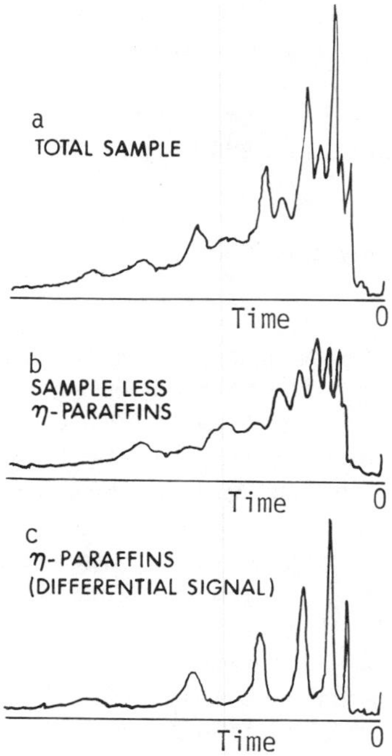

FIG. 4. Differential chromatogram of n-paraffins in kerosine.

Stuckey modified Folmer's method by adapting it for use with thermal conductivity detectors and by programming the column temperature. By doing this he was able to determine the distribution of n-paraffins in gas oils as well as in kerosenes. The advantages of speed and simplicity remain, but so do the problems of the initial adjustment of the equipment.

IV. BIOLOGICAL MARKERS

While there are still a few enthusiastic supporters of an inorganic origin for petroleum [25], the majority of petroleum chemists assent to the theory that crude oil is derived from biological sources. Among the evidence for a biogenic origin is the presence of compounds, such as the isoprenoid hydrocarbons, the steranes, and the pentacyclic triterpanes, whose precursors are commonly found in living matter.

A. Isoprenoid Hydrocarbons

At least some of the isoprenoid hydrocarbons are postulated to be derived from the phytol moiety of chlorophyll, thus offering strong evidence for the biogenic origin of petroleum. They have attracted the attention of petroleum chemists since the C_{19} species, pristane, was first isolated by Bendoraitis et al. [3]. In the chromatograms of the straight-run distillates of many crude oils, especially those from sediments of recent origin, the presence of these isoprenoids can be clearly observed as distinct peaks of an homologous series. The usually observed isoprenoids contain 13-21 C-atoms per molecule. There are two prominent members of this series; the C_{20} homolog, phytane, which elutes close to the n-C_{18} paraffin peak and the C_{19} homolog, pristane, which has a retention time somewhat offset to that of n-heptadecane. On an Apiezon column, these isoprenoids elute before the n-paraffins; on a nonpolar silicone column they elute after the peaks for the saturated hydrocarbons. The C_{21} isoprenoid is often masked by the C_{19} paraffin peak, and a high-resolution column is generally required to separate these two compounds. The C_{18} isoprenoid elutes between the C_{18} and C_{17} n-hydrocarbon peaks. For reasons not quite established, one finds none or very little of the C_{17} isoprenoid in crude oils. The C_{16} isoprenoid elutes close to the n-C_{15} paraffin peak and the C_{15} isoprenoid, farnesane, elutes before the n-C_{14} paraffin. The C_{13} and C_{14} isoprenoids elute between the n-C_{12} and n-C_{13} paraffin peaks. Figure 5 shows the isoprenoid and n-paraffin peaks from an Iraqi oil shale. In that particular chromatogram, which was obtained on a polar column, the isoprenoids are eluted much earlier than the n-paraffins as compared to a chromatogram obtained on a nonpolar column.

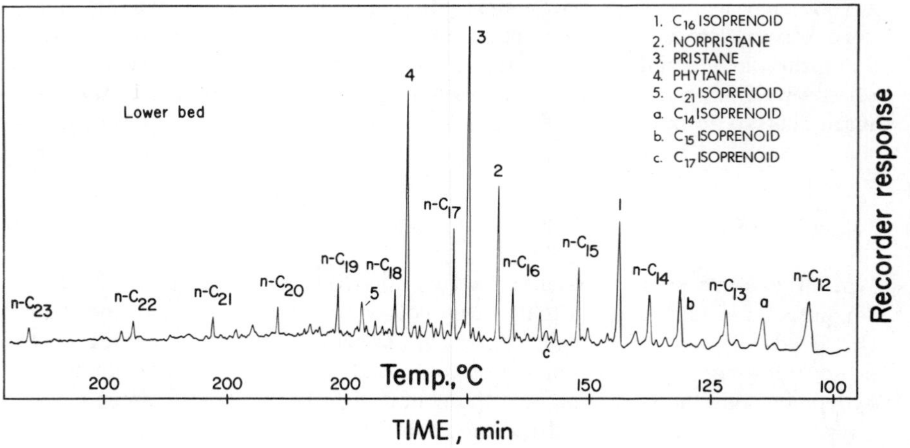

FIG. 5. Isoprenoid and other saturated hydrocarbons of Iraqi oil shale, Brazil.

Although in many cases these chromatograms are adequate to yield the required information one may obtain a better quantiation of these compounds by the use of a preseparation step. This would be the case when the isoprenoid level in the crude oil is fairly low.

The methods proposed for their analysis are, for the most part, basically similar. The isoprenoid hydrocarbons are concentrated by removing the aromatic portion of the sample on silica gel, and a chromatogram is obtained on the whole saturate fraction [13]. Further concentration is achieved by removing the n-paraffins by adduction with urea [12]. Open tubular columns coated with polyphenylether [12], Apiezon L [19], or squalane [13] are commonly employed, though in some studies a packed column was used [12, 34, 36]. In the author's laboratory, a 200-ft by 0.02-in. nickel capillary coated with Apiezon L is used for this analysis. This column has excellent resolution and shows sufficient temperature stability to perform the analysis in a reasonable time.

Gibert et al. determined the distribution of isoprenoid hydrocarbons in the oil shale of the Iraqi Formation of Brazil [12]. The shale is crushed and extracted with benzene-methanol (3:1) in a Soxhlet apparatus. After evaporating the solvent, the residue is fractionated on a silica gel column into pentane (or hexane), benzene and methanol eluates. The solvent is evaporated from the earliest fractions, and the recovered saturates are then separated on a 0.02-in. by 475-ft stainless steel capillary coated with Polysev, a 7-ring m-polyphenylether. The column temperature is programmed from 25 to 200°C at 1°C min^{-1}. A representative chromatogram is shown in Figure 5. A more complex approach is necessary if the isoprenoids are only present in trace quantities. MacLeod proposed a method for identifying hydrocarbons at trace levels in sediments [19]. The sample is extracted with 3:1 benzene-methanol in an ultrasonicator for 15 min. After the residue is allowed to settle, the supernatant liquid is decanted, washed with triply distilled water, and dried over benzene washed anhydrous sodium sulfate. The dried solution is reduced in volume by evaporation and separated by preparative thin layer chromatography. An 8-in. by 8-in. silica gel plate is prepared to a thickness of 0.25 mm from 5% silver nitrate and prewashed with benzene. The sample is streaked on the dried plate, developed with hexane, and the fraction corresponding to R_f 0.9 to 1.0 is extracted with hexane and evaporated almost to dryness in a glass tube fashioned from a melting point capillary. The resulting saturate sample is warmed, taken up into a microsyringe, and separated on a 0.02-in. by 100-ft SCOT column with decolorized Apiezon L as the liquid phase. The column temperature is programmed from 100 to 240°C at 2°C min^{-1}. The chromatograph was coupled through a Watson-Biemann separator to a Hitachi RMU-6E mass spectrometer to identify the peaks as they are eluted.

B. Steranes and Triterpanes

The steranes and pentacyclic triterpanes are structurally more complex and, because of their higher boiling range, less readily accessible. These compounds have, nevertheless, elicited no less interest from petroleum chemists. The presumed precursors, steroids and pentacyclic triterpenes, are widely distributed in living matter; and a study of the reduced species may allow a better understanding of the biomass from which kerogen and petroleum deposits are formed.

Pym et al. fingerprinted crude oils on the basis of the relative concentrations of 8 triterpanes isolated from the 300 to 540°C (572 to 1004°F) boiling range fractions [26]. Earlier work had shown that these compounds can be separated from the paraffins by elution chromatography over alumina [15]. A long column, 370 cm by 0.9 cm, and an eluent flow rate of 0.5 ml min^{-1} is required. Even then the separation is not complete, and minor amounts of n-alkanes persist in the pentacyclane fractions. A 1-g sample of distillate yields 10 to 20 mg of triterpane concentrate. The concentrate is taken up in 100 μl of cyclohexane and separated on a 0.25-mm by 50-m stainless steel column coated with Dexsil 300 maintained at 273°C.

Gallegos has described an analysis for steranes, terpanes, and branched paraffins in Green River shale [8]. The preliminary sample preparation is much less involved than that just described, more reliance being placed on the chromatographic column to achieve the desired separations. The shale sample is crushed to 2 to 3-mm particle size and extracted at reflux temperature with a 1:1 benzene-methanol mixture. A saturate fraction is isolated by making 4 passes through a fully activated alumina column with cyclohexane as the eluent. The analytical separation is carried out on a 0.02-in. by 150-ft capillary, coated with OV-17, programmed from 100-320°C at 4°C min^{-1}. A typical chromatogram is shown in Figure 6. The chromatograph is coupled directly to an AEI MS-9 mass spectrometer. The resulting spectra are used to establish the identity of components of the shale extract.

Gelpi et al. have investigated methods for concentrating steranes and triterpanes from Green River oil shale [10]. The shale is pulverized and the extractable portion is removed with benzene-methanol. The saturate fraction is chromatographically separated on alumina, and the normal paraffins are removed with molecular sieves. The remaining branched-cyclic fraction is treated with thiourea. In this process, the acyclic isoprenoid hydrocarbons are adducted almost completely; some cyclic compounds are also found in the crystalline matrix. For the most part, however, the steranes and triterpanes are concentrated in the supernatant portion.

The next step is to separate these two classes of compounds. Fractionation with long alumina columns alone is not adequate to obtain a clear cut separation. The best method found is by gel filtration chromatography with Sephadex LH-20. This methylated cross-linked dextran is useful in separating branched paraffins from cycloparaffins [33] and Gelpi et al. convincingly demonstrate its applicability to these mixtures. In a subsequent paper, this work was expanded to isolate milligram-size samples of steranes and triterpanes [10] from Green River shale by the use of automatic preparative gas liquid chromatography.

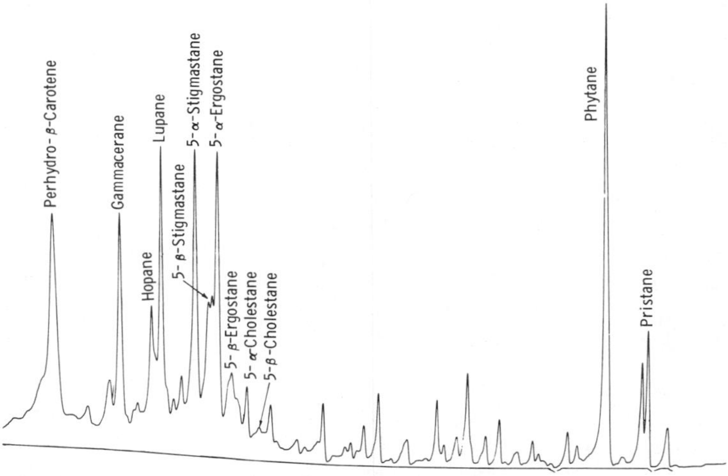

FIG. 6. FID chromatograph trace of the branched cyclic hydrocarbon fraction from Green River shale. (Reprinted with permission from Anal. Chem.)

V. SULFUR AND NITROGEN COMPOUNDS

A. Sulfur Compounds

Sulfur compounds in crude oils and in distillate fractions are of interest from a number of standpoints. As their concentration increases, the range of materials for the construction of processing units becomes more restricted and more expensive. A number of government agencies are now restricting the sulfur contents of distillate products, and effective processes for the removal of sulfur will be of increasing interest and necessity. Sulfur also tends to affect refining catalysts adversely, necessitating removal of these compounds prior to some processing steps. In all of these cases, gas chromatography can be used to determine the effectiveness of the pretreatment step for different types of sulfur compound.

Castex et al. have described an analytical method which affords a great deal of information about the types of sulfur compounds present and about their distribution in an oil [5]. The sample is distilled at 10 torr in an Oldershaw column to yield 200 to 250°C, 250 to 300°C, and 300 to 350°C fractions. Distillation is continued in a molecular still at 0.005 torr to obtain the 350 to 400°C and 400 to 450°C fractions. The aromatic portions of these fractions are isolated by silica gel chromatography using Oudin's procedure [24]. 40 g of silica gel activated at 180°C for 4 hr is used for 500 mg of oil. The column is saturated with n-heptane, the sample is introduced, and the saturates are eluted with a volume of n-heptane equal to 2.5 times the volume required to wet the column. Less than 1% of the aromatics are eluted with the saturates. The aromatics are eluted with benzene.

The aromatics, stripped of the benzene, are further fractionated on an alumina column 2 m long and 6 mm ID. The alumina is moistened with n-heptane, a 100-mg sample is placed on the column, and the solvents are introduced by means of a piston pump. The solvents, flow rates, and fractions eluted are given in Table 1. The optical transmittance of the eluate at

TABLE 1. Conditions for Separation of Aromatic Fractions on Alumina

Solvent (% diethylether in n-heptane)	Flow rate (ml min^{-1})	Compounds eluted
0	4.9	Monoaromatics
0.8	7	Diaromatics and benzothiophenes
8	7	Triaromatics and dibenzothiophenes
50	7	Polyaromatics and naphthobenzothiophenes

280 nm is monitored and cut points are made on that basis. A typical elution curve is shown in Figure 7. Each fraction is analyzed using a 0.5-mm by 25-m capillary column coated with OV-17 and programmed from 120 to 250°C at 2°C min^{-1}. A fraction of the effluent is sent to a flame ionization detector to produce a chromatogram of the total sample, and the balance is directed through a flame photometric detector set at 394 nm to produce a sulfur-specific chromatogram. From such chromatograms the distribution of benzothiophenes, dibenzothiophenes, and naphthobenzothiophenes can be found. After obtaining a high-resolution (4000) mass spectrum for each fraction from the alumina separation, the carbon number distribution for each class of sulfur compound can be plotted within each boiling range. This paper includes the results of the examination of two rock extracts and four crude oils of various ages [5].

Drushel has proposed a method for determining sulfur in distillate cuts which does not require any preliminary separations, uses packed columns for the gas chromatographic separation, and which features a microcoulometer for a sulfur-specific detector [6]. Despite its procedural simplicity and speed, it does yield a great deal of information concerning the types of sulfur compounds present. The chromatogram in Figure 8 is typical of a vacuum gas oil. The chromatographic separation is carried out on a 0.25-in. by 10-ft column of borosilicate glass packed with 10% Carbowax 20M on Chromosorb W. The column temperature is programmed from ambient to 300°C.

A Dohrmann microcoulometer equipped with a T-300-P titration cell serves as the sulfur-specific detector. A combustion tube is connected to the outlet of the GC column by a glass transfer line heated to 350°C. The inlet area is maintained at 550°C and the combustion zone, through which

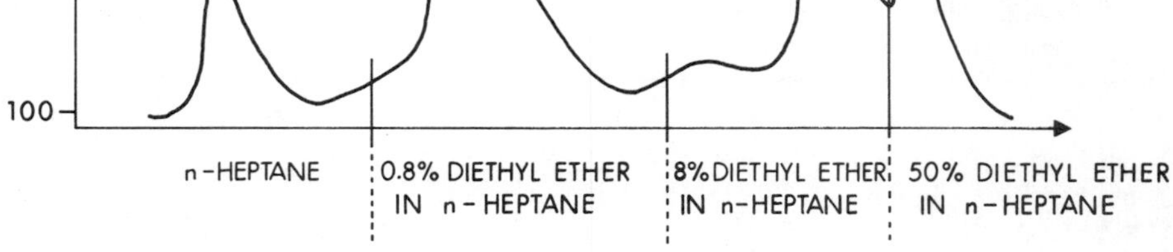

FIG. 7. Elution curve for sulfur compounds from activated alumina.

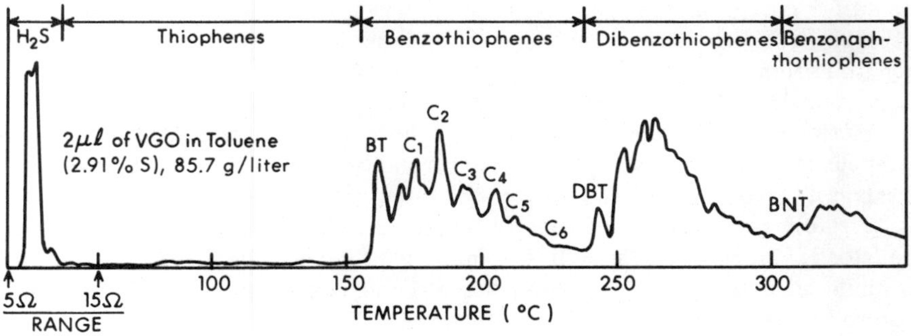

FIG. 8. Sulfur compounds in a vacuum gas oil. Reprinted with permission from Anal. Chem. 41:569 (1969).

oxygen flows, is maintained at 850°C to convert all forms of sulfur quantitatively to SO_2. This SO_2 is absorbed in the titration cell and oxidized to sulfate by electrogenerated iodine.

The gas chromatographic investigation of sulfur compounds in high boiling fractions requires the use of elevated column temperatures. This, in turn, leads to column bleed and column degradation. For this reason George et al. investigated the use of various inorganic materials as column substrates [11]. They report that 15% lithium chloride coated on acid washed Chromosorb W separates sulfur compounds. Unlike some inorganic materials, this column requires temperatures no higher than required for conventional column packings containing organic liquids, thus minimizing errors due to pyrolysis.

The column material is fired at 700°C for 30 min before being packed into 1/4-in. by 6-ft glass tubes. The columns are operated isothermally, and the flame ionization detector oven is kept at a 25°C higher temperature than the column to prevent condensation of the eluted compounds. The results of the investigation are reported in 15 tables of Kovat's indices for a large number of representative standard compounds. An interesting observation in this study was that if the columns are not in use for several hours with the carrier gas (nitrogen, at 53 ml min^{-1}) flowing, the retention time of these compounds are observed to increase. This is particularly noticeable at higher temperatures. The columns can be regenerated by flushing with oxygen at 400°C for at least 1 hr.

B. Nitrogen Compounds

The determination of the distribution of nitrogen compounds in crude oils and distillate products has received much less attention than similar efforts for sulfur compounds. Crude oils typically contain less than 0.3% total nitrogen. The presence of nitrogen-containing compounds does not impose any limitations on the selection of materials of construction for processing units, nor does it appear that for environmental purposes, governmental restrictions will be imposed on their levels in distillate products. Nitrogen compounds do adversely affect the performance of reforming and cracking catalysts. In both cases, however, the effect is temporary; and normal activity returns within a short time of the "change back" from a high nitrogen feedstock. More attention will undoubtedly be focused on nitrogen compounds as shale oil production increases, since the nitrogen contents of shale oils tend to be significantly higher than those of crude oils.

A gas chromatographic method for the analysis of nitrogen compounds in distillate fractions has been described by Martin [21]. The system employs a microcoulometer as a nitrogen-specific detector, thereby eliminating the need for chemical pretreatment to separate the nitrogen compounds from the hydrocarbon matrix. A block diagram of the apparatus is shown in Figure 9. The 0.25-in. by 20-ft stainless steel column is packed with 9% Bakelite DYLT (12,000 mol. wt. polyethylene) on Chromosorb W which has been precoated with 3% potassium carbonate. Column temperatures of 50 to 190°C are used for naphthas and 120 to 330°C for catalytic cycle oils.

The high molecular weight of the liquid phase contributes to keeping retention times short, thus making it easier to elute high boiling compounds. However, polyethylenes much above 12,000 mol. wt. are difficult to coat on the support and give poor separating efficiencies. The potassium carbonate and the high level of liquid phase serve to minimize any reaction of the nitrogen compounds with the solid support. Polyesters and silicone fluids

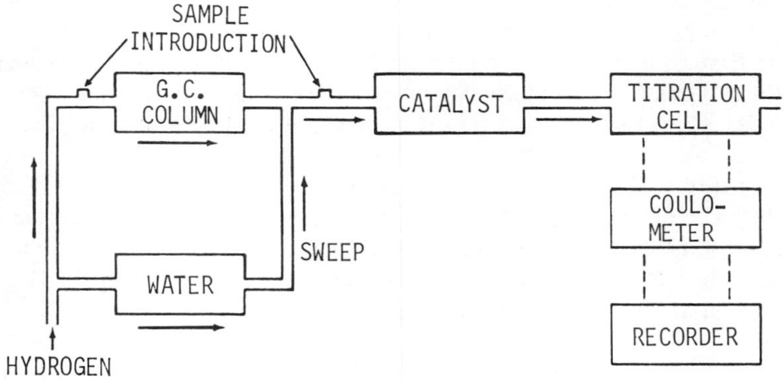

FIG. 9. Block diagram for the determination of nitrogen compounds in distillate fractions. Reprinted with permission from Anal. Chem. 38: 1210 (1966).

cannot be used as liquid phases because at elevated temperatures they decompose in the presence of potassium carbonate and their decomposition products cause an interfering response. Petroleum-derived oils and greases (Apiezon greases) contain nitrogen impurities and cannot be used. Microcrystalline waxes give excellent separations but catalyze the polymerization of pyrroles and indoles. Even though the polyethylene begins to decompose at 270°C, it is usable to 300°C since the decomposition products neither cause a detector response nor deactivate the hydrogenation catalyst.

The hydrogenation catalyst is placed in a 3-in. by 1/4-in. borosilicate tube in an oven maintained at 440 ± 10°C and mounted as shown in Figure 10. The catalyst is prepared by precipitating hydrous nickel oxide with a slurry of magnesium oxide, collecting and drying the precipitate, and reducing the nickel oxide with flowing hydrogen at 450°C.

Hydrogen is used both as the carrier gas for the gas chromatographic separation and as the sweep gas in the detection of the nitrogen compounds. That portion used as sweep gas is humidified by bubbling through water. This procedure minimizes the deposition of coke on the catalyst and the adsorption of ammonia on the coke. Without humidified hydrogen, ammonia adsorption increases as coke accumulates; and the peaks soon tail too much for accurate measurement. The catalyst remains active for at least 100 samples. Organic nitrogen is quantitatively converted to ammonia in the catalyst tube. Acidic compounds such as hydrogen sulfide or hydrogen chloride are irreversibly adsorbed by the magnesium oxide as long as the catalyst temperature is kept below 480°C. The ammonia is swept into the coulometric titration cell where it is titrated with electrogenerated hydrogen ions. A chromatogram of nitrogen compounds in a catalytic cycle oil is shown in Figure 11.

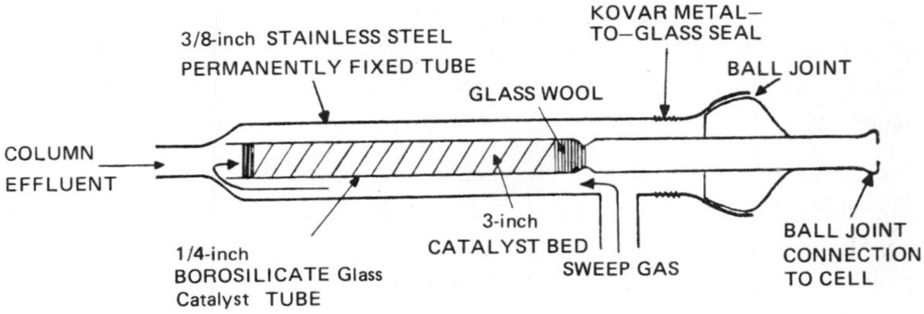

FIG. 10. Hydrogenation tube for the quantitative conversion of organic nitrogen to ammonia. Reprinted with permission from Anal. Chem. 38:1210 (1966).

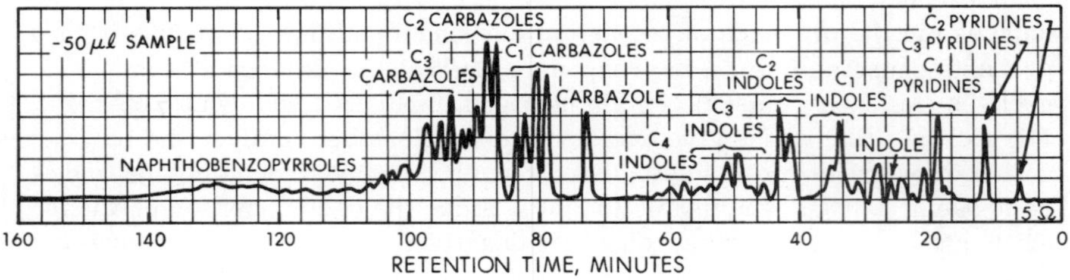

FIG. 11. Chromatogram of nitrogen compounds in a catalytic cycle oil. Reprinted with permission from Anal. Chem. 38:1213 (1966).

VI. HYDROCARBON TYPE ANALYSES

Knowledge of the distribution of saturates, olefins, and aromatics is of importance in describing the composition of gasolines and naphthas. Historically, this determination has been carried out by the fluorescent indicator analysis, ASTM D 1319-70. Work in several laboratories has been directed toward a chromatographic method which would yield answers similar to the FIA but with greater speed and, perhaps, greater accuracy.

Soulages and Brieva have described the apparatus shown schematically in Figure 12 [27]. The sample is split in three fractions which are introduced into the three parallel-placed columns. In the first column a mercuric perchlorate-perchloric acid mixture (MP-PA) is used to remove all olefins and aromatics. The third column contains a mercuric sulfate-sulfuric acid mixture (MS-SA) which will retain all olefins with three or more C atoms. The second column is a standard gas chromatographic column. The carrier gas is humidified by passing it through a 50-ml U tube packed with water-impregnated Chromosorb P. At 20 ± 2°C the packing imparts a water vapor pressure of 10 to 14 torr to the carrier gas, which is approximately the vapor pressure of water from the two absorber tubes in the chromatographic oven. Carrier gas humidification reduces variabilities in performance of the two absorber columns because of changes in the acid concentrations.

The amount of sample entering each branch of the analysis accessory is proportional to the flow through that branch. This is controlled by the length of the capillary restrictors, B_{1-3}. The retention times are further altered by the delayers in branches 2 and 3.

The flows from the three branches are combined at A_2 for a common exit. The different hydrocarbon types reach the detector in the following order:

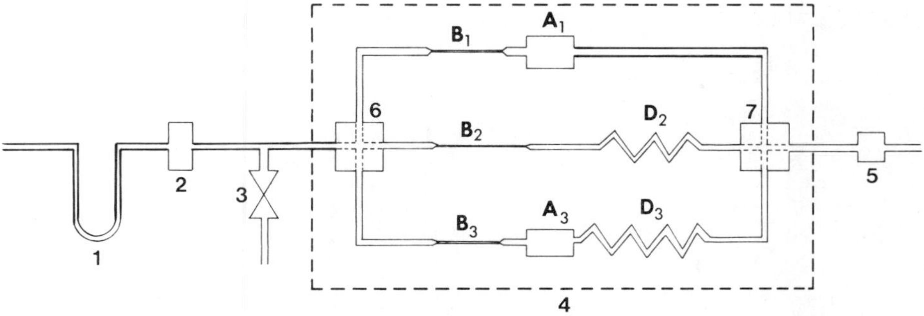

FIG. 12. An apparatus for determining saturates, olefins, and aromatics in gasolines. Key to figure is as follows: 1, carrier gas humidifier; 2, injector; 3, splitter; 4, column oven; 5, flame ionization detector; 6, 7, 4-way connectors; A_1, MP-PA absorber; A_3, MS-SA absorber; B_1, B_2, B_3, capillary restrictors; D_2, D_3 delayers.

1. Saturates from Branch 1.
2. Total sample from Branch 2.
3. Saturates and aromatics from Branch 3.

Typical chromatograms are shown in Figure 13.

The fraction of sample entering each branch can be determined by injecting a pure saturated hydrocarbon, such as n-heptane, which is not retained by any of the absorbers. The fraction of the total area contributed by each branch becomes the calibration factor for that branch.

The FID response is proportional to the carbon atom number, so the results are in weight percent. As it is customary to report these results in volume percent it is necessary to use an assumed, or experimentally determined, density for each hydrocarbon type. If the samples to be analyzed have a limited range of compositional variation, it is feasible to determine the required densities once experimentally and to use them in the above calculations. However, if the composition of the sample being analyzed is greatly different from the standard used to determine the densities, intolerable error may result. Typical results obtained by this method and those obtained by FIA are given in Table 2.

Stavinoha described a somewhat different approach for the isolation and determination of aromatics in gasoline both with [28] and without [29] an internal standard and for the determination of saturates, olefins, and aromatics in gasoline [29]. The apparatus shown in Figure 14 permits the determination of the three hydrocarbon types as well as the detailed analysis of the aromatics described in Reference 30.

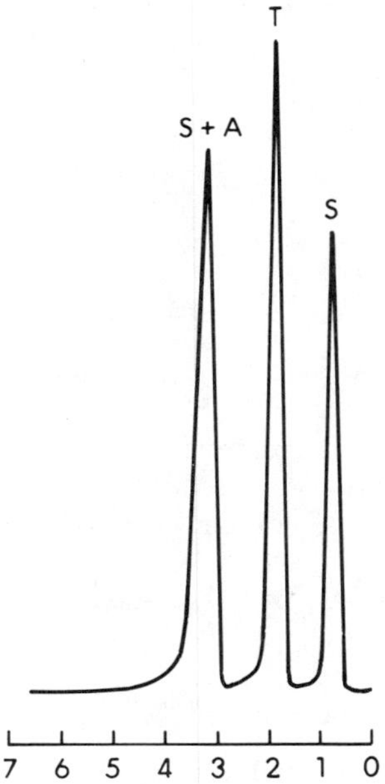

FIG. 13. Chromatogram from the determination of saturates, olefins, and aromatics in a gasoline. Reprinted with permission from the Journal of Chromatographic Science.

TABLE 2. Comparison of Hydrocarbon-Type Analyses by FIA and by GC [27]

Sample		Wt. %	
		FIA	GC
Synthetic gasoline 1	Saturates	54.4	55.4
	Olefins	26.7	26.0
	Aromatics	18.9	18.6
Commercial gasoline A	Saturates	36.1	34.7
	Olefins	44.1	43.9
	Aromatics	19.8	21.4
Commercial gasoline B	Saturates	52.3	50.1
	Olefins	23.2	26.0
	Aromatics	24.5	23.9

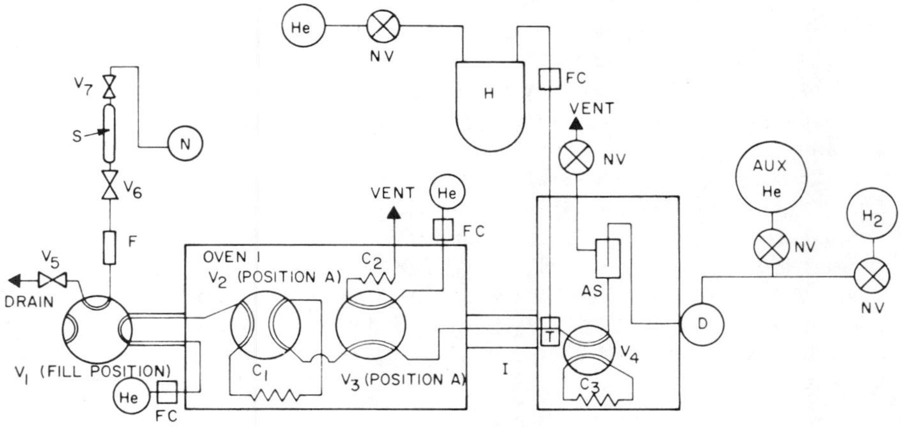

FIG. 14. An apparatus for determining saturates, olefins, and aromatics in a gasoline.

Oven 1.	Isothermal CEF column oven.	C_1.	CEF column.
Oven 2.	Absorber column oven.	C_2.	SE-30 column.
V_1.	Heated sampling valve in fill position; turn 90° CW to inject position.	C_3.	MP-PA absorber column.
V_2.	Backflush valve in position A; turn 90° CW to position B for backflush.	N.	Nitrogen.
V_3.	Column switching valve in position A; turn 90° CW to position B.	S.	High pressure cylinder containing sample.
V_4.	Dual column valve in position A; turn 90° CW to position B.	He.	Helium carrier gas.
V_5, V_6, V_7.	Shut-off valves.	D.	Flame ionization detector.
F.	7 microinch in-line sample filter.	I.	Heated interface, same temperature as oven 1.
		FC.	Flow controller.
		Aux He.	Auxiliary helium.
		H_2.	Hydrogen.
		NV.	Needle valve control.
		AS.	Annular splitter, removed for no split.
		H.	Humidifier vessel.
		T.	Stainless steel (316) cross-pattern needle valve.

(Reprinted with permission from the Journal of Chromatographic Science.)

The sample is introduced by means of a high-temperature, high-pressure liquid sampling valve at 175°C and 450 psig of nitrogen. The CEF column, 0.25-in. by 3-ft, is packed with 35% N, N-bis(2-cyanoethyl) formamide on 60-80 mesh Chromosorb P, AW-DMSC, and is maintained at 120°C. The absorber tube, 0.25-in. by 1-ft of 316 stainless steel, is pakced with mercuric perchlorate-perchloric acid and maintained at 60°C. Two helium streams, one dry and one humidified, are used. The saturates and olefins are eluted from Column C_1, while the aromatics are being retarded. After 5 min Valve V_4 is rotated 90 degrees clockwise and the aromatics are backflushed from Column C_1. 5 min is adequate for n-tetradecane to pass through the column but not long enough for the elution of benzene. For this pass, Valve V_4 is positioned to bypass the MP-PA absorber, C_3. A second pass is then made with Valve V_4 rotated 90 degrees clockwise so the sample flows through the MP-PA absorber where the aromatics and olefins are removed.

The equipment is calibrated using a synthetic standard mixture, the composition of which is given in Reference 29. The peak areas are obtained by electronic integration, and the percent of each hydrocarbon type is calculated as follows:

$$\frac{A_u}{A_s} C_s = C_u$$

where

A_s = aromatic area of standard

A_u = aromatic area of unknown

C_s = aromatic concentration (vol %) of standard

C_u = aromatic concent tion (vol %) of unknown

and

$$(100 - C_u) \frac{A_{sat}}{A_{sat} + A_{ole}} = C_{sat}$$

$$(100 - C_u) \frac{A_{ole}}{A_{sat} + A_{ole}} = C_{ole}$$

where

C_u = aromatic concentration, vol %

A_{ole} = area assigned to olefins

A_{sat} = area assigned to saturates

C_{sat} = saturate concentration, vol %

C_{ole} = olefin concentration, vol %

These calculations assume that detector response is linear with volumetric concentration and, therefore, independent of hydrocarbon type or density. This, of course, is not the case. The density range of olefins and aromatics is higher than that for saturates. Relative to the amount present this method shows saturates lower by 7%, aromatics higher by 7%, and olefins higher by as much as 40% than FIA results on the same sample. Typical values are given in Table 3.

Since the method is calibrated with an external standard, the greater the difference in composition between the standard and the sample, the greater the possibility of error in the results. Again, with the FID responding to carbon weight and the customary results being expressed in terms of volume, the same problems with assumed densities are encountered as described earlier.

A somewhat analogous method as described above has been reported for the analysis of hydrocarbon mixtures such as straight-run naphthas, catalytic reformates, and hydroprocessed streams with final boiling points of up to 200°C [4, 18]. This method, which is used to determine the levels of paraffins, naphthenes, and aromatics in the sample is based on the use of a polar, a nonpolar, and a Molecular Sieve 13X column connected to each

TABLE 3. Comparison of Hydrocarbon-Type Analyses by FIA and by GC [29]

Sample		Wt. %	
		FIA	GC
Calibration standard	Saturates	62.7	59.9
	Olefins	10.9	10.1
	Aromatics	26.4	30.0
Gasoline 7	Saturates	73.0	70.2
	Olefins	6.2	10.0
	Aromatics	20.8	19.8
Gasoline 9	Saturates	66.5	60.0
	Olefins	6.6	11.5
	Aromatics	26.9	28.5

other by 8-way switching valves [18]. The polar column described by Boer and Van Arkel was 3 m long, packed with tris-(cyanoethyl) nitromethane on 60-80 mesh Chromosorb P. There are now also available other comparable stationary phases showing extremely high paraffin aromatic selectivity with a relatively low paraffin-naphthene selectivity. OV-275 has also been used but with questionable results. The nonpolar column can be a 4 m long column packed with a nonpolar silicone or with UC CW982 [4].

Both these columns are maintained at 120°C. A cold trap is mounted between these two columns outside of the actual column oven. This trap can be chilled to -80°C with solid dry ice and heated up rapidly with an electric heater to about 250°C. The molecular sieve column is located in a separate compartment, the temperature of which can be programmed.

To start the analysis the sample is injected on the polar column. The paraffins and low molecular weight naphthenes are eluted into the molecular sieve column. The latter is maintained at ambient temperatures. Under these conditions, very little migration of the solutes takes place.

At the appropriate time, the valves are switched; and the balance of the sample in the polar column is flushed back into the cold trap maintained at -80°C. The trap is then heated rapidly to 250°C to flash its contents into the nonpolar column where the C_6-C_9 aromatics are separated, eluted in order of increasing molecular weight, and detected by a flame ionization detector. After most of the C_9 aromatic hydrocarbons have been detected, the switching valves are again activated. This point generally corresponds to the elution time of 1,3,5-trimethylbenzene. The portion of the sample still in the nonpolar column is backflushed into the cold trap and reinjected on the polar column. The high boiling naphthenes are now eluted and combined with the saturates already trapped on the molecular sieve column. This portion of the analysis is continued until all naphthenes are eluted while still retaining the C_9 aromatics on the polar column. By reactivating the switching valves at this moment, the aromatics are flushed back into the nonpolar column from where they are eluted and detected.

The final step consists of switching the valves such that the carrier gas stream only goes through the molecular sieve column, which is now rapidly programmed up in temperature. In this process, the paraffins and naphthenes are eluted from the column in order of their molecular weight and detected. For each carbon number, the paraffins and the naphthenes are eluted as two distinctly separated bands. The members of each species are grouped together. The separation between these two bands is quantitative through C_9, fairly good for C_{10}, and still quite useful for C_{11}. A typical chromatogram is shown in Figure 15. A commercial unit specifically designed for this analysis is also available from Packard Instrument Co.

Most of the analyses described in this chapter are performed by gas chromatography. Volatility considerations and lack of specificity with

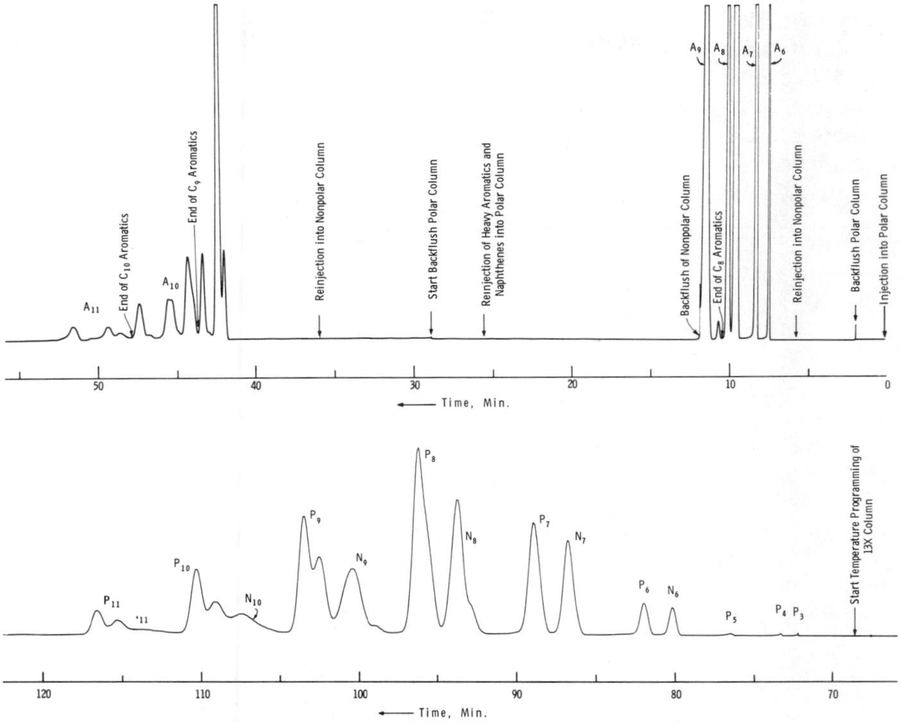

FIG. 15. Chromatogram of paraffins, naphthenes, and aromatics from a PNA analyzer.

increasing molecular weight limit the applicability of this technique to the gasoline and naphtha range. The rapid development of high performance liquid chromatography makes it now possible to carry out group analysis of materials boiling in this and higher ranges with a separation sharpness not attainable with the classical open column techniques. These techniques are of fairly recent origin and are described in more detail in Chap. 6 of this volume.

VII. CARBON NUMBER DISTRIBUTION

Even though a wealth of information can be obtained from high-resolution chromatograms as described in the previous chapter, the process of carrying out the analysis is time-consuming and expensive. In many cases, a more rapid method would be preferable to obtain the carbon number distribution. Garilli et al. found that the carbon number and hydrocarbon type of saturated compounds in naphthas and other mixtures with endpoints below 185°C can be determined using a gas chromatographic column of molecular sieve 13X [9]. This is equivalent to what has been described in the previous section. Because aromatic hydrocarbons are partially and irreversibly adsorbed by the molecular sieve, they must be determined by a separate method, such as by ASTM D 2267.

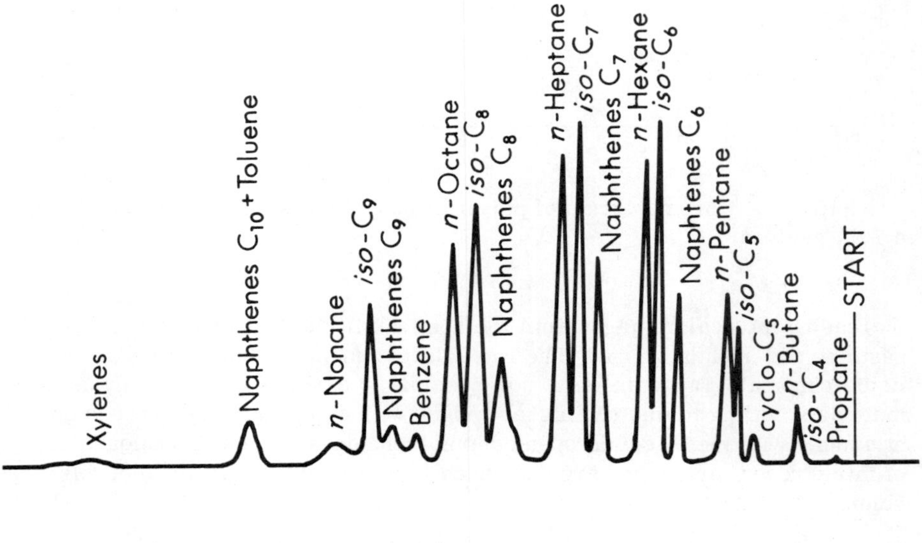

FIG. 16. Distribution of saturated hydrocarbons in a naphtha by carbon number and type.

The molecular sieve, 30-60 mesh, is treated with a 3% aqueous solution of NaOH, washed to neutral, and dried at 100°C for 24 hr. The sieve is heated at 200°C for 24 hr and then screened prior to use.

Virgin naphthas and the feed and liquid effluent products from a platformer have been analyzed by this approach. Comparison of the results with those from the FIA technique is difficult because the FIA values are in terms of volume percent and the GC values are in terms of weight percent. A chromatogram of a naphtha is shown in Figure 16.

VIII. LUBRICATING OILS

Polynuclear aromatic hydrocarbons (PAH) are characteristic products of all combustion processes. They build up in engine lubricating oils in a manner which is dependent on the type of vehicle, its condition, and its combustion parameters. Consequently, the distribution of PAH can be used to fingerprint lubricating oils. Lee et al. [17] have proposed a characterization based on the capillary column chromatography of PAH concentrates using both an FID and a nitrogen selective thermionic detector [16].

A sample of the oil in cyclohexane is extracted with nitromethane. The nitromethane portions in which the PAH are concentrated are combined and evaporated to dryness in vacuo. This residue is dissolved in methylene chloride and a 5-μl aliquot is transferred onto a precolumn of 2 mg of a special deactivated packing in a 1-mm ID glass tube [23].

This packing is prepared by treating ordinary diatomaceous earth supports, such as Chromosorb W or Gas-Chrom Q, with 6% Carbowax 20M or the nonpolar fraction of Apiezon L grease (obtained by silica gel chromatography). The mixture is then heated in a tubular furnace overnight in a slow flow of nitrogen and exhaustively extracted with a series of solvents of different polarities and dried prior to use. The glass wool used as plugs to hold the precolumn material is treated in a similar manner. Irreversible adsorption of the sample in the precolumn is thus minimized.

The precolumn with the sample is flushed with helium at room temperature to remove the solvent and then placed in a modified injection port of the gas chromatograph [35]. The modification consists of a device to water-cool the septum to minimize bleeding. The injection block is held at 250°C while the capillary column is held at room temperature. Helium is flowed at 1 ml min^{-1} for 30 min to sweep the sample off the precolumn and onto the first part of the analytical column, a 0.25-mm by 22-m glass capillary coated with SE-52 silicone gum. The column is programmed from 100 to 250°C at 20°C min^{-1}. The column effluent is monitored with both a flame ionization detector and a nitrogen-specific thermionic detector. Lee et al. used the method to compare the oils from three different automobiles and were able to unequivocally distinguish one from another.

REFERENCES

1. M. Baron, in "Physical Methods in Chemical Analysis" Vol. 4 (W. G. Berl, ed.), Academic Press, New York, 1961, p. 223.
2. E. M. Barrall and F. Baumann, J. Gas Chromatog. 2:256 (1964).
3. J. G. Bendoraitis, B. L. Brown, and L. S. Hepner, Anal. Chem. 34:49 (1962).
4. H. Boer and P. V. Arkel, Hydrocarbon Process 51(2):80 (1972).
5. H. Castex, J. Roucache, and R. Boulet, Rev. Inst. Fr. Petrole Ann. Combust. Liquides 29:3 (1974).
6. H. V. Drushel, Anal. Chem. 41:569 (1969).
7. O. F. Folmer, Jr., Anal. Chim. Acta 60:37 (1972).
8. E. J. Gallegos, Anal. Chem. 43:1151 (1971).
9. F. Garilli, L. Fabiani, U. Filia, and V. Cusi, J. Chromatog. 77:3 (1973).
10. E. Gelpi, P. C. Wszolek, E. Yang, and A. L. Burlingame, J. Chromatog. Sci. 9:147 (1971); Anal. Chem. 43:864 (1971).
11. A. E. George, G. T. Smiley, and H. Sawatzky, Can. Mines Br. Res. Rep. 1972, R249.
12. J. M. Gibert, I. M. R. DeAndrade Bruning, D. W. Nooner, and J. Oro, Chem. Geol. 15:209 (1975).
13. K. E. H. Gohring, P. A. Schenck, and E. D. Engelhardt, Nature 215:503 (1967).
14. T. H. Gouw, I. M. Whittemore, and R. E. Jentoft, Anal. Chem. 42:1394 (1970).
15. I. R. Hills and E. V. Whitehead, Nature 209:977 (1966).
16. B. Kolb and J. Bischoff, J. Chromatog. Sci. 12:625 (1974).
17. M. L. Lee, K. D. Bartle, and M. V. Novotny, Anal. Chem. 47:540 (1975).
18. N. G. McTaggart, L. A. Luke, and D. Wood, Proc. Int. Symp. (Dublin) 8:35 (1970) (Publ. 1971).
19. W. D. MacLeod, Jr., J. Gas Chromatog. 6:591 (1968).
20. J. R. Marquart, G. B. Dellows, and E. R. Freitas, Anal. Chem. 40:1633 (1968).
21. R. L. Martin, Anal. Chem. 38:1209 (1966).
22. J. V. Mortimer and L. A. Luke, Anal. Chim. Acta 38:119 (1967).
23. M. Novotny, M. L. Lee, and K. D. Bartle, J. Chromatog. Sci. 12:606 (1974).
24. J. Oudin, Rev. Inst. Fr. Petrole Ann. Combust. Liquides 25:3 (1970).
25. V. B. Porfir'ev, Am. Assoc. Petroleum Geologists Bull. 58:3 (1974).
26. J. G. Pym, J. E. Ray, G. W. Smith, and E. V. Whitehead, Anal. Chem. 47:1617 (1975).
27. N. L. Soulages and A. M. Brieva, J. Chromatog. Sci. 9:492 (1971).
28. L. L. Stavinoha, J. Chromatog. Sci. 11:515 (1973).
29. L. L. Stavinoha, J. Chromatog. Sci. 13:72 (1975).

30. L. L. Stavinoha and F. M. Newman, J. Chromatog. Sci. 10:583 (1972).
31. C. L. Stuckey, Anal. Chim. Acta 60:47 (1972).
32. T. L. Thomas and R. L. Mays, in "Physical Methods of Chemical Analysis" Vol. 4 (W. G. Berl, ed.), Academic Press, New York, 1961, p. 45.
33. P. C. Talarico, E. W. Albaugh, and R. E. Snyder, Anal. Chem. 40:2192 (1968).
35. W. Van Hoeven, P. Haug, A. L. Burlingame, and M. Calvin, Nature 211:1361 (1966).
36. J. F. J. van Rensburg, P. L. Mouton, and V. Pretorius, J. Chromatog. Sci. 10:580 (1972).
37. D. W. Waples, P. Haug, and D. H. Welte, Geochim. Cosmochim. Acta 38:381 (1974).
38. T. A. Washall, S. Blittman, and R. S. Mascieri, J. Chromatog. Sci. 8:663 (1970).

Chapter 6

HYDROCARBON GROUP-TYPE ANALYSIS BY HIGH PERFORMANCE LIQUID CHROMATOGRAPHY

Joseph C. Suatoni
Analytical Technology Department
Gulf Research and Development Company
Harmarville, Pennsylvania

I. INTRODUCTION

In the past, there has been a need in the petroleum industry for rapid hydrocarbon group analysis for evaluating feedstocks and products. The basic types of compounds in petroleum include saturates, olefins, aromatics, and polar compounds. Previous methods [2, 3] were time consuming and expensive to run on a routine basis. High-performance liquid chromatography (HPLC) has changed that. Techniques have been established that are rapid, accurate, and applicable to samples over a variety of boiling ranges. Many can be adapted to onstream analysis by means of a process chromatograph. Methods are available for the group-type analysis of gasolines, naphthas, furnace oils, middle distillates, gas oils, lubricating oils, residuals, and total crude oils [9,10,11,12]. These techniques with suitable modifications can be applied to synthetic hydrocarbons derived from coal liquids, shale oils, and tar sands.

HPLC can also be utilized as a nondestructive preparative technique to provide clean fractions of hydrocarbon classes for further characterization by other analytical techniques. Carcinogenic polynuclear aromatics can be isolated and identified rapidly by HPLC.

II. GASOLINES

An HPLC technique [9] has been developed as an alternative for ASTM D 1319 [1], the FIA (fluorescent indicator adsorption) method, which is the petroleum industry standard procedure for group analysis of gasoline range materials, 28 to 215°C (80 to 520°F). ASTM D 1319 has certain limitations: samples containing C_5 and lighter hydrocarbons must be depentanized or saturate values will be low, elapsed time for one sample is 4 hr, and human errors are introduced during measurement of dye boundaries. The advantages of the HPLC method include: (a) samples are analyzed as received, (b) total analysis time is only 10 min, (c) diolefins (except cyclic diolefins) are included with olefins, (d) results in general are more accurate and precise, and (e) the method can be adapted to onstream analysis for refinery applications.

A. Column and Mobile Phase

The HPLC gasoline separation is accomplished by means of a 10-μ silica column and a very nonpolar mobile phase, Fluorinert FC-78, which is an electronic testing fluid marketed by the 3M Company. FC-78 has an extremely low solvent strength [8] ($\epsilon^o = -0.25$) as compared to n-pentane ($\epsilon^o = 0.00$) [4]. The low refractive index (approximately 1.25) results in excellent sensitivities with hydrocarbons.

B. Calibration

The refractive indices of monoaromatics are relatively constant (± 2%) over the range C_6-C_{12} benzenes. Therefore, a blend of mononuclear aromatics typical of gasolines is used for calibration. Since the refractive indices vary greatly with structure and molecular weight for saturates and olefins, a direct measurement of the saturate and olefin values from area responses is not feasible. However, the ratio of delta refractive indices (differences between olefin and solvent divided by difference between saturate and solvent) was determined to be 1.10. Therefore, the saturate and olefin values can be determined by suitable calculation [9].

C. Procedure

Saturates are eluted first from the column (Table 1) followed by the olefins. The aromatics are eluted as a single peak by utilizing backflushing (Figure 1). The advantages of backflushing (flow reversal) are that all the aromatics are eluted as one sharp peak making the analysis time shorter and quantitations more accurate. The need for flow reversal is illustrated by the chromatograms of a heavy gasoline distillate shown in Figure 2. This HPLC method can be applied to naphthas, light catalytic gasolines, heavy catalytic gasolines, jet fuels, and motor gasolines.

III. PETROLEUM FRACTIONS BP 190 TO 360°C

The HPLC method established for gasoline boiling range samples can be adapted with suitable experimental modification to higher boiling fractions [11], 190 to 360°C (375 to 680°F). These petroleum fractions include virgin and FCC furnace oils, #2 heating oils, and middle distillates. Four basic hydrocarbon group types: saturates, monoolefins, diolefins, and aromatics can be determined in this boiling range.

The HPLC approach offers certain improvements over ASTM D 1319 in addition to extending the boiling range above 315°C (600°F); i.e., (a) total analysis time is 20 min, (b) sharp separations are obtained between olefins and saturates containing large amounts of naphthenes, (c) diolefins can be separated as a group, (d) the method can be adapted to onstream analysis, and (e) it will better analyze furnace oils derived from the FCC (fluid catalytic cracked) process.

A. Column and Mobile Phase

In this boiling range (190 to 360°C), hexane can be substituted for FC-78 as the mobile phase; however, to obtain the necessary separation of saturates

TABLE 1. Gasoline Boiling Range Model Compound Study

Compound	Retention volume (ml)
Saturates	
pentane	4.52
hexane	4.68
isooctane	4.81
cyclohexane	4.85
octane	5.05
decane	5.62
decalin	5.95
dodecane	6.22
Monoolefins	
pentene-1	6.01[a]
hexene-1	6.34
pentene-2	7.19
octene-1	7.38
2-methyl-butene-2	8.04
1-decene	8.67
1-methyl-cyclohexene-1	10.23
1-dodecene	10.48
Diolefins	
2-methyl butadiene-1,3	7.65
1-7-octadiene	11.86
1,5-cyclooctadiene	28.48[b]
Aromatics	
benzene	16.14]
toluene	21.37
p-xylene	27.88
t-butyl benzene	31.66
1,2,4-trimethyl benzene	40.23
triethylbenzene	50.82

[a]Pentene-1 will be included with the saturates.
[b]Cyclo-diolefins will be included with the aromatics.

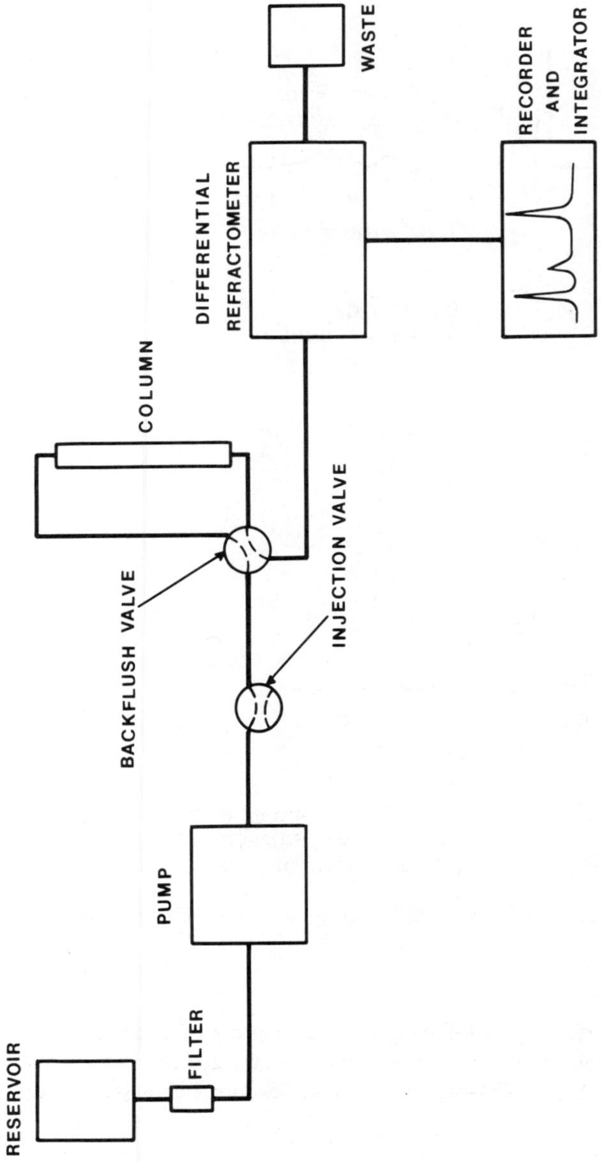

FIG. 1. Flow diagram of liquid chromatograph with flow reversal.

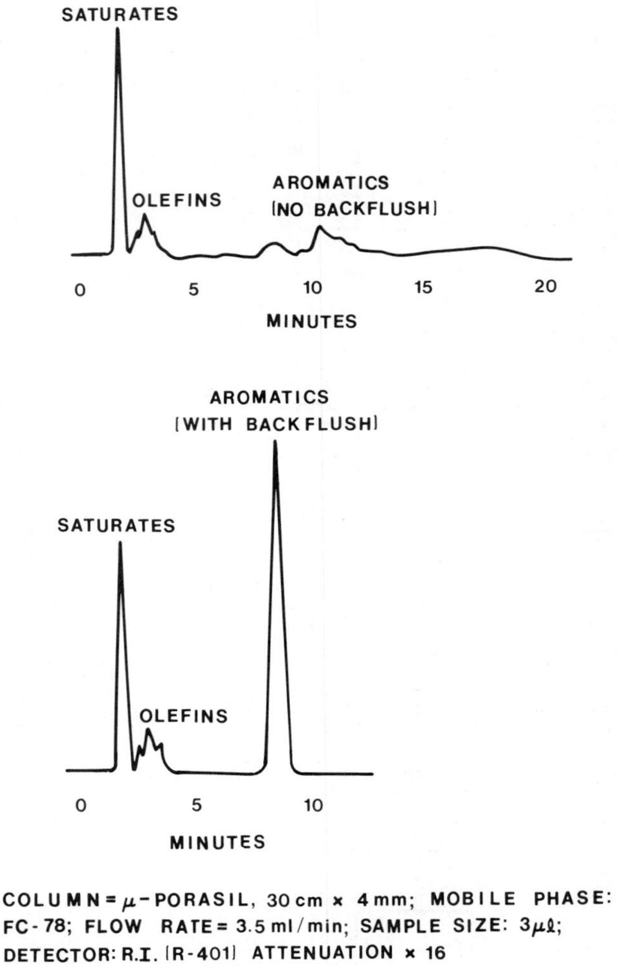

FIG. 2. Chromatogram of a heavy gasoline distillate with and without flow reversal.

from olefins a larger 10-μ silica column (10 mm by 50 cm) must then be employed. A differential refractometer can be utilized for detection since the delta refractive indices between C_{11+} hydrocarbons and hexane are large.

B. Calibration

Calibration in this boiling range is similar to the gasoline boiling range with the exception that the aromatics used for calibration are isolated from typical samples [11].

C. Procedure

The sample is injected without dilution into the chromatograph. As in the gasoline boiling range, the saturates (Table 2) are first to elute from the column, followed by the monoolefins and diolefins. The aromatics are back-flushed off the column as a single peak. A chromatogram of a furnace oil is shown in Figure 3.

TABLE 2. Model Compound Study (190 to 360°C)

Compound	Retention volume (ml)
Saturates	
dodecane	29.80
tridecane	29.62
pentadecane	29.35
hexadecane	29.32
octadecane	29.00
eicosane	28.90
tricosane	28.61
heneicosane	28.61
decylcyclohexane	29.50
decalin	29.50
Monoolefins	
decene-1	35.09
dodecene-1	35.05
hexadecene-1	34.20
heptadecene-1	33.96
octadecene-1	33.77
eicosene-1	33.25
docosene-1	32.80
Diolefins	
1,7-octadiene	47.51
1,11-dodecadiene	45.78

TABLE 1 (continued)

Compound	Retention volume (ml)
Diolefins (cont.)	
1,13-tetradacadiene	44.06
1,5-cyclooctadiene[a]	77.00
Aromatics	
1,3,5-triethylbenzene	62.50
octylbenzene	60.00
hexaethylbenzene	75.90
tetralin	75.84
benzene	60.60
biphenyl	143.30
thiophene[b]	62.60
p-dicyclohexylbenzene	61.21

[a]Cyclodiolefins will be included with the aromatics.
[b]Thiophenes will be included with the aromatics.

Results from this technique and from the previously discussed gasoline analysis are reported in volume percent as is the custom for petroleum fractions of these boiling ranges.

IV. PREPARATIVE HPLC

HPLC separations can be preparative as well as analytical. Preparative separations yield four classes of petroleum fractions: saturates, aromatics, polar compounds and hexane insolubles (asphaltenes). Fractions with little or no overlap are obtained for further identification by mass spectrometry, infrared analysis, gas chromatography, nuclear magnetic resonance, elemental analysis, and so on. The same preparative HPLC techniques can be applied to synthetic hydrocarbons except that some polar compounds are irreversibly adsorbed on the silica column. When separating samples which contain large amounts of polar compounds, coal liquids, and shale oils, it

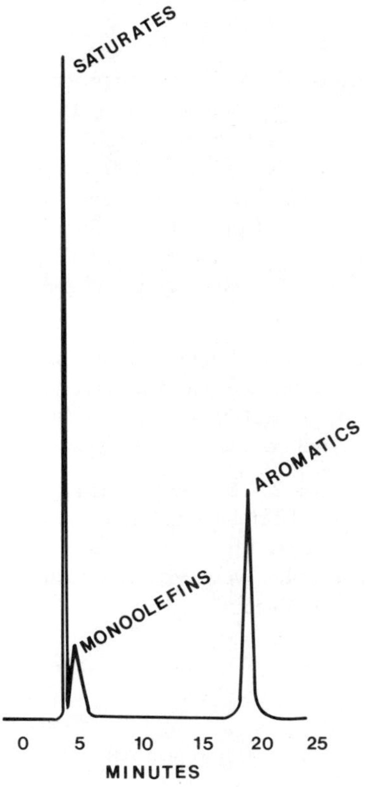

COLUMN: TWO 10μ LICHROSORBS, 10 mm × 250 mm;
MOBILE PHASE: HEXANE; FLOW RATE: 6.0 ML./MIN.;
SAMPLE SIZE: 3μℓ; DETECTOR: R.I. (R-401);
ATTENUATION: × 16.

FIG. 3. Chromatogram of a furnace oil with aromatics backflushed.

is advisable to insert a clay column [3] before the silica column to remove the polar compounds. The SARA (saturates, aromatics, resins, and asphaltenes) technique [3] is widely used for preparative group-type separations. It is a very useful technique, but time-consuming. Fractions derived from the HPLC and SARA techniques are similar but not identical: the saturates are the same, the aromatics from the HPLC technique will contain some very weak polar heterocompounds, and the resins from the SARA will contain the bulk of the higher polynuclear aromatic compounds, whereas the polynuclear aromatics are in the HPLC aromatic fraction.

A. Experimental

Before injecting a sample into the liquid chromatograph (Figure 4), the hexane insolubles are removed by means of a millipore filter and determined gravimetrically. The saturates (as a group) are eluted from the column (Bio-Sil A, 20-44 micron) with hexane. The aromatics (as a group) are eluted with hexane by backflushing. Polar compounds are eluted from the column with a 50:50 mixture of methylene chloride and acetone. The solvents are evaporated and the residue determined gravimetrically. The HPLC column is quickly regenerated by using a methylene chloride wash followed by a hexane wash.

If this technique is applied to hydrocarbon samples with initial boiling points >500°F, no loss of sample occurs during the solvent removal step. With lower boiling samples (initial boiling points of 400 to 420°F and end-points of 800 to 850°F), we normally see a 1 to 4% loss due to volatility.

Normally, one charges 200 to 250 mg [12] of hydrocarbon sample since this will provide a sufficient quantity of each fraction for further analytical study; however, larger amounts can be separated by substituting a larger column. For example, we have successfully charged and separated 15 g of middle distillate in 45 min.

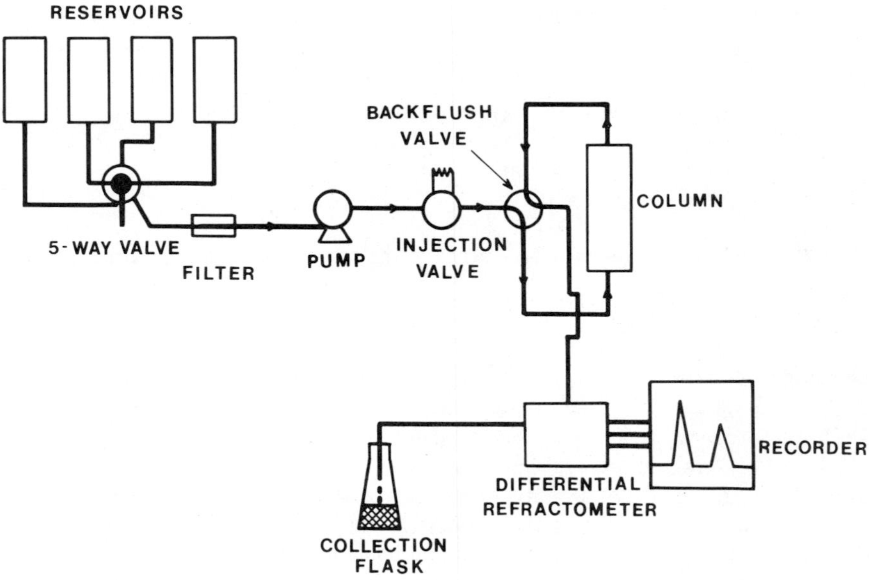

FIG. 4. Flow diagram of a prep liquid chromatograph.

A number of workers [2,3,5,6, and 7] have previously established preparative techniques for isolating hydrocarbon types; however, most of them involve time-consuming open-column separations. The HPLC separation is rapid (approximately 5 min) and columns are regenerated for further usage in an additional 10 min. The technique has been applied to a wide variety of samples: middle distillates, gas oils, lubricating oils, residua, and synthetics derived from coal, shale oil, and tar sands.

V. RAPID ANALYTICAL GENERAL TECHNIQUE

An HPLC rapid approach has been applied [10] to determine analytically hydrocarbon group-types: saturates (includes olefins), aromatics, polar compounds, and hexane insolubles in all liquid and solid hydrocarbons (natural or synthetic).

The basis for the method is a prior establishment of calibration factors [10] for the particular species. Within a given boiling range, the corresponding fractions (saturates, aromatics) of different crude oils usually had very close calibration factors, except for some highly naphthenic and highly paraffinic samples. A very efficient column packing (10 micron Porasil) is employed, which insures that all the saturates (normal, branched and cyclic) elute before the aromatics (Table 3). The aromatics are eluted as a single

TABLE 3. General Technique Model Compound Study

Compound	Compound type	Retention volume (ml)
Eicosane	Saturate	3.75
C_6-C_{36} blend	Saturate	3.75
Decylcyclohexane	Saturate	3.75
Cyclohexane	Saturate	3.75
Cholestane	Saturate	3.80
Adamantane	Saturate	3.80
Decalin	Saturate	3.80
Eicosene-1	Olefin	4.00
Octene-1	Olefin	4.20
Hexene-1	Olefin	4.25

TABLE 3 (continued)

Compound	Compound type	Retention volume (ml)
4-Methylcyclohexene	Olefin	4.50
Cyclohexene	Olefin	4.60
Octylbenzene	Aromatic	6.30[a]
2-Methyl 1,3-butadiene	Diolefin	6.35
2,5-Dimethyl 2,4-hexadiene	Diolefin	6.40
Thiophene	Sulfur compound	6.70
Benzene	Aromatic	6.80
Hexa-ethylbenzene	Aromatic	6.82
p-Xylene	Aromatic	7.55
Tetralin	Aromatic	7.75
1,2,3,4,5,6,7,8 Octahydro-anthracene	Aromatic	7.75
1,5-Cyclooctadiene	Diolefin	7.90
p-Dicyclohexylbenzene	Aromatic	8.33
Naphthalene	Aromatic	8.65
1,2,3,4,5,6,7,8,13,14,15,16 Dodecahydrochrysene	Aromatic	8.82
Dibenzothiophene	Sulfur compound	11.0
Pyrene	Aromatic	12.7
Phenanthrene	Aromatic	13.5
Anthracene	Aromatic	13.5
Fluoranthene	Aromatic	14.9
Crysene	Aromatic	21.0
Coronene	Aromatic	30.3
p-Diphenylbenzene	Aromatic	33.9
Diethylene glycol dimethyl ether	Oxygen compound	Does not elute

TABLE 3 (continued)

Compound	Compound type	Retention volume (ml)
Ethyl disulfide	Sulfur compound	Does not elute
p-Cresol	Oxygen compound	Does not elute
Aniline	Nitrogen compound	Does not elute
Acetone	Oxygen compound	Does not elute
Dioxane	Oxygen compound	Does not elute
Acetic acid	Oxygen compound	Does not elute

[a]In an actual analysis, the column will be backflushed before octylbenzene starts to elute so that all the aromatics are eluted as one peak.

peak with backflushing. Monoolefins, if present, are eluted with the saturates, diolefins with the aromatics. Polar compounds are retained on the column as shown in Table 3 and determined by difference.

This general technique has been applied to total crude oils, coal liquids, shale oils, tar sands, and their various distillate fractions [10].

VI. AROMATIC RING TYPE SEPARATIONS

A rapid HPLC procedure [13] has been developed for analytically determining aromatics by ring number: mononuclear, dinuclear, trinuclear, and polynuclear (PNAs). The technique with suitable modification can also be a preparative one with the above fractions isolated and available for further study by other analytical techniques. The preparative approach offers a very quick way to isolating PNAs (some of which are carcinogenic) for further separation and identification by reversed-phase HPLC procedures.

This analytical procedure was developed to provide an inexpensive means of generating FCC feedstock characterization data. However, the same method can be applied to other hydrocarbon samples as well: middle distillates, residua, synthetic fuels, and aromatics from exploration cores.

A. Apparatus

The liquid chromatograph is equipped with valves for precise sample injection and solvent backflushing. Detection is made by means of a 254 nm uv detector. The mobile phase is a 91:9 mixture of hexane and methylene chloride. The column (9.4 mm by 50 cm) is packed with Partisil PAC-CN. For analytical separations a 20-μl sample loop is used; for preparative separations usually 125 mg of sample are charged through a 500-μl loop.

B. Procedure

After sample injection the monoaromatics are first to elute followed by the dinuclear and trinuclear compounds. The PNAs are eluted as a single peak by backflushing. Retention times are established with model compounds (Table 4) and quantitation is accomplished by previously established calibration factors.

C. Discussion

Separation on the Partisil PAC column offers certain advantages for aromatic ring separations as compared to silica (Table 4). With silica, the noncondensed ring compounds elute very late (p-terphenyl); in addition, cycloparaffinic rings attached to aromatic rings cause late elution (fluorene). This is not the case with Partisil PAC.

VII. FUTURE APPLICATIONS

A. PONA Analysis

There is a need for a rapid PONA (paraffins, olefins, naphthenes, aromatics) group analysis for the gasoline boiling range. The present HPLC technique [9] does not split the naphthenes from other paraffins (branched and normal). It should be possible to separate naphthenes from the paraffins by means of a size separation.

B. Preparative Olefin Separation

The HPLC preparative method [10] does not separate olefins from saturates. This is usually not a problem since olefins are not present in high boiling naturally occurring petroleum fractions. However, shale oils contain large amounts of olefins. HPLC utilizing 10 micron silica or a silver nitrate coated column should be capable of preparatively separating olefins as a class.

TABLE 4. Aromatic Model Compound Study

Compound	Number of aromatic rings	Peak elution time (sec)	
		CN column[a]	Silica column[b]
Benzene	1	223	176
p-Xylene	1	225	190
Tetralin	1	229	212
n-Octylbenzene	1	196	162
Dodecylbenzene	1	193	163
1,2,3,4,5,6,7,8-Octa-hydroanthracene	1	226	215
Thiophene	1	224	188
Naphthalene	2	303	243
Biphenyl	2	303	337
Acenaphthalene	2	304	285
alpha-methylnaphthalene	2	303	260
Diphenylmethane	2	297	485[c]
Benzothiophene	2	305	245
Fluorene	2	416	400[c]
Anthracene	3	495	370
Phenanthrene	3	497	391
Dibenzothiophene	3	497	367
P-Terphenyl	3	496	860[c]
Fluoranthene	3	661	480
Pyrene	4	646[c]	400[c]
Crysene	4	905	672
1,2-Benzanthracene	4	908	678
Dibenz(a,h) anthracene	5	1,232	1,220
Coronene	6	1,430	—

[a]Partisil PAC, 9.4 mm × 50 cm. Mobile phase, 9% methylene chloride in hexane at 8.8 ml/min.
[b]Two Partisil 10 micron silica columns, 4 mm × 20 cm. Mobile phase, hexane at 2.0 ml/min.
[c]Abnormal elution times.

C. Carcinogenic PNAs

A combination of two HPLC techniques should result in a rapid determination of two known carcinogenic PNAs, benz(a)pyrene and 1,2-benzanthracene. The 4 and 5 ring aromatics can be isolated very quickly [13] from any hydrocarbon sample by employing the nitrile stationary phase. This fraction can then be separated into individual compounds by employing reversed-phase gradient elution.

D. Polar Compounds

HPLC is very useful in the characterization of polar compounds providing cuts for identification by mass spectrometry and infrared analysis, among others. Future work will include separations of the polar compounds into definitive subclasses.

REFERENCES

1. ASTM D 1319, Hydrocarbon Types in Liquid Petroleum Products by Fluorescent Indicator Adsorption, 17:476 (1973).
2. ASTM D 2007, Characteristic Groups in Rubber Extender and Processing Oils by the Clay-Gel Adsorption Chromatographic Method, 18:465 (1970).
3. D. M. Jewell, E. W. Albaugh, B. E. Davis, and R. G. Ruberto, I.&E.C. Fundamentals 13:278 (1974).
4. L. R. Snyder, "Principals of Adsorption Chromatography," Marcel Dekker, New York, 1968.
5. L. R. Snyder, Anal. Chem. 36:128 (1964).
6. L. R. Snyder, Anal. Chem. 36:774 (1964).
7. L. R. Snyder, Anal. Chem. 37:713 (1965).
8. L. R. Snyder, G. Chromatog. 36:476 (1968).
9. J. C. Suatoni, H. R. Garber, and B. R. Davis, J. Chromatog. Sci. 13:367 (1975).
10. J. C. Suatoni and R. E. Swab, J. Chromatog. Sci. 13:361 (1975).
11. J. C. Suatoni and H. R. Garber, J. Chromatog. Sci. 14:546 (1976).
12. J. C. Suatoni and R. E. Swab, J. Chromatog. Sci. 14:535 (1976).
13. J. C. Suatoni, unpublished work.

Part III ANALYSIS OF HEAVY DISTILLATES, RESIDUES, AND CRUDES BY GAS CHROMATOGRAPHY

Chapter 7

THE IDENTIFICATION OF OIL SPILLS BY GAS CHROMATOGRAPHY

E. R. Adlard
Shell Research Ltd.
Thornton Research Centre
Chester, England

I. INTRODUCTION

Any article on the identification of oil spills must be preceded by a number of general remarks. The first, and perhaps the most important, is that in the majority of cases analytical work is undertaken solely for forensic reasons, that is, to identify the immediate source of the spill and not its original source. It is useless to apply more and more elegant analytical techniques if other circumstantial evidence is not available to link the scientific evidence to a suspect ship or a land installation such as a tank farm. Thus, in the absence of other evidence it is futile to collect a tar ball from a beach and ask for the identity of the ship it came from, although the identity of the oil-field from which it came may be easy to determine. The ideal case from both the scientific and forensic point of view occurred in the author's laboratory in connection with an oil spill found in the estuary of a river. The river had a refinery belonging to a different company and processing a different crude oil on each bank and the question to be answered was, simply, did the spill resemble the oil processed by refinery A, the oil processed by refinery B, or neither? Quart quantities of the three samples were available and it was a straightforward task to show that the spilt oil was the same as that processed by one of the refineries.

The second point is that most samples received for identification will have undergone changes in composition, or "weathering." The most immediate effect of weathering is the loss of lower boiling compounds by evaporation. The nature and extent of this loss depends on the original composition of the spill and the climatic conditions prevailing at the time. Thus, a gasoline spill on water is quite likely to have disappeared within 48 hr, but cases have been known where this time scale has been doubled because of emulsion formation. A crude oil will gradually lose its more volatile components but even after several months hydrocarbons down to nC_{12} may still be present in tar balls, and material boiling as low as nC_{14} is frequently present even in samples more than a year old unless they have been exposed as thin films on rock surfaces [1].

Figure 1 shows a set of curves obtained for Kuwait crude oil artificially weathered in the laboratory. Although the experimental conditions used (22°C, air agitation and illumination by uv light) were designed to represent severe natural conditions, it can be seen that in 96 hr there has been little or no loss of material boiling above nC_{16}. The conclusion from these laboratory experiments and the analysis of numerous pollutant samples is that frequently much diagnostic information may be derived from compounds with boiling points as low as nC_{16}. These results are in general agreement with recently published quantitative studies [2, 3].

Other weathering processes such as biodegradation (the ultimate weathering process), oxidation, and solution will also cause considerable changes in the composition of the oil, and the oil itself may act as a solvent for other pollutants and detergent emulsifiers which may be present [4]. In the case of the river spill just described it was found that the iron content of the spilt oil had increased from its normal value of 12 ppm to 156 ppm in a period of 1 or 2 days, because the river received the effluent of a number of iron and steel works.

A third generalization is that, although gas chromatography is perhaps the most important single technique for pollutant identification, there are a number of other techniques available and some of these complementary techniques should always be used if possible before a positive identification is made [26, 27].

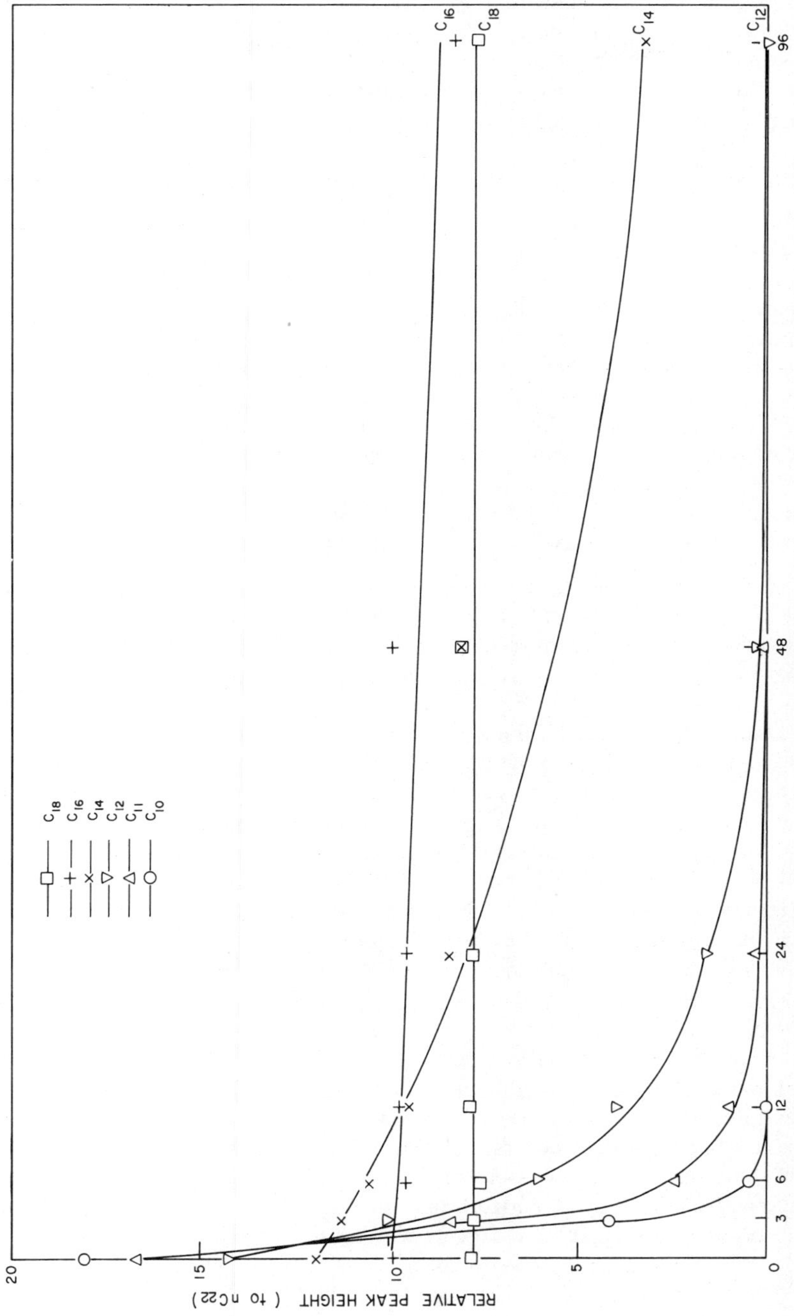

FIG. 1. Artificial weathering Kuwait crude oil.

Finally, the analyst must decide on a scheme of identification based on the available facilities and the lengths to which it is necessary to proceed. A large number of gas chromatographic (GC) and other analytical methods exist for identification purposes but some of them are so tedious and time-consuming or demand such specialized equipment that their use on a routine basis is most unlikely.

II. LOW-RESOLUTION GC METHODS

The objects of low-resolution GC methods for pollutant identification are, first, to provide a rough boiling range distribution of the compounds in the sample and, second, to provide a fingerprint chromatogram of limited diagnostic ability. If it were not for the changes produced by weathering, the simulated distillation method described in ASTM D 2887 could be applied. Since weathering will have caused an unknown loss of lower boiling compounds there is little point in trying to obtain an accurate simulated distillation curve but the column and apparatus described in D 2887 may be applied to spill identification with minor modifications.

Ramsdale and Wilkinson [5] wrote one of the first papers on the application of GC to spill identification in 1968, and a version of their method has now been adopted by the Institute of Petroleum as method IP 296-73T. This uses dual columns 1 m long by 3.1 mm OD (0.125 in. OD × 0.100 in. ID) packed with 5% silicone elastomer OV1 on 80-100 mesh Chromosorb G, acid washed and treated with dimethyldichlorosilane (DMCS). Nitrogen or helium is used as carrier gas but care should be taken to remove traces of oxygen which would decrease column life and stability. For high-boiling oils the columns are programmed from 100 to 300°C at 5°C min^{-1}; for lighter products the initial and final temperatures can be lowered. The upper temperature limit is fixed at about 325°C by the stability and volatility of the stationary phase. If very high-boiling residual materials are to be chromatographed (e.g., tank bottom residues) it might be better to use a column of only 50 cm in length and obtain the necessary resolution of the lower boiling part of the sample by starting the analysis at a lower initial temperature—say 30°C instead of 100°C. In order to do this, forced cooling facilities using liquid nitrogen or liquid CO_2 may be required to reach the initial temperature in a reasonable time after the columns have been up to their maximum temperature. Dual differential flame ionization detectors (FID) are used and the only nonstandard part of the apparatus is the sample introduction system. This consists of a double ball-valve gas lock which allows an open-ended ampule containing the sample to be applied to the inlet of the GC column (Figure 2). In this way it is possible to analyze "as received" samples without any prior clean-up to remove sand, water, and so on. This gives a considerable saving in time which is important if there are many samples to be examined as the result of a major spill.

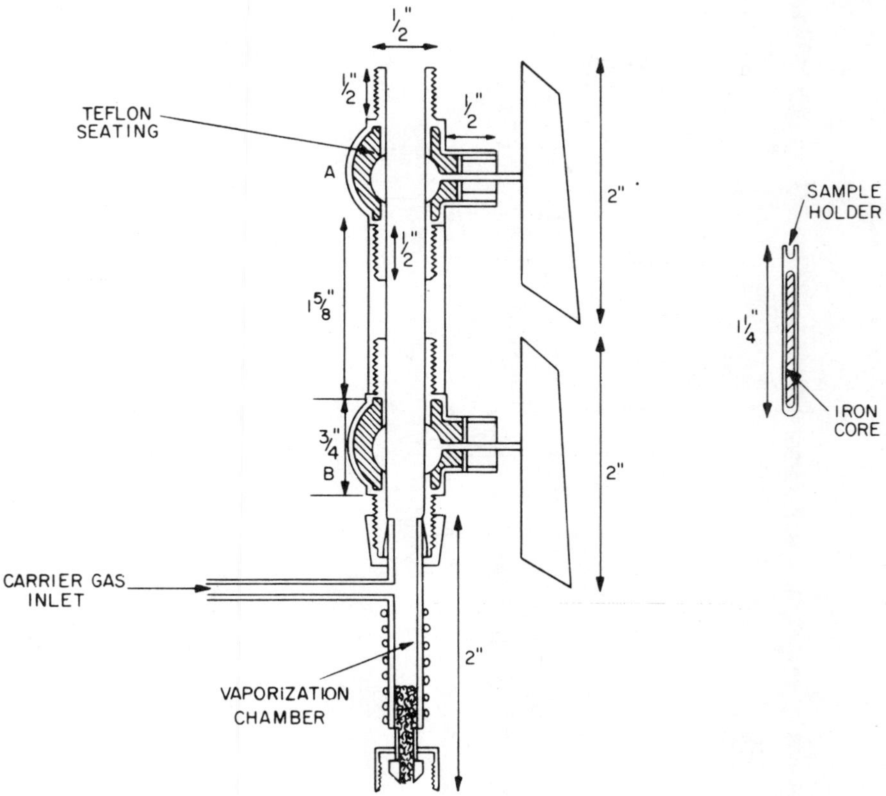

FIG. 2. Injection system for direct analysis of pollutant samples.

Figures 3 to 8 show some typical chromatograms obtained with the apparatus described above. Figure 3 is the chromatogram of a fuel oil showing clear evidence of blending and Figure 4 is the chromatogram of another fuel oil, the sharp fall-off at the rear of the trace indicating that the sample has been distilled (see Figures 5, 6, and 7 showing chromatograms of weathered crude oils). The crude oils in Figures 5-7 range from paraffinic to naphthenic in composition. Figure 8 is a chromatogram of a Kuwait crude oil tank bottom residue, showing a considerable increase in the proportion of higher boiling material stretching into the C_{40}-C_{50} region.

The chromatograms shown in Figures 3-8 illustrate the three main categories of persistent oils recognizable by low-resolution GC, i.e., fuel oils, crude oils, and tank residues; additionally, oils of nonmineral origin may be recognized. It may occasionally be possible to identify tentatively specific crude oils which have characteristics such as the abnormally large

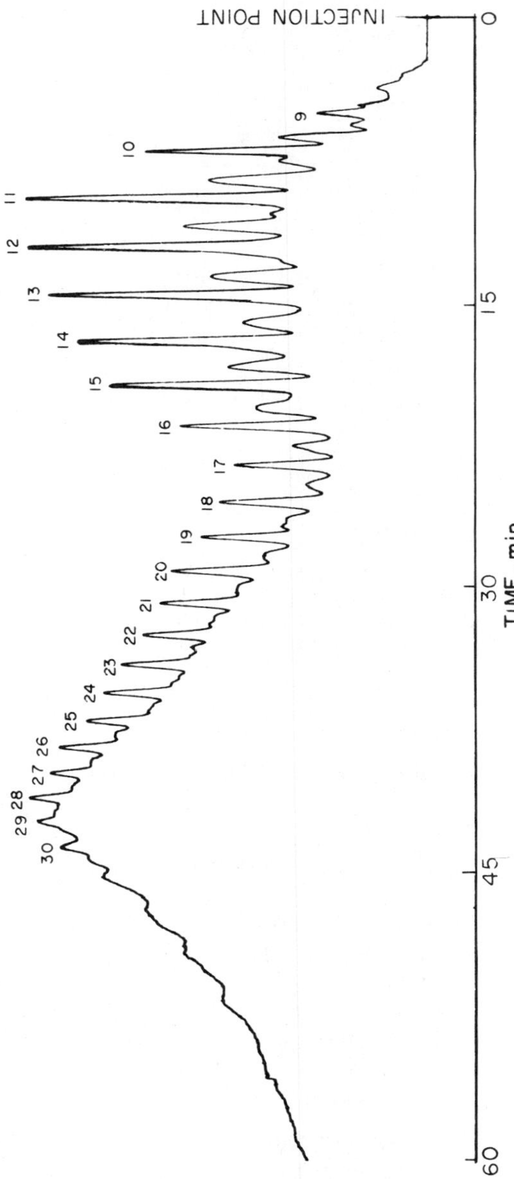

FIG. 3. Low-resolution chromatogram of Bahrein fuel oil.

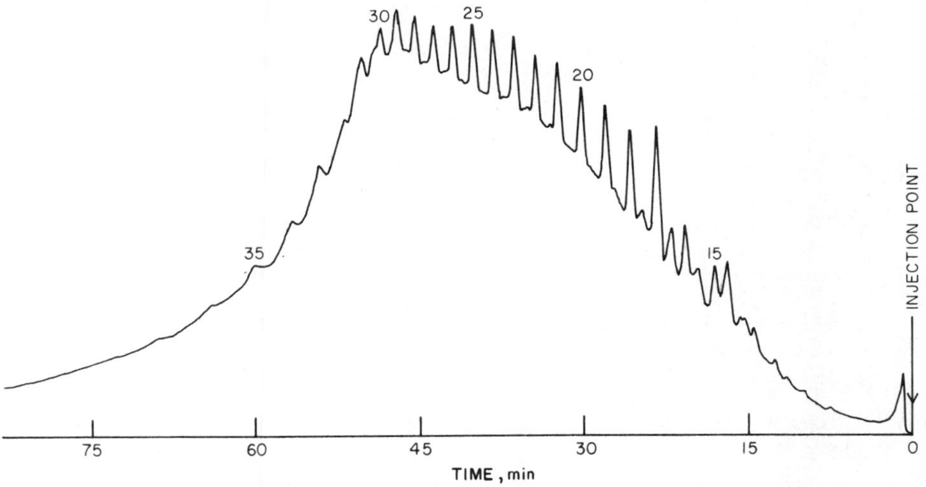

FIG. 4. Low-resolution chromatogram of Marine diesel fuel.

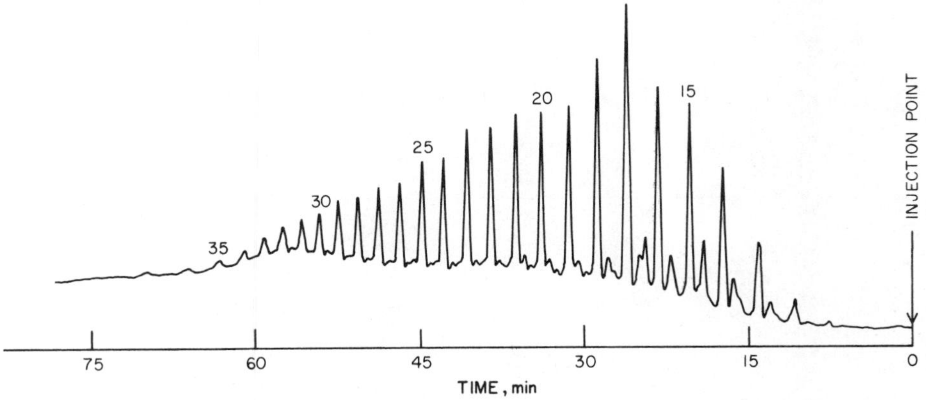

FIG. 5. Low-resolution chromatogram of weathered Sarir crude oil.

"C_{17}" peak in Nigerian crude oil (Figure 6) due, in fact, to the presence of a large amount of pristane and not due to a large amount of the normal paraffin. On the debit side, it is frequently impossible to differentiate between a weathered fuel oil and a weathered crude oil, and indeed this can be a difficult problem no matter what analytical method is used. Since fuel oils have been subjected to heating during distillation it has been claimed that the asphaltene content of a fuel oil is higher than that of the parent crude, but the

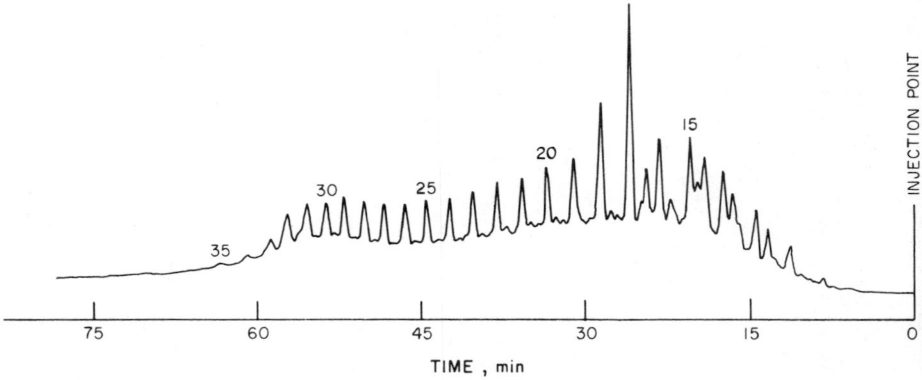

FIG. 6. Low-resolution chromatogram of weathered Light Nigerian crude oil.

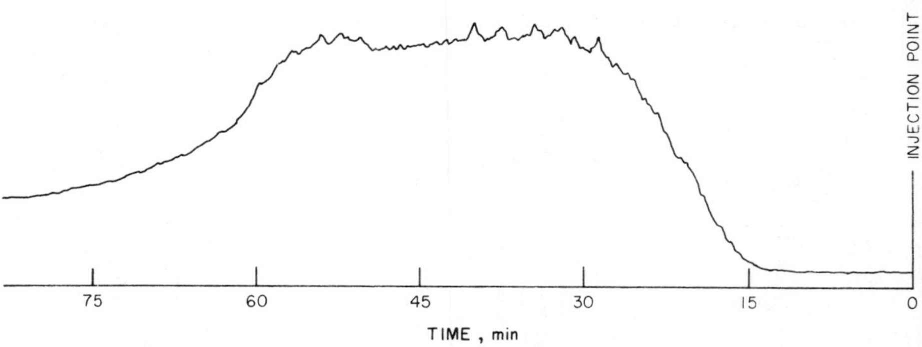

FIG. 7. Low-resolution chromatogram of weathered Bachaquero crude oil.

measurement of asphaltene contents has a poor precision, so there must inevitably be a region of considerable doubt between the two types of oil.

In addition to the Institute of Petroleum, the American Society for Testing and Materials (ASTM) has also standardized methods for oil spill identification and D 3328-74aT gives two GC methods. The first uses a 3-m by 3-mm column packed with 10% silicone OV101 on 60-80 mesh Chromosorb W, acid washed and DMCS treated. The longer column gives better resolution and therefore a more detailed fingerprint than the column specified by the IP method. This results in some improvement in the ability to identify an oil but no attempt is made in the ASTM method to analyze samples

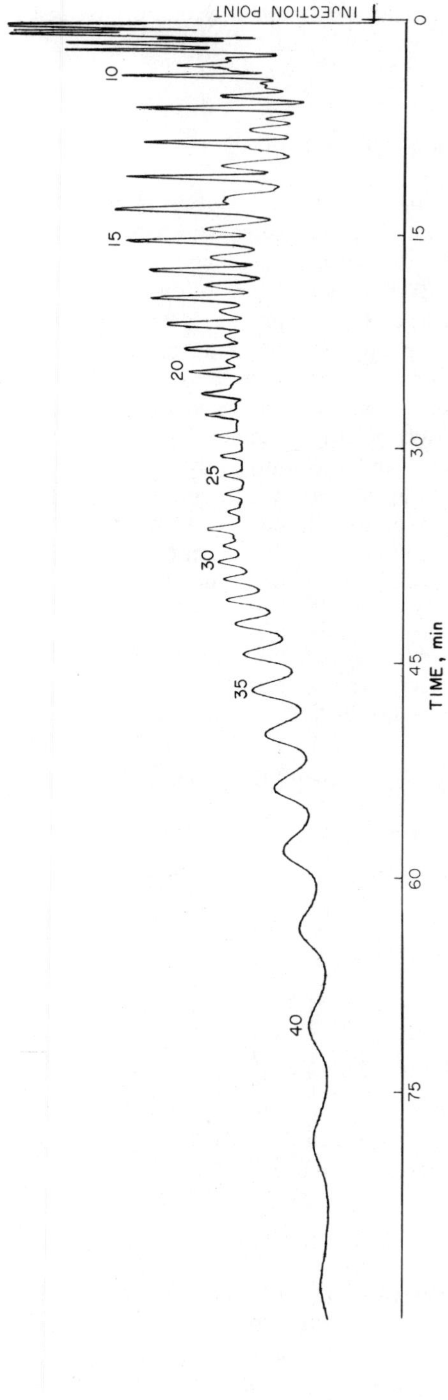

FIG. 8. Low-resolution chromatogram of a Kuwait crude oil cargo residue.

"as received" and they must be cleaned by dissolving in chloroform followed by centrifuging, or by refluxing with toluene.

III. HIGH-RESOLUTION GC METHODS

As pointed out above the higher the resolution of the GC column, the more detailed will be the fingerprint obtained from pollutant samples and the more detailed the fingerprint the better the chance of unambiguous proof of similarity (or proof of dissimilarity which is often of equal importance). It is obvious, therefore, that at first sight the use of wall-coated open tubular (capillary) columns for oil spill identification is extremely attractive [6]. There are, however, considerable practical difficulties in employing narrow-bore capillary columns to analyze crudes and heavy fuel oils. Many crudes and fuel oils contain high-boiling, high-molecular-weight compounds, the so-called "asphaltenes," and it is impossible to get asphaltenes to vaporize in the inlet of the GC, which would then soon fail to function correctly. With a packed column and on-column injection, a considerable number of 1-μl samples may be tolerated, even if 10% of each one stays on the column, but the inlet splitter of a capillary GC could become blocked after the injection of only one such sample.

To avoid this difficulty several different approaches have been proposed, each with advantages and disadvantages. Zafiriou [7] recommends a 50-ft support-coated open tubular (SCOT) column coated with silicone OV101 with a readily interchangeable glass liner in the GC inlet, and Method B in ASTM D 3328-74aT recommends the same type of column but the separation is

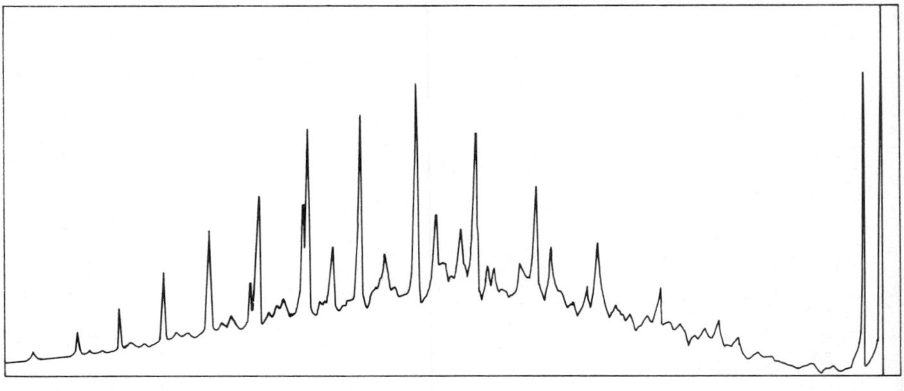

FIG. 9. Chromatogram of a number 2 fuel oil on a 50-ft OV101 SCOT column (by courtesy of Analytical Chemistry [7].

carried out on a saturated hydrocarbon fraction of the sample obtained by liquid adsorption chromatography over silica gel and alumina. Zafiriou's method needs no prior sample manipulation apart from solution of the oil in CS_2 and is attractive if large numbers of samples are to be examined. On the other hand, the 50-ft SCOT column gives only partial resolution of the isoprenoid compounds, pristane and phytane,* from the adjacent C_{17} and C_{18} normal paraffins (Figure 8).† This has the disadvantage, as exemplified by the results from Zafiriou's paper shown in Table 1, that it is difficult to obtain good reproducibility from column to column and laboratory to laboratory for the peak height ratios nC_{17}:pristane, nC_{18}:phytane.

Unless some resolution criterion is given it is not possible for one laboratory to know if its results are comparable with those from another location. In fairness to the ASTM method, it does not suggest that the ratios nC_{17}:pristane and nC_{18}:phytane should be measured, and it is proposed as a comparative, not an absolute, method.

TABLE 1. n-Paraffin:isoprenoid Ratios on Different SCOT Columns [7] (Courtesy of Analytical Chemistry)

Column number	Number of runs averaged	Sample	nC_{17}:Pristane	nC_{18}:Phytane
218	11	No. 2 fuel oil	1.41	1.90
219	11	No. 2 fuel oil	1.41	1.90
216	13	No. 2 fuel oil	1.29	1.65

*Isoprenoid hydrocarbons are naturally occurring compounds which can be considered to be derived from the isoprene molecule (2-methyl-1,3-butadiene) by addition and hydrogenation. Phytane, 2,6,10,14-tetramethylhexadecane, can be built up (on paper at least) by the head-to-tail addition

and saturation of four isoprene units. Odd-numbered compounds such as pristane, 2,6,10,14-tetramethylpentadecane ($C_{19}H_{40}$), are also isoprenoids but their derivation appears to be more complex.

†See also Figure 5 on page 100.

The Institute of Petroleum method IP 318-73T specifies a 50-m by 0.5-mm ID stainless steel capillary column coated with OV101 to separate a less-than-343°C distillation cut. (343°C is the normal boiling point of nC_{20} and is specified since the fraction boiling above this figure is required for other IP characterization methods such as nickel and vanadium contents. The data obtained from the >343°C material are regarded as not being influenced by weathering effects.) Helium is used as the carrier gas and the column is programmed from 100 to 220°C at 5°C min^{-1}. Under these conditions nearly complete resolution of pristane, phytane, and the adjacent normal paraffins is achieved and it has been found that satisfactory reproducibility of the n-paraffin:isoprenoid ratios can be obtained if the column resolution requirements given in the method are attained. Figures 10 and 11 show the chromatograms obtained by this method from two typical crude oils, one paraffinic and the other naphthenic in character. Table 2 lists the paraffin:isoprenoid ratios for some common crude oils. From Table 2 it can be seen that although the ratios may not provide an unambiguous identification, they will often eliminate a number of possibilities.

TABLE 2. n-Paraffin:isoprenoid Ratios Determined by IP 318-75T

Crude oil[a]	nC_{17}:pristane Peak height ratio	nC_{18}:phytane Peak height ratio
Amal	3.2	5.5
Amposta	3.4	2.0
Arabian Light	6.2	3.8
Bachaquero	1.2	2.4
Boscan	1.4	0.7
Castellon	1.2	2.6
Es Sider	2.5	2.8
Huntington Beach	0.36	0.43
Iranian Light	2.3	2.0
Khafji	5.7	3.2
Kuwait	4.9	3.0
Nigerian Light	0.61	1.9
Qatar Marine	4.8	3.6
Sarir	2.7	4.9
Tia Juana	1.5	2.0

[a]<343°C fraction in every case.

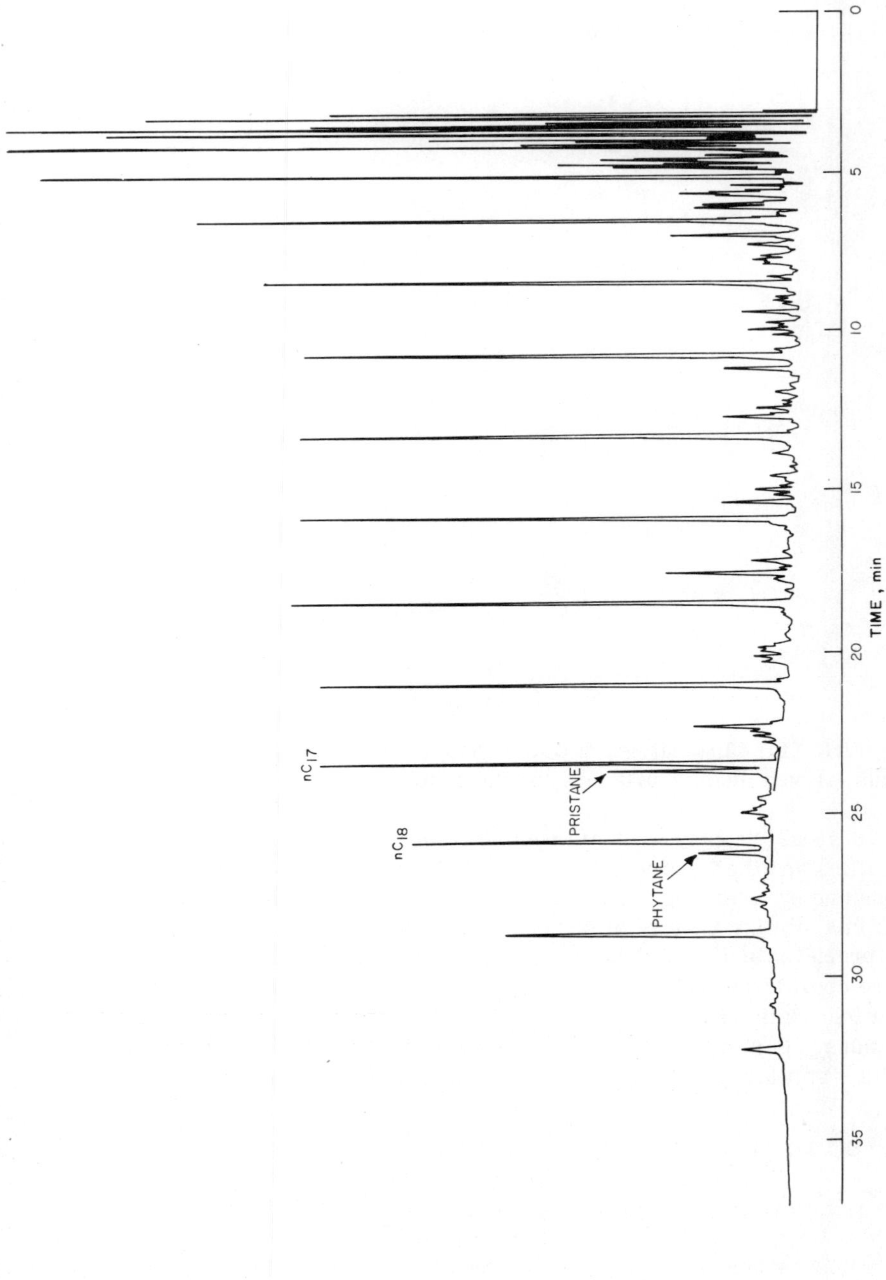

FIG. 10. Chromatogram of the <343°C fraction of Sarir crude oil on a 50-m × 0.5-mm column coated with OV101.

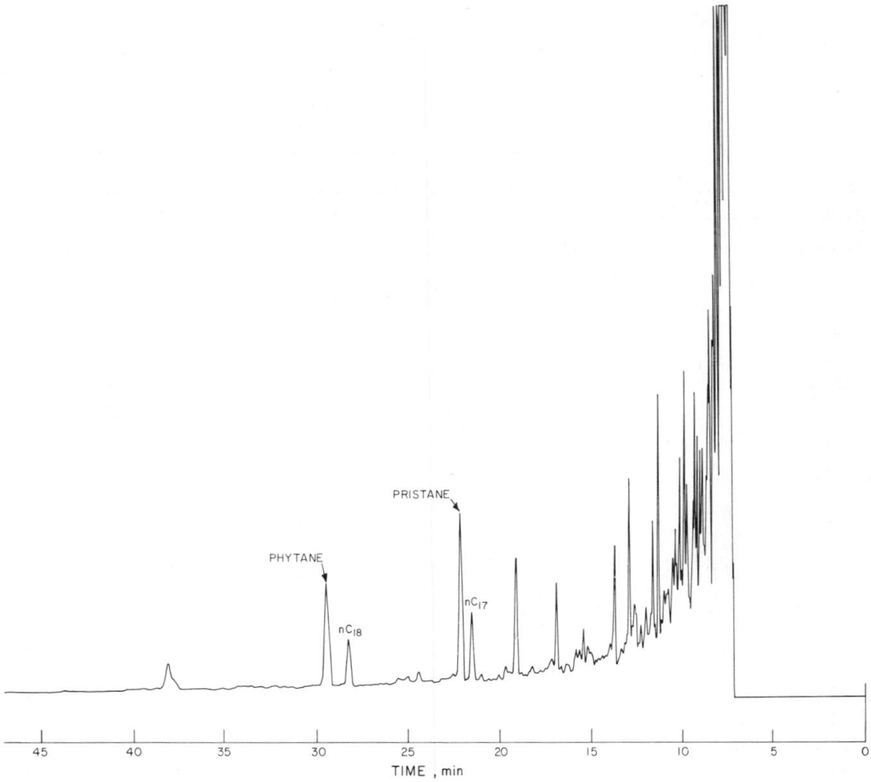

FIG. 11. Chromatogram of the <343°C fraction of Huntington Beach crude oil on a 50-m × 0.5-mm column coated with OV101.

It should be easy to obtain complete separation between the normal paraffins and the isoprenoids by using longer, narrower-bore capillary columns and by analyzing the saturated hydrocarbon portion of the <343°C fraction in a similar manner to ASTM D 3328. It does not follow, however, that the precision and reproducibility of the peak height ratios would improve since considerable difficulties have been encountered, for example, in correlation trials on the analysis of gasolines with high-resolution capillary columns. Part of these difficulties may be attributed to the malfunctioning of inlet splitters, especially if large split ratios are required.

IV. GC METHODS EMPLOYING SELECTIVE DETECTORS

The last 10 years have seen the development of a number of selective detectors for gas chromatography and several are available from most of the instrument manufacturers including detectors selective for sulfur and nitrogen [8].

Sulfur is present in crude oils in widely varying amounts; there are crudes with sulfur contents as high as 14% but the sulfur contents of commercially important crudes range from about 0.06% to 6% (Table 3), i.e., about a 100:1 range of concentration. Nitrogen is also present in most crude oils but usually in smaller quantities than sulfur and with a smaller variation in concentration (about 40:1). There are many other elements present in small quantities in crude oils, e.g., metallic elements, in particular nickel and vanadium, and nonmetallic elements such as fluorine, bromine, and iodine, but these elements are currently determined by techniques other than GC [9, 10].

In addition to being present in substantial amounts, the sulfur and nitrogen compounds have the advantage of being present largely in the higher boiling fractions of crude oils and are, therefore, relatively less affected by evaporative weathering. On the other hand, since they are associated with the higher boiling part of the oil, they belong to the most chemically complex part of the oil and the part least amenable to GC.

Of all the selective GC detectors the flame photometric sulfur detector (FPD) has proved to be one of the best, with a high sulfur-to-hydrocarbon selectivity ($>10^5$:1), a high sensitivity (<1 ng s^{-1} S) and a low dead volume. The latter two features make it particularly amenable to use with capillary

TABLE 3. Sulfur and Nitrogen Contents of Crude Oils

Crude oil	Sulfur[a] (wt. %)	Nitrogen[a] (wt. %)
Zarzaitine, Algeria	0.06	0.02
Nigerian Light	0.16	0.09
Brega } Libya	0.24	—
Es Sider } Libya	0.42	0.16
Abu Dhabi	0.80	0.03
Qatar Marine	1.50	0.07
Tia Juana, Venezuela	1.61	0.24
Kuwait	2.59	0.16
Bachaquero, Venezuela	2.64	0.41
Arabian Heavy	3.00	0.17
Boscan, Venezuela	5.1	0.78
Amposta, Spain	5.7	0.18

[a]These values are only approximate and may vary slightly from batch to batch.

columns. The detector operates by burning the effluent from the GC column in a hydrogen-rich flame and observing the chemiluminescence produced by sulfur at 394 nm by means of an optical interference filter and photomultiplier tube. The quantitative response of the detector is complex but this is not important for the fingerprinting of oils provided that reproducible chromatograms can be obtained which differ from one oil to another. Some FPDs have an FID incorporated into the design but the author's experience is that is is preferable to split the effluent from the GC column, taking part to a conventional FID and the rest to the FPD. The split ratio can be adjusted by using a micro needle valve of the type employed in GC mass spectrometry for similar purposes. Whichever method is used it is possible to obtain both an FID hydrocarbon fingerprint and an FPD sulfur fingerprint simultaneously. The two sets of fingerprints obtained in this way often have a considerably greater diagnostic value than that of the FID trace alone [11-15]. It must be emphasized, however, that the FPD is not the universal panacea for pollutant identification but does frequently yield different sulfur fingerprints even from oils which have very similar hydrocarbon fingerprints. Figures 12 and 13 show the chromatograms of a Zuetina crude oil, for which the hydrocarbon and sulfur fingerprints are different.

The FPD can also prove useful if biological degradation has proceeded to the point where most of the normal paraffins, which are the first type of compound to be attacked, have disappeared [13]. It is possible, of course, for degradation to proceed much further, and the author has encountered oil samples of known provenance so biodegraded as to be unrecognizable by any GC method.

The chromatogram shown in Figure 13 was obtained with a Tracor FPD coupled to a 50-m long, 0.75-mm ID stainless-steel capillary column coated with silicone OV101 and programmed from 60 to 300° C at 5° C min^{-1}. These columns have proved very satisfactory, with good resolution and a sample capacity of about 0.5 μl without a splitter, but SCOT columns and high-efficiency packed columns should be equally effective. If the sulfur contents of the oils under examination are high (say greater than 1%) then there is no reason why 50-m long 0.5-mm ID columns of the sort recommended for n-paraffin:isoprenoid ratios should not be used. If low-efficiency packed columns are employed, however, the diagnostic power of the FPD chromatogram is very much reduced and the technique loses much of its advantage.

The possibility of obtaining "nitrogen" chromatograms analogous to "hydrocarbon" and "sulfur" chromatograms is rendered difficult because of the generally lower nitrogen concentrations in crudes and fuel oils and because, up to recently at least, the performance of nitrogen-selective detectors has not equaled that of the FPD for sulfur. It is possible that the new

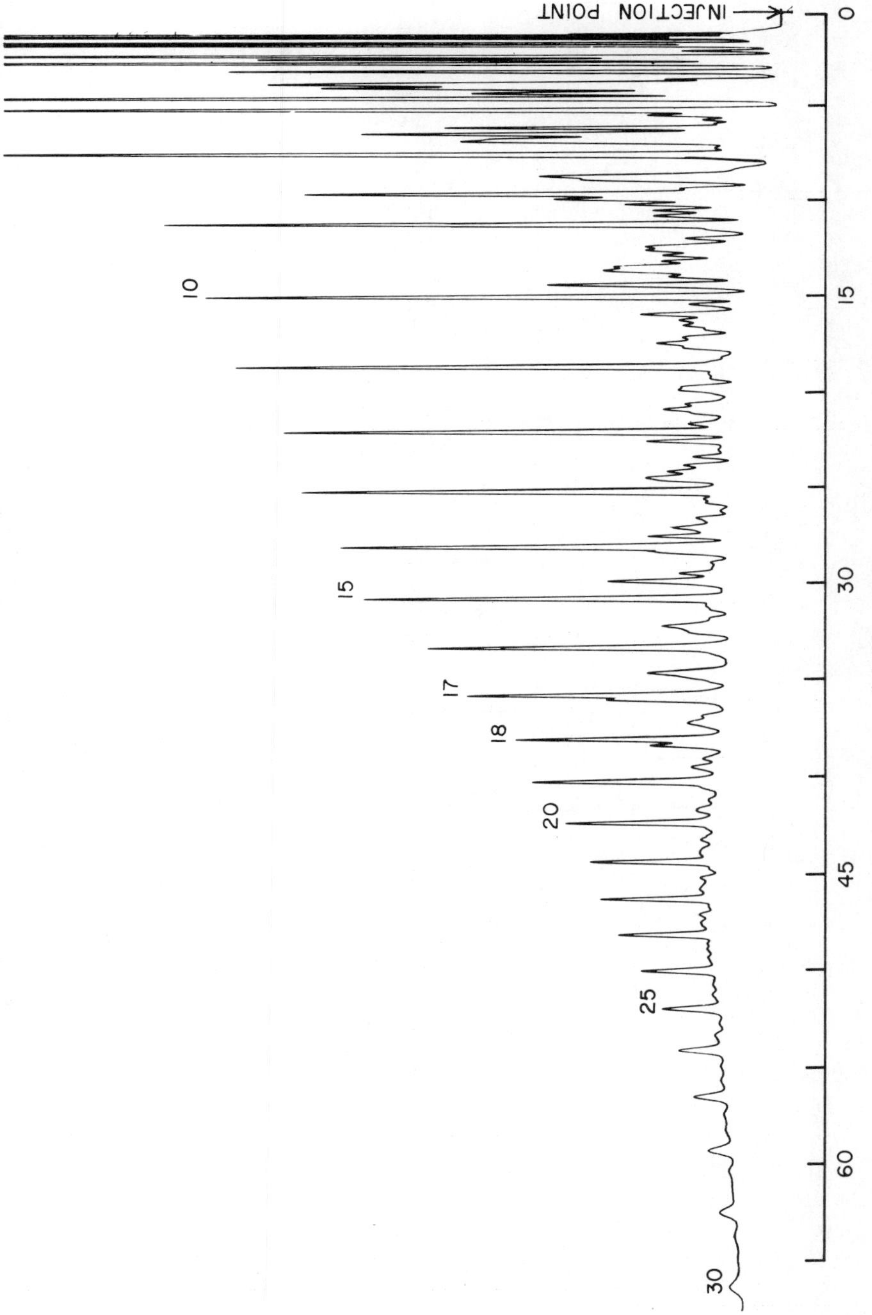

FIG. 12. Chromatogram of Zuetina crude oil with FID.

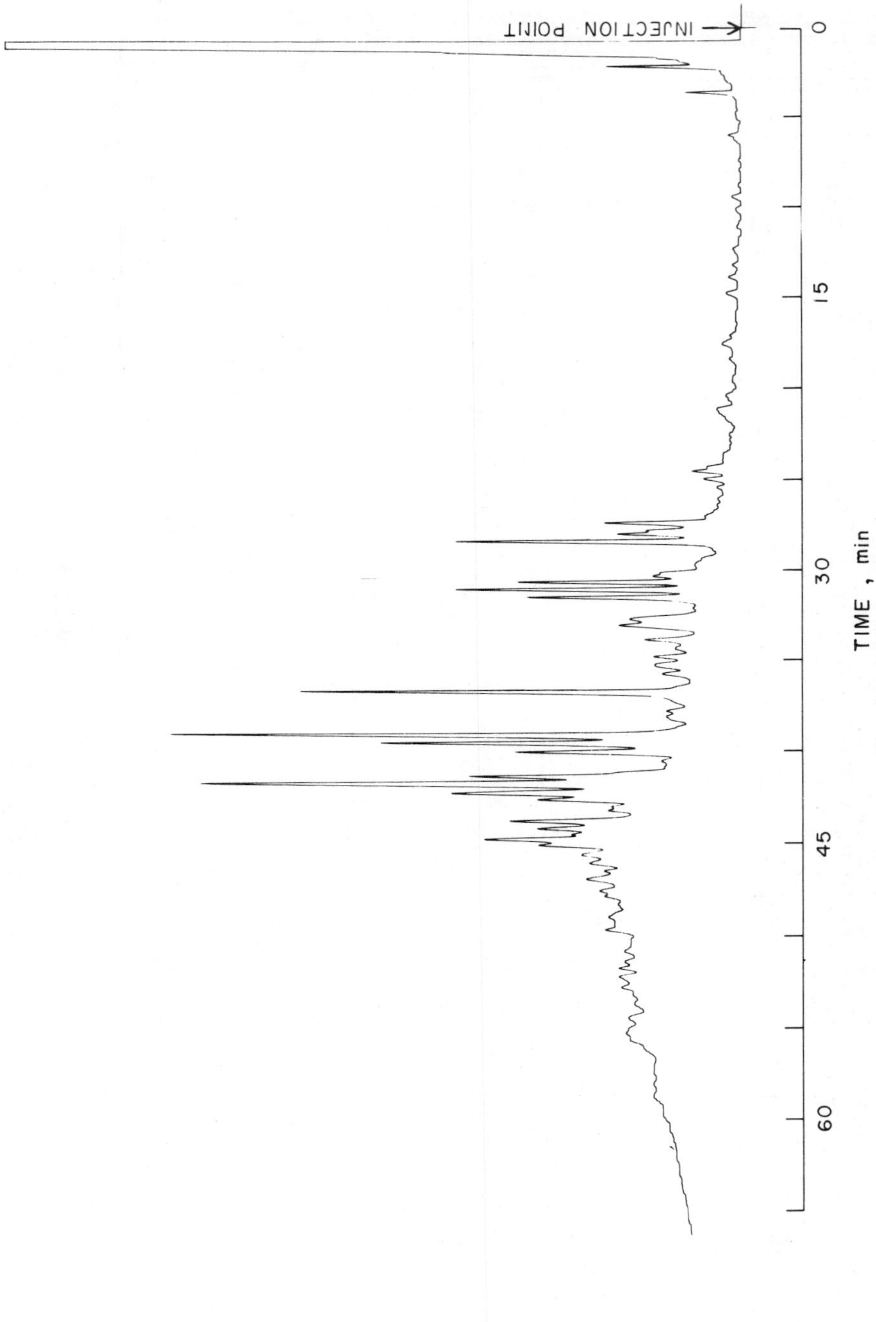

FIG. 13. Chromatogram of Zuetina crude oil with FPD.

generation of selective nitrogen detectors will prove considerably better in performance and the problem of low concentration can be overcome by preconcentration by liquid adsorption chromatography. There is, however, a more fundamental difficulty in that the nitrogen compounds are associated with the highest boiling fractions of crudes and as such are not very amenable to GC separation. Nitrogen chromatograms have recently been published by Flanigan and Frame [28].

Finally, it is probable that characteristic fingerprints could be obtained with an electron capture detector (ECD), especially from the aromatic portion of crudes and fuel oils obtained by liquid adsorption chromatography. The ECD response to different types of aromatic compounds can vary over a range of up to 10^6 and, just as the same sulfur compounds are found in most oils but in differing relative proportions, it is possible that a similar situation will apply to aromatic compounds of high electron affinity.

V. OTHER GC METHODS

The methods described so far are suitable for routine application in that the equipment required is more or less standard, the minimum of sample preparation is needed and operator time is relatively small. There are at least two other GC methods which do not fall into this category since they require considerably more skilled operator time and also longer elapsed time. These are methods which have been developed for geochemical studies but which can be applied to oil spill identification if the cost is justified and if the necessary expertise is available.

The first of these methods is the determination of normal paraffin profiles [16]. A sample of the oil, to which has been added a known amount of a normal paraffin outside the range of interest, is passed in a stream of nitrogen over a bed of powdered molecular sieve 5A heated from 20 to 300° C in 20 min. The normal paraffins are retained by the sieve and the rest of the hydrocarbons in the oil are swept away. After cooling under nitrogen, the sieve is dissolved in a 20% HF solution and the normal paraffins released are dissolved in a small volume of iso-octane. Finally, the iso-octane solution is analyzed on a short GC column to determine the amount of each normal paraffin. From the results obtained it is possible to plot a graph of percentage of each normal paraffin against carbon number, which is characteristic for many crude oils. Examples of such graphs comparing the higher boiling part of crude oils with pollutant samples are shown in Figures 14 and 15; Figure 16 shows a similar graph for a whole crude oil [18]. The method described takes 4 hr or more for a duplicate analysis and is therefore too time-consuming for use if many samples have to be examined but is can be used if extra evidence is required in specific instances.

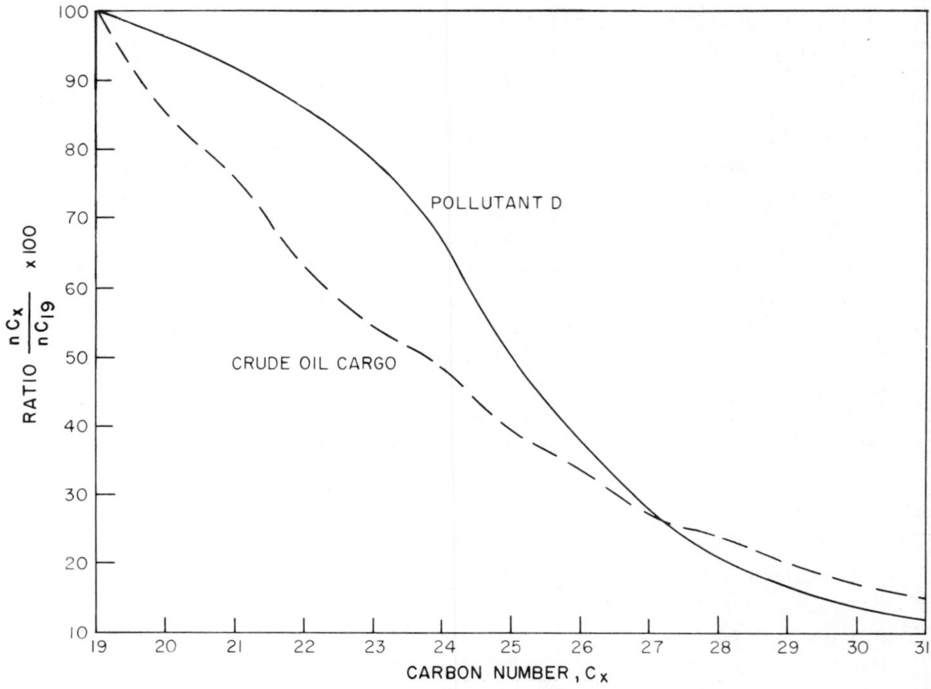

FIG. 14. Normal paraffin ratios for a crude oil and a pollutant sample (by courtesy of the Institute of Petroleum [17]).

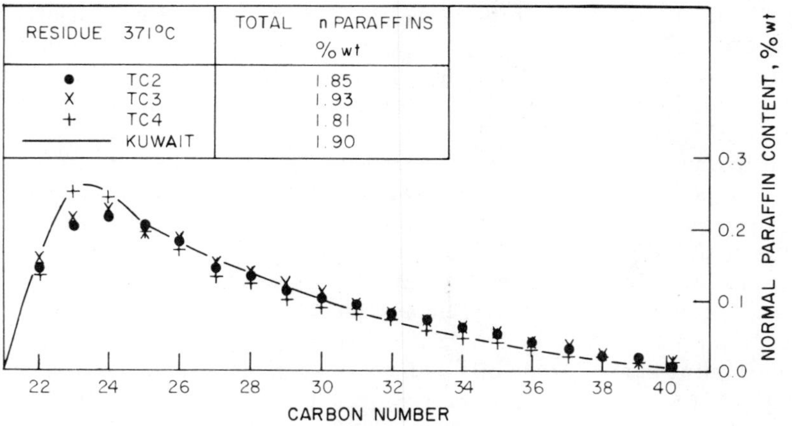

FIG. 15. Normal paraffin distributions of residues above 371°C from Kuwait crude and weathered "Torrey Canyon" cargo samples (by courtesy of the Institute of Petroleum [17]).

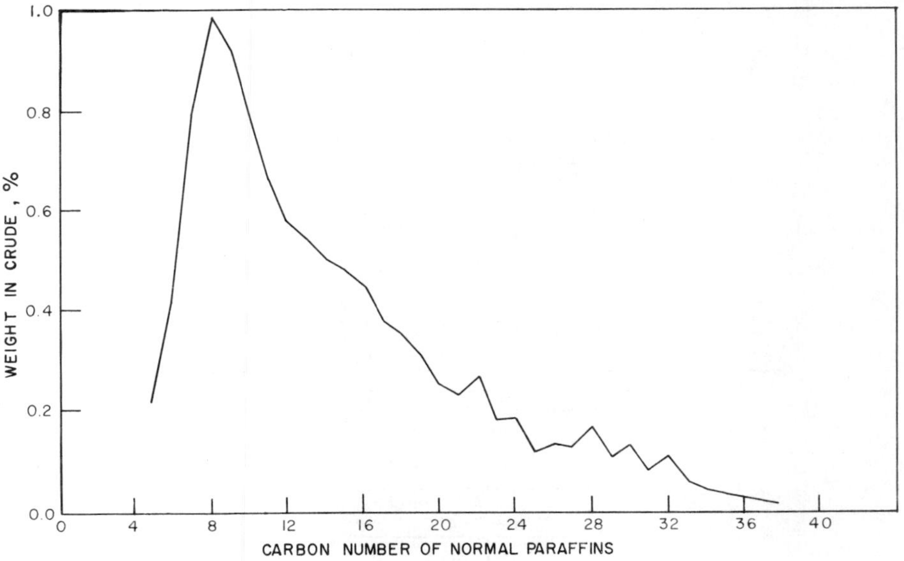

FIG. 16. Distribution of n-paraffins in Amposta-marino crude oil (by courtesy of Nature [18]).

In the second method [19] a 450 to 500° C distillation fraction of the oil is further separated by means of a liquid chromatography column packed with alumina with pentane as the eluent. The fractions from this separation are examined by GC on a 50-m by 0.25-mm ID stainless-steel column coated with Dexsil 300 operated isothermally at 273° C to give chromatograms of the C_{27}-C_{32} triterpanes in the oil. Further simplification of the samples by removal of the normal paraffins with molecular sieve 5A may be required. A schematic diagram of the method is shown in Figure 17. It is apparent that the method is so time-consuming in its present form that it is unlikely to be used for oil spill identification. It is probable, however, that a simpler version using GC combined with selective ion mass spectrometry could be developed specifically for identification purposes [29].

VI. METHODS FOR THE IDENTIFICATION OF OIL IN THE WATER COLUMN

All the methods described so far assume that at least a few milligrams of oil are available as a phase separate from water. If the oil is present in low concentration, as it may well be in the ocean some time after a spill, then the analyst is presented with an extremely difficult problem of identification.

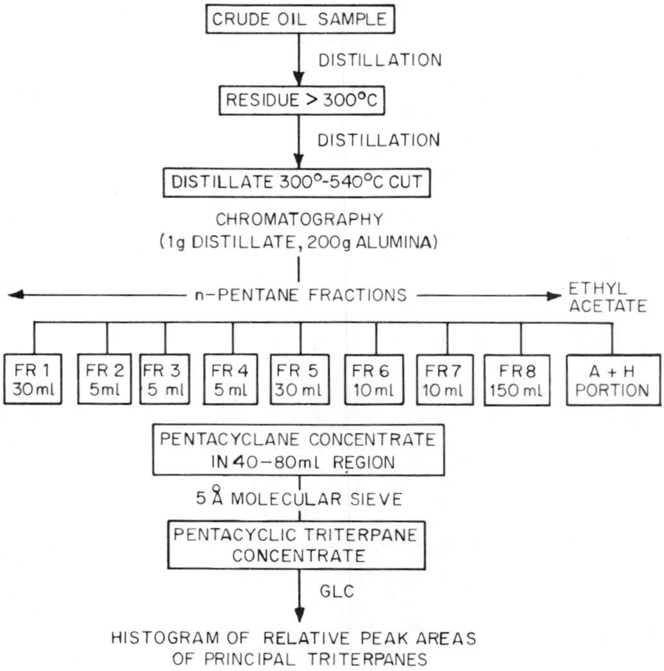

FIG. 17. Schematic diagram of procedure for triterpane separation (by courtesy of Analytical Chemistry [19]).

Firstly, some concentration technique must be used to obtain enough sample for analysis by GC. Secondly, even when this has been achieved, identification can only taken place if a considerable amount of prior knowledge is available since the fingerprints obtained from dissolved oil will bear no resemblance to those from the oil itself. The reasons for this are twofold: first, only the lower hydrocarbons up to about C_{20} are likely to be present since solubility* decreases with increasing molar volume and, second, because aromatic hydrocarbons are much more soluble in water than are aliphatic hydrocarbons of similar molecular weight [20].

*There is considerable doubt as to whether the concentrations of oil found in the water column represent oil in true solution, in colloidal suspension or in microparticulate form. It is probable that some of the values reported represent a combination of all three.

Concentration from a large volume (1-10 liters) of seawater can be achieved by solvent extraction, absorption by a resin [21], or by a cyclic degasification as described by Grob [22-24]. The disadvantages of solvent extraction are, first, the difficulty of obtaining a sufficiently pure solvent and, second, the possibility of incomplete extraction of the more hydrophilic compounds. On the other hand, the method is well established and easy to use. Absorption by a resin also requires relatively large volumes of ultrapure solvents for the purification of the resin prior to use. The most attractive method is that described by Grob in which an inert gas such as helium is recirculated continuously through the water sample for periods of 24 hr or more. After bubbling through the water the helium plus any organic compounds are passed through a very small charcoal filter. The organic compounds are removed by the charcoal and the helium passes back into the water for another cycle. The organic material is removed from the charcoal by a few microlitres of CS_2 and the mixture is examined by high-resolution GC. The whole extraction equipment must be made of glass and metal, although very restricted use of P.T.F.E. is permissible in some parts of the apparatus.

Whatever method of extraction and concentration is used the net result of GC capillary column analysis of the concentrate is a complex chromatogram showing several hundred peaks. Identification of a particular oil from such a chromatogram is extremely difficult in the present state of knowledge and the information obtained is of more use to the biologist than to the forensic chemist. As more background information is obtained it is possible that analysis of the hydrocarbons in the water column will attain equal importance with other methods for spill identification. The topic is of growing importance because of the large amount of offshore drilling now taking place in many parts of the world.

VII. CONCLUSION

Gas chromatography is undoubtedly the most powerful single method of pollutant identification. It requires only a few milligrams of sample and is relatively inexpensive in terms of both capital and operating costs. For spill identification, however, it is largely a qualitative technique which relies heavily on the skill and judgment of the person interpreting the chromatograms. It is for this reason that attempts to obtain quantitative data such as n-paraffin:isoprenoid hydrocarbon ratios are to be welcomed.

Computer pattern recognition techniques are beginning to be applied to the identification of oil spills from elemental composition [25] and it is likely that such methods will be applied to GC chromatograms in the next few years. However, it cannot be overemphasized that positive identification of oil spills should use as many analytical techniques as are available to the investigator and that reliance on gas chromatography (especially low-resolution GC) alone is likely, sooner or later, to lead to erroneous conclusions.

REFERENCES

1. P. R. Mommessin and J. C. Raia, Conference on Prevention and Control of Oil Pollution, American Petroleum Institute, San Francisco, 1975, p. 155.
2. W. Harrison, M. A. Winnik, P. T. Y. Kwong, and D. Mackay, Environ. Sci. Tech. 9:231 (1975).
3. Z. R. Regnier and B. F. Scott, Environ. Sci. Tech. 9:469 (1975).
4. O. C. Zafiriou, J. Myers, and F. Freestone, Marine Polln. Bull. 4:87 (1973).
5. S. J. Ramsdale and R. E. Wilkinson, J. Inst. Petrol. 54:326 (1968).
6. R. D. Cole, Nature 233:546 (1971).
7. O. C. Zafiriou, Anal. Chem. 45:952 (1973).
8. E. R. Adlard, Crit. Rev. Anal. Chem. 5:13 (1975).
9. J. Wilson and C. Z. Marczewski, Anal. Chem. 45:2409 (1973).
10. D. E. Bryan, V. P. Guinn, R. P. Hackleman, and H. R. Lukens, U.S. Atomic Energy Commission Report GA-9889, 1970.
11. E. R. Adlard, L. F. Creaser, and P. H. D. Matthews, Anal. Chem. 44:64 (1972).
12. J. W. Miller, Conference on Prevention and Control of Oil Spills, American Petroleum Institute, Washington, D. C., 1973, p. 195.
13. M. E. Garza and J. Muth, Environ. Sci. Tech. 8:249 (1974).
14. C. E. Davis, A. E. Krc, J. J. Szakasits, and R. L. Hodgson, Conference on Prevention and Control of Oil Pollution, American Petroleum Institute, San Francisco, 1975, p. 93.
15. J. S. Warner, Conference on Prevention and Control of Oil Pollution, American Petroleum Institute, San Francisco, 1975, p. 97.
16. J. V. Mortimer and L. A. Luke, Anal. Chim. Acta 38:119 (1967).
17. D. F. Duckworth, Water pollution by oil (P. Hepple, ed.), Institute of Petroleum, London, 1971, p. 165.
18. J. Albaiges and J. M. Torradas, Nature 250:567 (1974).
19. J. G. Pym, G. E. Ray, G. W. Smith, and E. V. Whitehead, Anal. Chem. 47:1617 (1975).
20. C. McAuliffe, J. Phys. Chem. 70:1267 (1966).
21. G. A. Junk, J. J. Richard, M. B. Grieser, D. Witiak, J. L. Witiak, M. D. Arguello, R. Vick, H. J. Svec, J. S. Fritz, and G. V. Calder, J. Chromatog. 99:745 (1974).
22. K. Grob, J. Chromatog. 84:255 (1973).
23. K. Grob and G. Grob, J. Chromatog. 90:303 (1974).
24. K. Grob, K. Grob, and G. Grob, J. Chromatog. 106:299 (1975).

25. D. L. Duewer, B. R. Kowalski, and T. F. Schatzki, Anal. Chem. 47: 1573 (1975).
26. A. P. Bentz, Anal. Chem. 48:454S (1976).
27. E. R. Adlard, J. Inst. Petrol. 58:63 (1972).
28. G. A. Flanigan and G. M. Frame, Research/Development 28:28 (1977).
29. J. Albaigés and P. Albrecht, Internat. J. Environ. Anal. Chem., in press.

Chapter 8

PYROLYSIS GAS CHROMATOGRAPHY

Emilio J. Gallegos
Research Services Department
Chevron Research Company
Richmond, California

I. INTRODUCTION

A very thorough discussion of pyrolysis gas chromatography (PGC) appeared in 1966 in Chromatographic Reviews [1]. In this presentation, Levy [1] gave a comprehensive treatment of the terminology, historical development, apparatus, applications, and many other aspects of this technique. PGC may be defined as gas chromatography of products formed from a substance that has undergone thermal transformation. In many ways, PGC is comparable to analysis by mass spectrometry. Levy [2] showed this diagramatically in Figure 1.

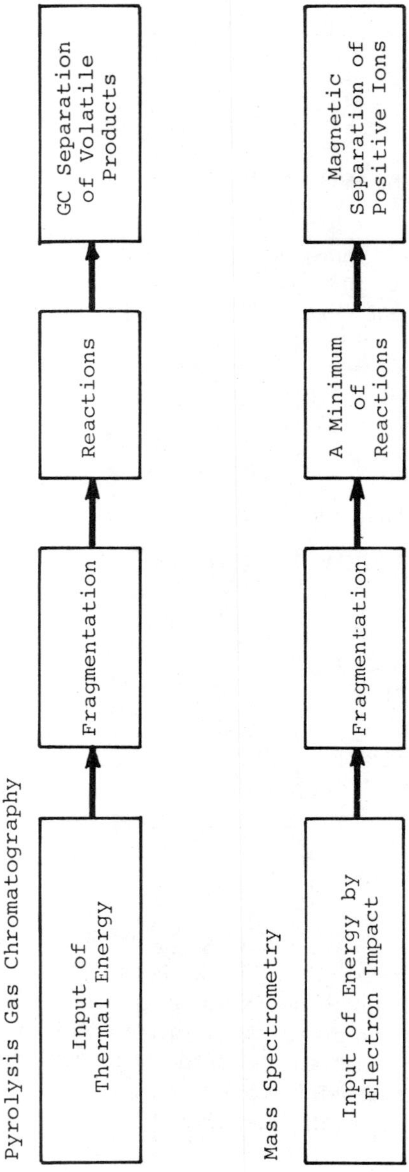

FIG. 1. Comparison of the principal events in pyrolysis gas chromatography and mass spectrometry.

This chapter compares briefly the basic differences in performance of the more important types of pyrolyzers that are presently in use, followed by a short review of vapor phase pyrolysis of pure hydrocarbons. The final portion of this chapter reviews and discusses the recent applications of PGC to the analysis of petroleum residua, coals, shales, and recent sediments.

II. INSTRUMENTATION

There are two basic modes of operation for pyrolysis units, the pulse mode and the continuous mode. A third alternative, vapor phase flow-through pyrolysis, can be considered as a combination of the first two basic forms.

In the pulse mode, the sample is first placed in the pyrolysis region, e.g., coated onto a wire. Then, ideally, the wire is heated instantaneously to the desired temperature. In the continuous mode, the pyrolysis chamber is held at a desired temperature, say 800°C; the sample is then conveyed from an area at room temperature into the pyrolysis region. The pulse mode of operation is considered to be the superior technique for interlaboratory comparison of data. This is because the pulse mode parameters more closely resemble those described by a theoretical model, and it is easier to describe and reproduce the conditions of heat transfer to the sample in the pulse mode than in the continuous mode. Figures 2 and 3 depict idealized models of the pulse and the continuous modes used in PGC.

For a comparative study of the performance characteristics of different pyrolyzers, we refer to a publication by Walker and Wolf [3]. In this paper, the PGC behavior of two model compounds, 2,6,10- and 2,6,11-trimethyldodecane, was investigated using three different pyrolyzers:

1. A vapor phase flow-through tube reactor
2. A Curie-point pyrolyzer based on radiofrequency induction heating of a ferromagnetic wire
3. A capacitive boosted filament heater

The studies on vapor phase PGC showed that the product distribution obtained by this method compared favorably with that predicted by a free radical chain process similar to that proposed by Kossiakoff and Rice [4]. Walker and Wolf [3] concluded that, although there are substantial similarities between the products from the reference system (vapor phase pyrolysis) and from the filament and Curie-point pyrolyzers, significant differences exist. Because of the presence of heat transfer and secondary reactions, pyrolysis in the condensed phase is far more complex than in the vapor phase so that the three basic assumptions made by the Kossiakoff-Rice [4]

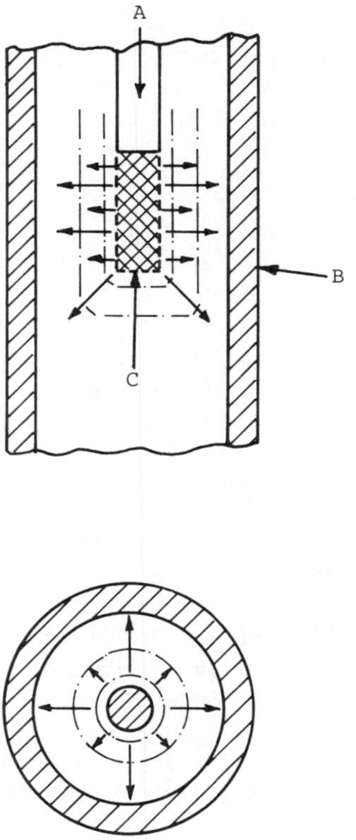

FIG. 2. Cross-sections of an idealized model of a pulse mode unit. A, metal rod; B, enclosure; C, area coated with sample; →, direction of heat transfer; ⟶, direction of mass transfer; and -·-·, points of equal temperature.

theory may not be applicable to the Curie-point or filament pyrolyzer. Hence, with the latter systems: (a) the chain length of the radical can be small so that reactions other than chain propagating steps may be important, (b) the rapid thermal pulse used with these pyrolyzers may not allow the system to attain complete thermal equilibrium, and (c) secondary reactions, which change the net product distribution, may occur.

III. PURE HYDROCARBONS

Recently, several investigators have instigated fundamental studies by comparing pyrograms obtained from pure hydrocarbons. Fanter et al. [5] reported the pyrograms of 83 different C_8-C_{10} hydrocarbons. These comprised

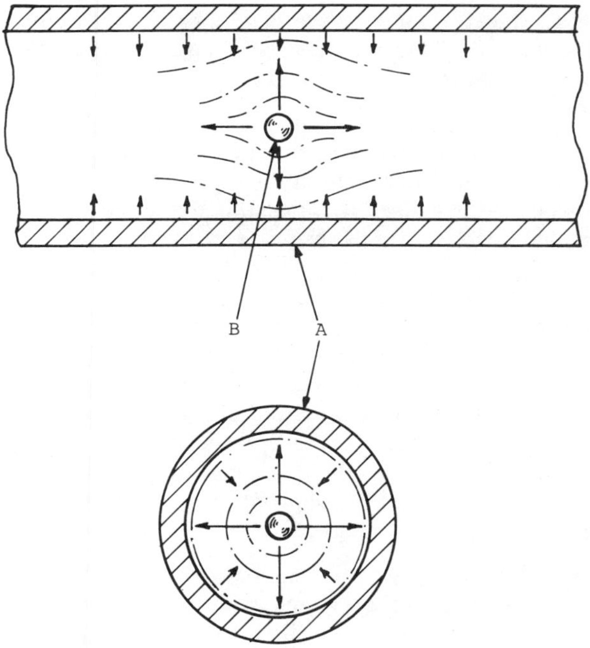

FIG. 3. Cross-sections of an idealized model of a continuous mode unit. A, metal, glass, or quartz tube; B, milligram quantity of solid sample in the form of a sphere; →, direction of heat transfer; ⟶, direction of mass transfer; and -·-·, points of equal temperature.

24 alkanes, 10 alkynes, and 49 straight-chain, branched-chain, and cyclic alkenes. The results were reported in terms of relative mole response as a function of specific retention index interval. These investigations showed that, with the exception of cis-trans isomers, most of the alkenes of the same molecular weight produce significantly different pyrograms. This is an interesting observation because alkene isomers are difficult to distinguish by mass spectrometry. This work was performed in a vapor phase pyrolysis gas chromatographic system. It was also shown that in the temperature range from 575 to 625°C patterns were quite repeatable.

The data from these studies have been used to compare pyrograms with mass spectra. The first step in these correlations has been the tabulation of the relative peak intensity from the pyrograms as a function of retention index intervals of one carbon number. Mass spectral intensity is tabulated as a function of carbon number rather than for the separate m/e, i.e., the intensity of the mass spectral fragments having a given number of carbons are summed together to give the total intensity for that number of carbons. From Figure 4 we perceive that, as a general rule, the mass spectra tend to show the largest amount of fragments in a carbon number range one number higher than that observed for the pyrograms. There are, of course, some notable exceptions.

Kelly and Wolf [6] observe the existence of an obvious but superficial analogy between PGC and mass spectrometry. Based on consideration of the physical processes involved in thermal and electron impact fragmentation, they conclude that it is unlikely that any general correlation can be developed in which existing mass spectral data can be used to correlate with pyrograms. However, in a recent study, Green et al. [7] observed that two different electron impact-induced hydrogen abstraction elimination reactions appeared to be superficially analogous to certain thermal reactions. They compared the stereochemical prerequisites in the transition states of thermal and electron impact reactions by determining the relative transfer of competitive diastereotopic hydrogens. For both electron impact reactions and thermal reactions the stereochemical course developed exactly parallel to each other. This suggests that in some cases PGC and MS will continue to enjoy mutual support.

Brown [8] made a study of the compositional data from the thermal degradation products of alkanes as reported by Cramers and Keulemans [9] and by Cramers [10]. Using a simplified representation of the Rice theory [11], Brown [8] demonstrated that the monoolefins in the pyrolysis products originate primarily from simple cleavage of the parent molecule. From this simple concept, correlations and working rules were developed to predict alkane structures accurately from the pyrograms. Most of the C_6-C_8 alkanes could be reconstructed exactly by this process. Of the others, it was possible to indicate the principal basic structure. It is interesting to note that all pyrograms of the n-alkanes show the presence of 1-olefins, the heaviest having the same C-number as the parent paraffin. The modified Rice theory predicts for the C_6-C_8 alkanes that fragmentation is favored at bonds α and β to branch points. In gem structures, γ cleavage produces two mono-olefins; the position of the double bond is favored to occur at a methylene rather than at a methyl carbon. Actually, Keulemans and Perry were the first to propose the use of pyrograms for structural interpretation [12]. A pyrolysis gas chromatograph, they suggested, can be used as a "poor man's mass spectrometer."

In a series of papers entitled "Pyrolysis of Liquid Hydrocarbons," Szepesy et al. [13] and Illés et al. [14-18] describe the pyrolysis of individual hydrocarbons, hydrocarbon mixtures, and naphtha fractions using a vapor phase flow-through pyrolyzer. The first paper in the series treats the characteristics of the pyrolysis reactors and the experimental technique. Their following three papers describe the overall decomposition rates for n-hexane, 2-methylpentane, 2,2-dimethylbutane, cyclohexane, n-heptane, 2,4-dimethylpentane, n-octane, isooctane, and the distribution of the pyrolysis products obtained.

	C_1	C_2	C_3	C_4	C_5	C_6	C_7	C_8	C_9	C_{10}	
	38	100	89	68	27						TF
	(2)	(36)	(100)	(63)	(22)	(10)					(MS)
	85	100	16	89	[95]	8					TF
	(3)	(53)	(95)	(100)	(14)	(23)					(MS)
	85	17	2	12	100	6					TF
	(3)	(34)	(100)	(58)	(18)	(18)					(MS)
	20	100	53	57	18	12					TF
	(2)	(59)	(100)	(83)	(40)						(MS)
	20	100	52	37	28	13					TF
	(1)	(40)	(100)	(55)	(38)	(36)	(5)				(MS)
	39	100	66	44	27	50					TF
	-	34	(100)	(62)	(30)	(9)	(7)				(MS)
	39	100	63	50	19	60					TF
	-	(42)	(100)	(92)	(52)	(22)	(12)				(MS)
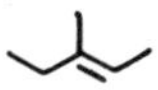	100	-	-	-	95						TF
	(4)	(46)	(100)	(65)	(73)	(32)					(MS)
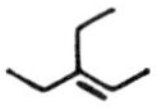	100	24	12	16	89	43					TF
	(4)	(51)	(100)	(68)	(53)	(24)					(MS)
	74	100	58	33							TF
	(3)	(57)	(100)	(78)	(6)	(9)					(MS)
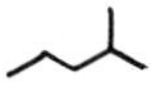	60	60	100	50	10						TF
	(3)	(22)	(100)	(10)	(17)	(2)					(MS)
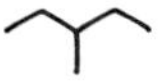	100	85	28	64	50						TF
	(3)	(58)	(58)	(100)	(4)	(2)					(MS)
	100	80		90							TF
	(6)	(42)	(100)	(67)	(51)	-					(MS)
	33	100	35	25	12	6					TF
	(2)	23	(100)	(70)	(1)	(8)					(MS)
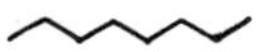	32	100	35	20	13	17					TF
	(23)	(37)	(100)	(39)	(25)	(27)	-	(4)			(MS)
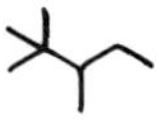	100	19	50	87	82	12	19				TF
	(2)	(21)	(34)	(100)	(2)						(MS)
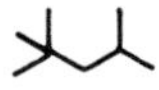	50	2	30	100	3	19					TF
	(3)	(16)	(35)	(100)	(1)	-	(5)	-			(MS)
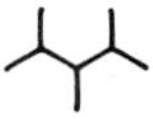	95	12	100	49	72	36					TF
	(2)	(22)	(100)	(22)	(78)						(MS)
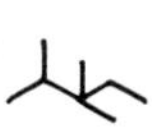	100	64	75	53	52	53					TF
	(2)	(21)	(100)	(39)	(60)	(25)					(MS)
	34	100	35	16	10	16					TF
	-	(33)	(100)	(64)	(18)	(22)	(4)	-	(3)		(MS)

FIG. 4. Thermal fragmentation (TF) patterns from 600°C pyrolysis compared to carbon number mass spectra (MS).

The thermal decomposition kinetics of binary and tertiary mixtures of n-heptane, isooctane, and cyclohexane, and the experimental as well as the calculated pyrolysis product distributions are covered in the last two papers in this series. One major purpose of these extensive investigations is to find overall kinetic expressions suitable for describing both decomposition rate as well as product distribution as a function of feedstock composition.

The findings can be summarized as follows: With the exception of cyclohexane, the overall decomposition rates of the hydrocarbons investigated do not obey first-order kinetics. In the given temperatures and reaction times, the yield of the main reaction products from single compounds are simple curves which are independent of the temperature. In addition, for mixtures, the yield curves can be calculated additively from the yield curves of the individual components in the initial mixture.

This section would not be complete without reference to studies of thermal unimolecular reactions of pure hydrocarbons in the gas phase. In this field, it is possible to report precise quantitative thermodynamic data, which are indispensable for an understanding of the decomposition process. A review covering this subject was recently published in Chemical Reviews by Frey and Walsh [19].

IV. NONVOLATILES FROM PETROLEUM

Leplat [20] applied PGC to the study of nonvolatile petroleum fractions. His intention was to establish standard programs for petroleum residues and related products and to work out a PGC technique for the rapid assay of samples without preliminary treatment. Such treatment is generally necessary when petroleum fractions are analyzed; for routine determination, this aspect can be quite onerous and time-consuming. His studies were also directed towards the application of PGC to problems such as: (a) determination of a possible correlation among various petroleum residues, (b) determination of the characteristic surface indices of petroleum, (c) soil evaluation, and (d) possible correlations between various properties of street asphalt. PGC was chosen because the method is simple and rapid and requires only micrograms of sample. Both the filament method and the furnace method were investigated. He noted that excellent results can be obtained by the latter approach; the program can be reproduced and the differences noted were significant enough to be useful in geochemical prospecting. The results on his other goals were not as conclusive.

The applicability of PGC to the resinous-asphaltic components of petroleum was the subject of a paper by Martynov et al. [21]. These workers evaluated their pyrograms by surveying the light ends evolved.

The distribution of C_1-C_{12} pyrolysis products are actually quite similar for widely different bitumens and derived fractions. This was observed not only in the work discussed above but also in the studies by Knotnerus [22] described further below. Even though Martynov et al. [21] correctly claim the presence of differences for different samples, these variations tend to be subtle, limiting somewhat the applicability of their method.

Based on the information obtained from pyrolysis (hydrogenation)-GC of a number of model compounds, Knotnerus [22] studied the possibility of using this technique to characterize bitumens from various sources. Correlations and differences were determined by looking specifically at the ratio of cyclic saturates to paraffins in the C_4-C_{10} range, the yield of C_4-C_{10} paraffins, the ratio of 5-membered rings to 6-membered rings, and the aromaticity factor F, the ratio of the weight percent of aromatic cyclanes in the pyrolyzate to total yield of pyrolyzate. He concluded from his results that bitumen may be characterized on the basis of PGC. A recent paper by Ramljak et al. [23] describes the use of this technique to determine the amount of maltenes and asphaltenes in bitumens. Pyrolysis was carried out in a furnacelike microreactor. The determining parameter in this study is the ratio between the sum of the peak areas of the unsaturated C_3-C_5 hydrocarbons and of their saturated counterparts. For good results, very strict control of the pyrolysis conditions is required; the complete chromatographic resolution of the lower molecular weight degradation products is also essential. The authors claim that, in general the method allows the determination of asphaltenes in bitumens with an accuracy of 1%.

V. COAL

Surprisingly little has been published on the analysis of coal by modern analytical techniques. Only recently have Romováček and Kubát described the characterization of coal substances by PGC [24].

Coal samples were weighed into small tin capsules which, at a preset time, were automatically dropped into a bath of molten tin maintained at a suitable temperature. Pyrolysis was considered to be complete in less than 30 sec. To minimize secondary reactions, the pyrolysis products were quickly removed from the hot region by a carrier gas. Gas chromatography was used to analyze these products.

The highest molecular weight product detected under the conditions used was about that of the C_8 aromatics, i.e., the xylenes. This does not imply the absence of higher molecular weight material, but under the conditions of the analysis, these were at too low a level to be detected.

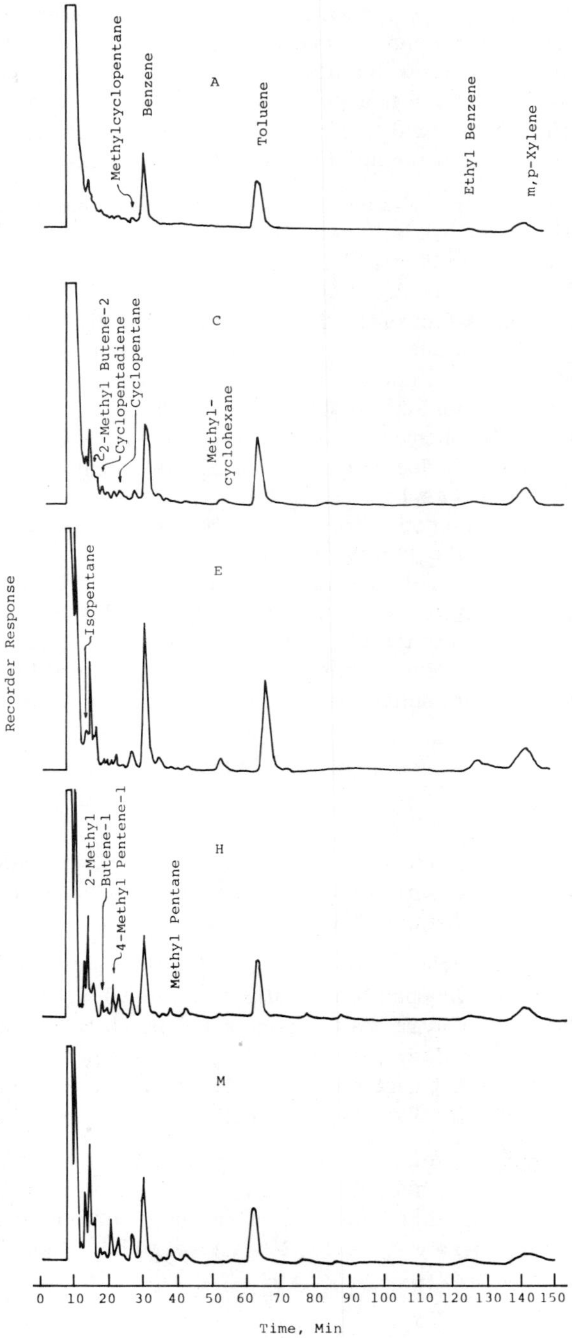

FIG. 5. Chromatograms obtained from coals of different ranks.

At 700° F, the optimum pyrolysis temperature, the greatest differences between relative peak heights for coals of different rank were observed. This is illustrated in Figure 5. Chromatogram A in this figure is obtained from semianthracite, a high-rank coal; Chromatograms C and E are from intermediate-rank low-volatility bituminous coals, and Chromatograms H and M are derived from low-rank subbituminous coals. The pyrograms of the five different coals show a great deal of similarity. Quantitation of the differences is established by determining the ratios of specific reproducible peaks in the pyrogram. For all coals studied, the most important C_5 and higher components are pentane, benzene, toluene, and m- and p-xylene.

An interesting approach to emphasize the variations in the chromatograms obtained is to subject coal macerals to PGC. A maceral is a petrological unit in a microscopic section of coal. Macerals taken from exinite, inertinite, and vitrinite (a bituminous high-volatility coal) produced pyrograms with greater differences among one another than were observed among the original coals.

Tamaki et al. [25] used pyrolysis GC-MS to analyze 16 different domestic and foreign coals. The relationship between pyrolysis products of each sample and its degree of coalification were compared by analyzing the levels of carbon monoxide, carbon dioxide, methane, ethane, and residue in their pyrolysis products. A summary of their results on 16 different coals is shown in Table 1. In each grouping, the coals are listed in increasing maturity (higher coalization).

The pyrolysis arrangement used by these investigators consisted of a furnacelike pyrolysis chamber with a nichrome wire coil heater. The sample was placed in a glass boat, which was moved into the furnace region by a magnet. Two iron pieces surrounded the sample boat. The uninterrupted inert gas flow presumably reduced secondary reactions in this design.

VI. SHALE

Simmonds et al. [26] published the results of their pyrolysis-GC-MS studies on three samples; viz., Fig Tree shale, an ancient South African sediment ($3.0 \cdot 10^9$ years), the Murray meteorite, and desert soil from the Mojave desert in Thermal, California. This work was performed to demonstrate the ability of the method to distinguish material of recent biological origin from fossil or meteoritic organic matter. This, in turn, was done in anticipation of using this technique for the analysis of organic material in Martian soil.

A summary of their results is shown in Figure 6. Naphthalene, toluene, and benzene are indicated as guides on the chromatograms. The

TABLE 1. Pyrolysis Gas Products and Residue Content of Coal Samples

Name of coal	Dry, ash-free basis (%)	Ash* and fixed carbon (%)	Pyrolysis gas products peak area					Residue, (%)
			CH_4	CO	CO_2	C_2H_4	C_2H_6	
Tenhoku	72.6	54.2	184	277	111	27	12	59.0
Chikubetsu	76.8	51.1	234	181	87	41	15	49.9
Taiheiyo	77.8	49.1	311	217	60	50	43	53.2
Kashima	78.1	49.6	332	222	43	47	27	51.7
Takamatsu	79.7	56.9	300	229	56	40	30	57.9
Mihoro	81.1	51.4	392	180	62	81	60	59.8
Akadaira	83.5	55.8	417	138	32	56	65	50.1
Miike	84.5	56.1	425	106	43	112	58	62.3
Yubayi	86.2	59.3	503	66	32	90	39	62.5
Hashima	37.2	65.4	504	16	32	63	38	73.6
Ohmine	93.1	91.7	65	21	43	38	26	84.3
Southern Bell	86.2	63.7	447	82	42	80	64	62.1
Moura	87.6	79.9	444	66	23	67	58	71.7
Michel	88.4	73.7	458	58	17	40	53	80.8
New River	88.9	75.1	409	47	12	43	43	75.6
Itmann	90.8	82.9	308	26	47	50	11	87.7

*Dry basis.

numbers shown are identified in the original paper. Identification of the eluted peaks from the chromatograph is carried out by mass spectrometry. This approach, as illustrated by these workers, demonstrates the tremendous amount of detailed information that can be obtained by this technique.

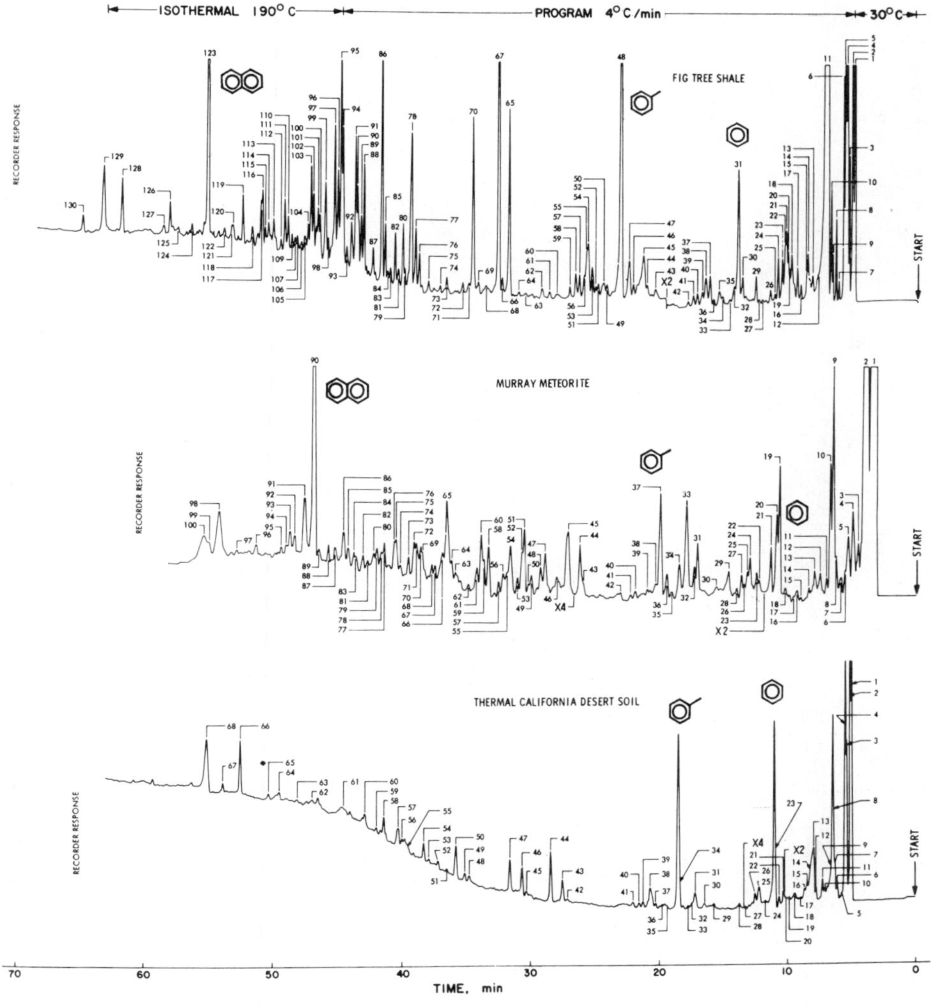

FIG. 6. Pyrolysis GC-MS of Fig Tree shale, Murray meteorite, and Thermal desert soil.

More recently, Gallegos [27] analyzed Green River shale kerogen and the saturate fraction of the pyrolysis oil from kerogen by pyrolysis-GC-MS. Figure 7 depicts the PGC-MS results for kerogen from solvent-extracted Green River shale. The chromatogram at the bottom is obtained from the total of the ions collected by the mass spectrometer. The trace, which is, therefore, equivalent to that obtained with an FID, is called a reconstructed

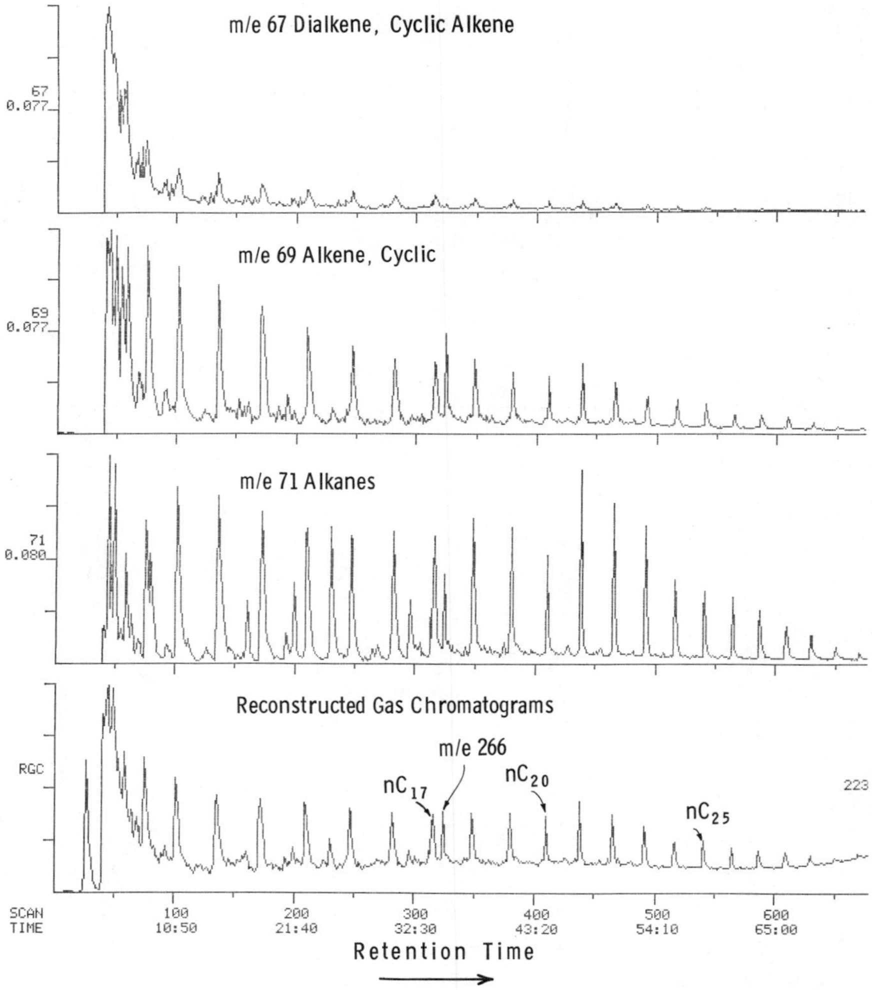

FIG. 7. Pyrolysis GC-MS mass chromatograms of kerogen in solvent extracted Green River shale.

gas chromatogram. The horizontal axis shows the scan number and the retention time. The temperature of the chromatograph oven ranged from 150-330°C. The upper three spectra, which are obtained concurrently with the signal from the total ion monitor, are the signals observed for specific m/e channels from the mass spectrometer. Their use is based on the fact that each class of compound will yield specific m/e fragments. Hence, the m/e 69 channel will monitor the emergence of olefins and 1-ring saturates. In the experimental configuration used, a minimum of secondary reactions presumably occur so that many of the peaks in the m/e 69 mass chromatogram are probably due to monoolefins; more specifically, they are probably derived from a homologous series of α-olefins. One-ring saturates are also detected by this channel so they may not be discounted in the evaluation of the results. The m/e 71 mass chromatogram indicates the presence of the paraffins whereas the m/e 67 channel monitors diolefins and/or cyclic olefins.

A GC-MS analysis of the saturate fraction of the pyrolysis oil is shown in Figure 8. Here, alkanes, steranes and terpanes are characterized by their mass chromatograms at m/e 197, m/e 217 and m/e 191.

An interesting approach was described recently by Reed and Warren [28] who analyzed Green River shale by thermal chromatography. This technique, although closely related to PGC, is different in some important aspects. In PGC, secondary reactions are minimized by sweeping the pyrolysis products as quickly as possible out of the pyrolysis area onto the chromatographic column with an inert carrier gas. Thermal gas chromatography, however, first vaporizes that part of the sample which is vaporizable. The residue is then pyrolyzed by controlled thermal heating. This results in a thermogram which carries with it information similar to that obtained by thermal gravimetric analysis. The vaporized components and products of pyrolysis are then condensed in a cold trap, followed by rapid vaporization of the sample into the column for GC analysis.

Condensation of the pyrolysis products encourages secondary reactions. The material which has been pyrolyzed, condensed, and then introduced into the GC column will not have the same composition as the pyrolyzate, which is immediately introduced into the GC column. The "thermal chromatography" apparatus can actually also be used to perform PGC by bypassing the condensation stage.

An example of the thermograms which Reed and Warren [28] and Scrima et al. [29] obtained on Green River shale is shown in Figure 9. The corresponding chromatogram of the condensate, i.e., the condensed pyrolysis products from bitumen and kerogen, is shown in Figure 10. The authors also demonstrated that the recoverable oil in Green River shale determined by this technique is comparable to results obtained by the classical Fischer assay. The Fischer assay is a retort technique for determining the amount of gas, liquids, and residue of coals, shales, and so on.

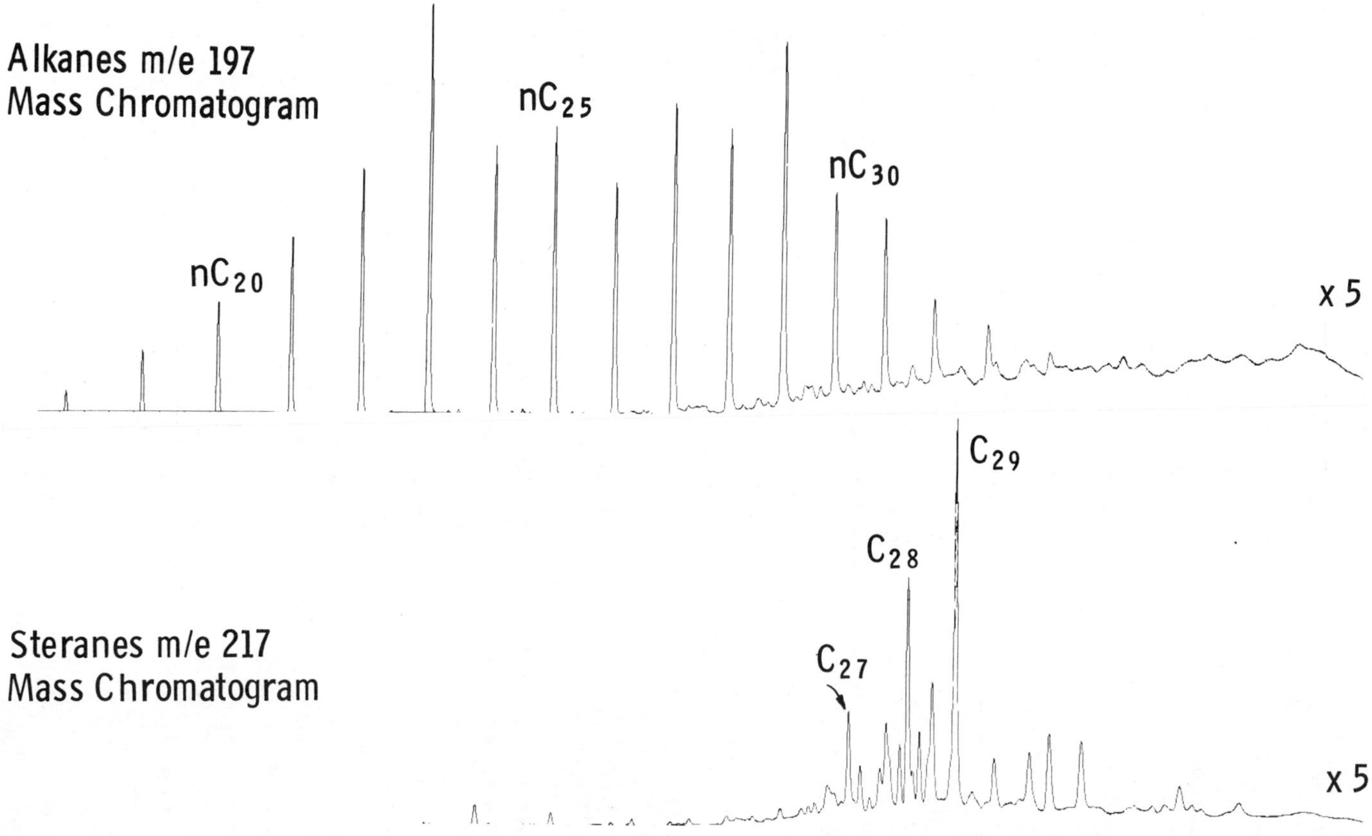

Alkanes m/e 197
Mass Chromatogram
nC20
nC25
nC30
x 5
Steranes m/e 217
Mass Chromatogram
C27
C28
C29
x 5

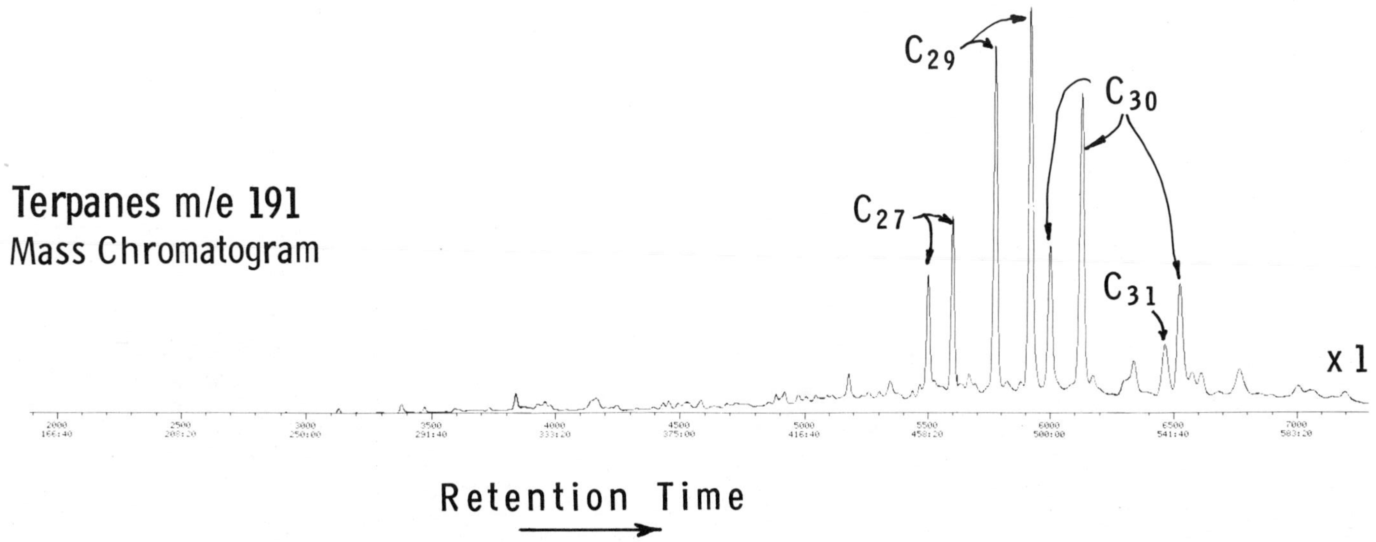

FIG. 8. Sterane, terpane analysis by GC-MS of the saturate fraction of Green River shale kerogen pyrolyzate.

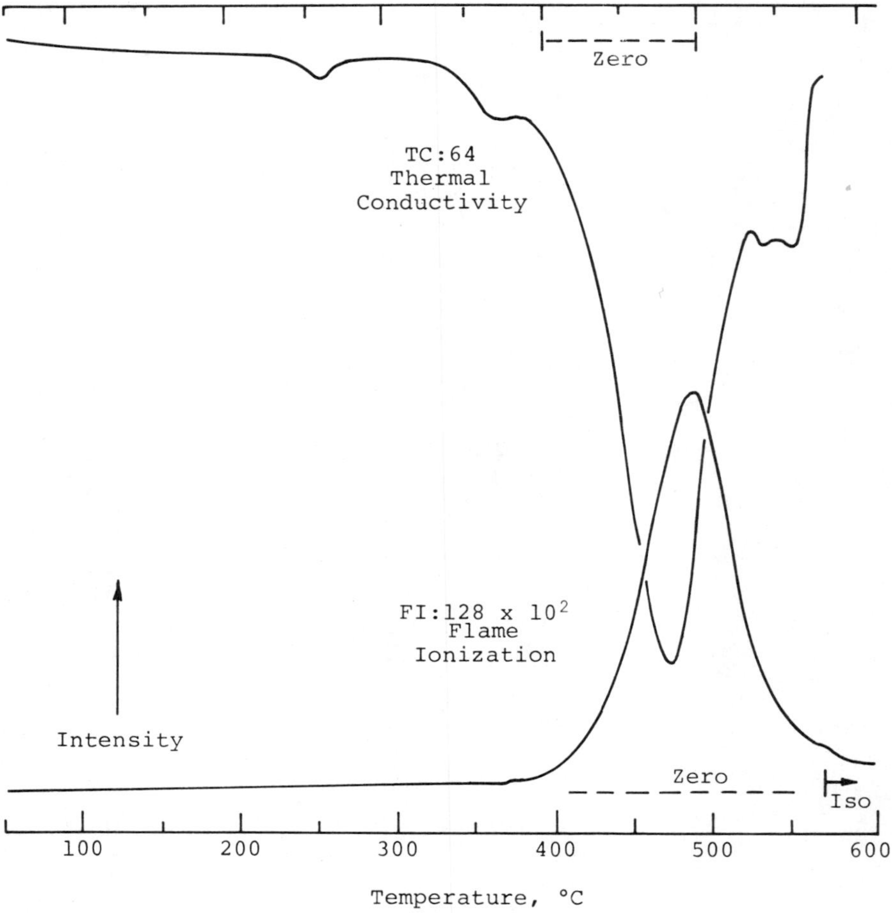

FIG. 9. Thermograms of oil from Green River shale—yield of oil, 34.7 gallons per ton. The upper and lower traces were recorded simultaneously with a TCD and a FID, respectively. The TCD "sees" everything and the FID "sees" only organic materials. Additional peaks in the upper trace are probably due to water, CO_2, CO, NH_3, and other inorganics. Sample heating rate was 28°C min^{-1}.

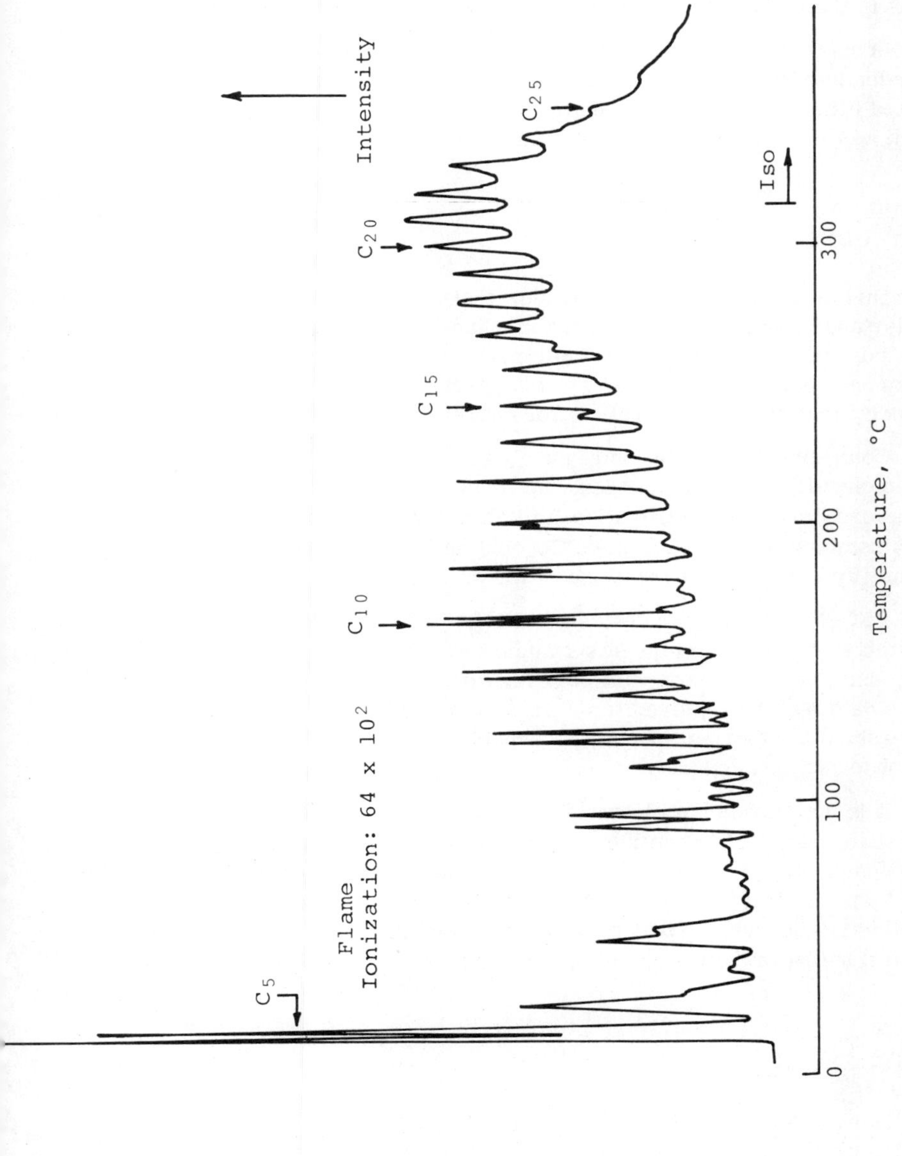

FIG. 10. Chromatogram of condensate from Green River shale—materials observed here were formed by thermalyzing oil shale. The temperature program rate was 8°C min^{-1}.

VII. HUMIC SUBSTANCES

Gomez et al. [30], using a Curie-point pyrolyzer, pyrolyzed humic acids extracted from coal. Their purpose was to use PGC as a vehicle of identification. They found that the pyrograms were reproducible and that the variations in the history of the sample were reflected in the pyrograms.

In another study, Marzakov and Gostenkov [31] used PGC to distinguish between humic substances from soil and from peat. A resistance wire coil heated furnace was used for the pyrolysis. A major observation from the work was that peat gave richer pyrograms than coals.

VIII. CONCLUSIONS

It should be apparent from this review that the analysis of petroleum residua, coals, and shales by PGC is still in its infancy. Results so far indicate that it is possible to obtain reproducible pyrograms from the pyrolysis of complex condensed systems. Most of the pyrograms are currently used in fingerprint identification as well as for correlation studies.

Component identification of pyrolysis product has been limited to the low molecular weight components up to about $\sim C_{10}$. This is due to sample complexity and the identification limitations in GC. The addition of a mass spectrometer to the outlet of a PGC column improves the identification capability of PGC by at least one order of magnitude.

PGC-MS, or, better yet, PGC-HRMS (high-resolution mass spectrometry) should therefore offset somewhat the effect of sample complexity and, thus, improve pyrogram peak identification capabilities. Even then, more data may be required to provide the information needed to completely understand the thermal cracking mechanism and role of secondary reactions in condensed phase pyrolysis.

It is not only in the classical fields of petroleum residues, bitumen, and shale that PGC is contributing to a better knowledge of the basic composition of these substances. The current interest in synthetic fuel production by pyrocatalytic conversion of petroleum residua, coals, and shales should also provide a great impetus to developing better PGC techniques and understanding of the pyrolysis processes involved.

REFERENCES

1. R. L. Levy, Chromatog. Rev. 8:48 (1966).
2. R. L. Levy, J. Gas Chromatog. 5:107 (1967).

3. J. Q. Walker and C. J. Wolf, J. Chromatog. Sci. 8:513 (1970).
4. A. Kossiakoff and F. O. Rice, J. Am. Chem. Soc. 65:590 (1943).
5. D. L. Fanter, J. Q. Walker, and C. J. Wolf, Anal. Chem. 40:2168 (1968).
6. J. D. Kelly and C. J. Wolf, J. Chromatog. Sci. 8:583 (1970).
7. M. M. Green, J. M. Moldowan, and J. G. McGrew II, J. Org. Chem. 39:2166 (1974).
8. R. A. Brown, Anal. Chem. 43:900 (1971).
9. C. A. M. G. Cramers and A. I. M. Keulemans, J. Gas Chromat. 5:58 (1967).
10. C. A. M. G. Cramers, Thesis, Technological University, Eindhoven, The Netherlands, 1967.
11. F. O. Rice, "Free Radical (Collected Papers of F. O. Rice)," Catholic University Press, Washington, D.C., 1958.
12. A. I. M. Kuelemans and S. G. Perry, "Gas Chromatography" (N. van Swaay, ed.), Butterworth, Washington, D.C., 1962.
13. L. Szepesy, V. Illés, K. Welther, and J. Simon, Acta Chim. Acad. Sci. Hung. 78:341 (1973).
14. V. Illés, K. Welther, and I. Pleszkats, Acta Chim. Acad. Sci. Hung. 78:357 (1973).
15. V. Illés, I. Pleszkats, and L. Szepesy, Acta Chim. Acad. Sci. Hung. 79:259 (1973).
16. V. Illés, K. Welther, and L. Szepesy, Acta Chim. Acad. Sci. Hung. 80:1 (1974).
17. V. Illés and I. Pleszkáts, Acta Chim. Acad. Sci. Hung. 80:247 (1974).
18. V. Illés, I. Pleszkáts, and L. Szepesy, Acta Chim. Acad. Sci. Hung. 80:267 (1974).
19. H. M. Frey and R. Walsh, Chem. Rev. 69:103 (1969).
20. P. Leplat, J. Gas Chromatog. 5:128 (1967).
21. A. A. Martynov, P. M. Usacheva, V. G. Khasanov, G. P. Karbskiĭ, and M. S. Vigdergauz, Upsekhi Gaz Kromatografii Vy P. 3:133, Kazan, 1973.
22. J. Knotnerus, I&EC, Prod. Res. Dev. 6:43 (1967).
23. Z. Ramljak, D. Deur-Šiftar, and A. Šolc, J. Chromatog. 119:445 (1976).
24. J. Romováček and J. Kubát, Anal. Chem. 40:1119 (1968).
25. S. Tamaki, Y. Mashiko, and H. Honda, Kogyo Kagaku Zasshi 70:884 (1967).
26. P. G. Simmonds, G. P. Shulman, and C. H. Stembridge, J. Chromatog. Sci. 7:36 (1967).
27. E. J. Gallegos, Anal. Chem. 47:1524 (1975).
28. P. R. Reed and P. L. Warren, 7th Oil Shale Symposium, Golden, Colorado, April 1974.
29. D. A. Scrima, T. F. Yen, and P. L. Warren, Energy Sources 1(3), Crane, Russack, New York, 1974.

30. A. V. Gomez, F. J. Oscar, and B. G. Revuelta, Anales de Quimica 68:1407 (1972).
31. B. G. Marzakov and V. F. Gostenkov, Khimiia Tverdogo Topliva 6:145 (1971).

Part IV ANALYSIS OF HEAVY DISTILLATES, RESIDUES, AND CRUDES BY LIQUID AND SUPERCRITICAL FLUID CHROMATOGRAPHIC TECHNIQUES

Chapter 9

SEPARATION SCHEMES

Klaus H. Altgelt
Fuels and Asphalts Research Department
Chevron Research Company
Richmond, California

Douglas M. Jewell
Gulf Research and Development Company
Pittsburgh, Pennsylvania

DeWitt R. Latham
U.S. Department of Energy
Laramie Energy Center
Laramie, Wyoming

Milan L. Selucky
Hydrocarbon Research Center
Department of Chemistry
University of Alberta
Edmonton, Alberta, Canada

I. BACKGROUND (D. M. Jewell)

The complexity of petroleum increases rapidly with increasing boiling point not only because of the (exponential) increase in the number of isomers of any given carbon number but also because of the prevalence of heteroatomic compounds. In some crude oils, nearly all molecules containing more than 50 carbon atoms also contain one or more heteroatoms (O, N, S). When this situation occurs, separation and isolation techniques known to work for purely hydrocarbon frameworks should not be expected to be applicable (and usually are not). The presence of any heteroatom in a molecule present in petroleum gives rise to the trivial term "nonhydrocarbons." Current and future separation techniques are being directed to the selective and predictable isolation of specific and structurally significant groups of nonhydrocarbons, as well as hydrocarbons, from petroleum and other fossil fuels.

Historically, most separation techniques for petroleum have been developed for hydrocarbon mixtures and have ignored nonhydrocarbon constituents. This practice is acceptable for the lower molecular weight components in which nonhydrocarbons may be $<1\%$. However, when nonhydrocarbons are in excess of several percent ($>5\%$), they can exert a negative influence on chromatographic systems designed and calibrated for hydrocarbons.* The nonhydrocarbons most responsible for this deleterious influence are "strongly polar." They contain functional groups which may be strong Brønsted acids (-OH, $-CO_2H$), strong Brønsted bases ($-NH_2$), or strong Lewis bases (N:, $-\ddot{\underset{..}{S}}-$, $-\ddot{\underset{..}{O}}-$). Prior removal of these strongly polar nonhydrocarbons maximizes the efficiency of the chromatographic system for the remaining hydrocarbons. Numerous techniques employing mineral acids and bases have been developed over the years for this purpose, especially by natural product chemists. Although many of these techniques offer high selectivity, [15, 17, 38], aqueous extractions have serious limitations and experimental complications when applied to petroleum:

1. Removal of acids or bases is dependent on complete solubility of the corresponding salt in the aqueous phase; but solubility of these salts diminishes rapidly with increasing molecular weight.

2. Aqueous extractions are not truly quantitative; acidity or basicity, when measured by nonaqueous titrimetry [18] is always greater than that determined by aqueous techniques.

3. Emulsion problems become almost insurmountable with residuals.

Many of these limitations were overcome by use of modern macroreticular ion exchange resins. In these resins strongly selective acidic or basic groups are part of an insoluble, rigid framework. Extractions are heterogeneous rather than homogeneous and can be performed by traditional or modern chromatographic procedures. Macroreticular ion exchangers are

---

*This point of view is not universally accepted; see pp. 200 ff. and 207 ff.

commercially available in many forms (very weak to very strong, various counterions) and offer exceptionally useful applications to petroleum owing to their porosity, their very reactive functional groups, and their inertness to organic solvents. The modern nonaqueous ion exchangers remove all acids and bases from hexane or cyclohexane solutions of petroleum, as defined by nonaqueous titrimetry, and are easily regenerated. However, as in aqueous ion exchangers, their capacity is limited and will vary with the molecular weight (size) of the petroleum components.

In their present usage, ion exchangers can remove all Brønsted acids and bases from n-pentane or cyclohexane solutions of petroleum. The remaining nonhydrocarbons contain at least one Lewis basic functional group (N:, —Ö—, or —S̈—). Of these three, only nitrogen compounds can be quantitatively removed from petroleum because of their unique ability to form hexane-insoluble coordination compounds with ferric chloride; certain carbonyl compounds (e.g., ketones), if present, are also removed with this reagent. If the ferric chloride can be supported as a thin film on an inert support (ideally kaolin clay), then this extraction can also be performed by common chromatographic procedures [18]. Experimental details for ion exchange and coordination chromatography can be found in Chapter 11. Techniques to remove selectively and quantitatively all ether and thioether nonhydrocarbon constituents from petroleum have not yet been developed.

II. THE USBM-API METHOD (D. R. Latham and K. H. Altgelt)

A. Background

The USBM-API (U.S. Bureau of Mines, American Petroleum Institute) method combines ion exchange and coordination chromatography with adsorption chromatography to separate heavy petroleum distillates or residues into seven broad fractions: acids, bases, neutral nitrogen compounds, saturates, and mono-, di-, and polyaromatic compounds. The acid and base fractions are isolated by ion exchange chromatography, the neutral nitrogen compounds by complexation chromatography on ferric chloride, the saturates and aromatics by adsorption chromatography on activated alumina [19] or on a combined alumina-silica column [16, 20].

The method separates the total sample into chemically significant fractions that are suitable for analysis according to compound type. The acid fraction contains primarily carboxylic acids, phenols, amides, and carbazoles [20]; the base fraction, pyridine benzologs, amides, carbazoles, and diaza compounds [25]. The composition of the hydrocarbon fractions as well as that of the polar fractions of several petroleum and tar sand bitumens and of shale oil distillates has been published [7, 9, 10, 26, 27].

The USBM-API method and its applications have recently been described in detail [31]. The scheme is outlined in Figure 1. The results obtained for some selected, widely differing petroleum distillates are displayed in Table 1.

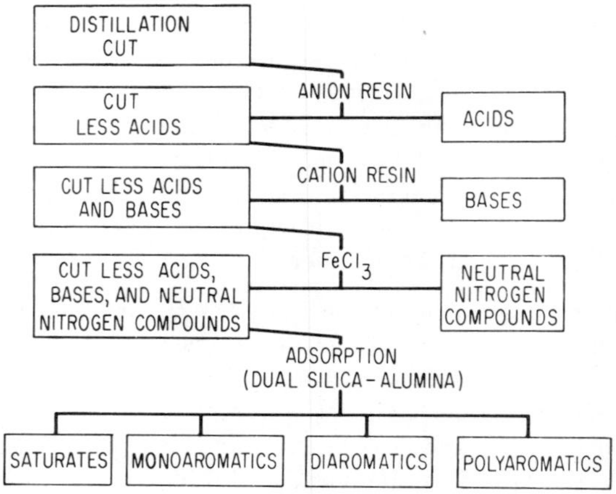

FIG. 1. USBM-API separation scheme [19]. (Courtesy ACS, Anal. Chem.)

B. Separation Procedure

The type of chromatographic column used to remove (separately) the acid, base, and neutral nitrogen fractions is illustrated in Figure 2. It is water jacketed and contains a recycling arrangement that permits the continuous elution of sample without need for large quantities of solvent. The chromatographic column used to obtain the saturate and aromatic fractions is a 2.4-cm OD by 224-cm glass tubing of 960-ml capacity, fitted with a 600-ml eluant reservoir at the top. All separations are carried out under nitrogen.

The scale of the procedures described here is that commonly used in the USBM laboratory at Laramie. Each procedure can be scaled up or down if the ratio of sample to resin or adsorbent remains constant. Details of the preparation of the resins and adsorbents are reported in Chapter 11 and in Ref. 31.

1. Acid Fraction

The sample (10-20 g) is dissolved in cyclohexane if it contains asphaltenes or large amounts of wax; otherwise, n-pentane is used. The solution is charged to Amberlite IRA 904 anion exchange resin (40 g) that had previously been dry packed into the column and prewet with solvent. Entrained, unreactive material is washed from the resins with cyclohexane or n-pentane for approximately 42 hr, using the recycling arrangement of the column. The reactive compounds (acids) are recovered by successive 24-hr

TABLE 1. Examples of Characterization Data from USBM-API Separation

Crude oil	Boiling range (°C)	Weight percent of distillate							
		Acids	Bases	Neutral compounds	Saturates	Aromatic compounds			Recovery
						Mono-	Di-	Poly-	
Wilmington	370–535	5.6	6.8	4.2	36.9	16.8	12.4	17.3	100.0
	535–675	9.3	12.7	21.3	20.8	11.8	8.9	15.3	100.1
Gach Saran	370–535	1.7	2.1	2.3	48.5	16.8	11.7	18.0	101.1
	535–675	5.4	8.7	8.9	31.6	12.1	8.2	25.0	100.0
Swan Hills	370–535	1.8	2.2	1.9	65.9	12.6	6.3	10.8	101.5
	535–675	3.5	4.5	5.6[a]	39.2[a]	9.7	5.0	14.2	99.9
Recluse	370–535	1.4	1.1	0.9	74.1	11.3	5.1	6.1	100.0
	535–675	2.9	3.3	3.0[a]	68.8[a]	8.1	3.7	10.1	99.9

[a]Values corrected from wax-free distillate to whole distillate basis. Waxes considered entirely saturate and essentially free of nitrogen and sulfur.

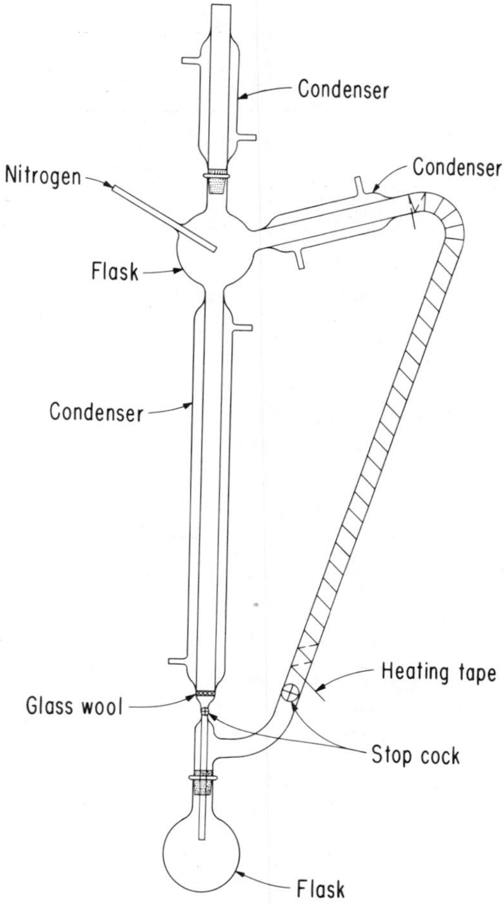

FIG. 2. Recycling chromatographic column [19]. (Courtesy ACS, Anal. Chem.)

elutions with benzene, followed by methanol, and finally with methanol saturated with carbon dioxide. These three solvents remove compounds of increasing acid strength, and the fractions can be analyzed separately or combined to give a total acid fraction.

2. Base Fraction

The acid-free sample from the above separation is charged to a column packed with Amberlyst 15 cation exchange resin. The amounts of acid-free sample and resin are the same as those for the acid fraction removal, as is the procedure of introducing sample and eluting unreactive material. The reactive material is removed from the resin by successive 24-hr elutions with benzene, followed by methanol, and finally with 8% v/v isopropyl amine in methanol. After the benzene elution is completed, nitrogen is forced up through the column until the resin bed is loose. This is a precautionary procedure to prevent column breakage which otherwise might result from swelling of the resin in methanol and from exothermic reaction of the resin with isopropyl amine. As with the acids, the bases can be removed as three fractions and analyzed separately, or they can be combined to give one fraction.

3. Neutral Nitrogen Fraction

The acid- and base-free sample is charged to a column that had been dry packed with 50 g of Amberlite IRA 904 anion at the bottom and 70 g of the ferric chloride-Attapulgus clay packed on top. A layer of glass wool is used to separate the two materials. The sample, dissolved in cyclohexane or n-pentane, is slowly percolated downward through the column that has been prewet with solvent. The color changes from yellow to green or blue as the distillate passes through the column, indicating the formation of iron complexes on the surface of the clay. Entrained oil is removed by cyclohexane or n-pentane elution. The colored complexes are desorbed from the clay by elution with 1,2-dichloroethane and broken upon contact with the anion-exchange resin. The metal salt is retained on the resin, and the nitrogen compounds are recovered in the eluate.

4. Saturates, Mono-, Di, and Polyaromatics Fractions

Distillate samples with a boiling range of about 330 to 535°C are separated on a chromatographic column that had been dry-packed with 440 g of fully activated F-20 Alcoa alumina gel in the bottom half and 310 g of fully activated Grade 12 Davison silica gel in the top half. The column is prewet with n-pentane and an amount of acid-, base-, and neutral nitrogen-free sample containing no more than 1 g of aromatics per 100 g of gel (usually 10 to 25 g of sample) is charged to it. The four compound-type fractions are eluted with the solvent sequence listed in Table 2 at a flow rate of approximately 200 ml hr^{-1}. Each fraction can either be collected in total by the use of a continuous solvent stripper attached to the bottom of the column or in small subfractions which are individually analyzed. An example of the results of this separation is shown in Figure 3. Spectroscopic examination of fractions and radioisotope studies with model compounds showed that the fractions contain predominantly the designated compound types (saturates, monoaromatics, etc.).

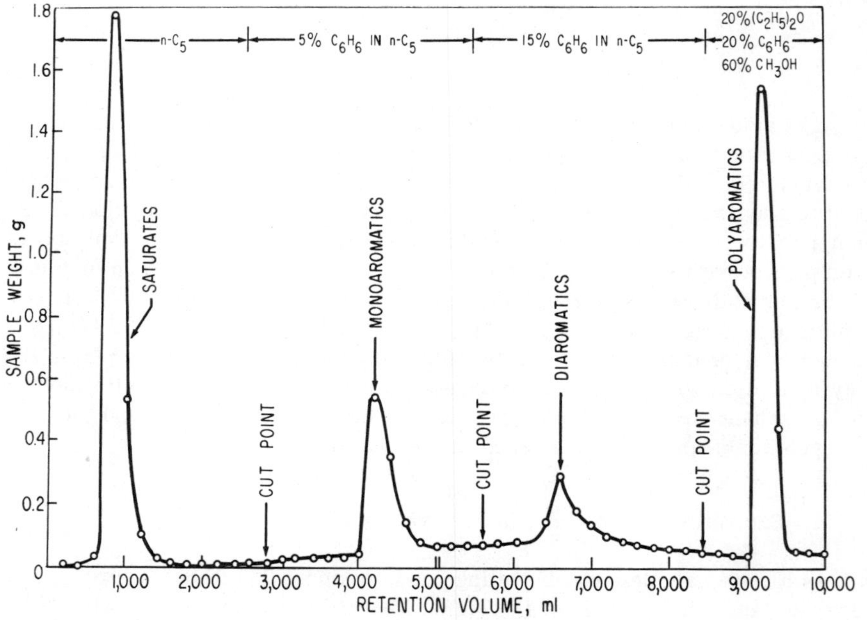

FIG. 3. Chromatogram of the hydrocarbon fraction of Gate Ridge 370 to 535°C distillate obtained on a silica-alumina column [16]. (Courtesy ACS, Anal. Chem.)

TABLE 2. Solvents Used for Eluting Hydrocarbons from Silica-Alumina Column for Samples Boiling Below and Above 535°C

	ml	Eluent
	Distillates boiling below 535°C	
Saturates	2500	n-Pentane
Monoaromatics	3000	n-Pentane + 5% benzene
Diaromatics	3000	n-Pentane + 15% benzene
Polyaromatics	500	Methanol + 20% benzene + 20% ether
	+1000	Methanol
	Distillates boiling above 535°C and residuals	
Saturates	2500	n-Heptane
Monoaromatics	4000	n-Heptane + 5% benzene
Diaromatics	4500	n-Heptane + 15% benzene
Polyaromatics	2000	Methanol + 25% benzene + 25% ether
	+1500	Methanol

Separation of distillates and residues boiling above 535°C (1000°F) requires the larger pore size Grade 62 Davison silica gel in place of Grade 12, n-heptane instead of n-pentane, and slightly adjusted eluent volumes and cut points as indicated in Table 2. A jacketed 8-ft column of 1500-ml gel volume is used at a temperature near 40°C. It is cooled before and while introducing the ether-benzene-methanol mixture.

III. THE SARA METHOD (D. M. Jewell and K. H. Altgelt)

A. Background

The SARA method [14, 21] is an extension of the API method to permit faster and more standardized separations. This was achieved by placing the two ion exchange resins and the $FeCl_3$-clay-anion exchange resin packing into a single column, the SARA column, and by using medium to high pressure LC equipment. The adsorption chromatography of the nonpolar part of the sample is still performed in a separate operation. In one version of the SARA technique, usually employed for analytical runs, both columns are integrated into a single medium- or high-pressure LC system. In another version, usually for preparative runs, the SARA column is operated separately at hydrostatic pressure conditions and only the adsorption LC column is run at higher pressure.

Another feature of the SARA method is that it separates the asphaltenes as a group. The asphaltene content is a common and often important parameter of a bitumen or coal liquid. The name SARA was derived from the main fractions determined in its analytical application: saturates, aromatics, resins, and asphaltenes.

The method is flexible, reproducible, and applicable to a large variety of the most difficult sample types, such as vacuum residuals, tar sand bitumens, shale oils, and coal liquids. Examples of these applications can be seen in Ref. 21 as well as Chap. 16 and Ref. 5 of Chap. 16.

The SARA column is packed with H^+ and OH^- ion exchange resins and ferric chloride-clay in the sequence and ratios shown in Figure 4 (10 ml, 10 ml, 70 ml, 3 ml). This particular column could accommodate 20- to 50-mg samples. Others may vary from microscale (a few grans packing and μg-size samples) to preparative-scale (kg of packing and g of sample). However, the sequence and ratios of the reagents should not be changed for reasons discussed in Chap. 11. The recycle feature of the column conserves solvent and provides a means for the exhaustive extraction of resins and oils from the asphaltenes.

B. Separation Procedures

1. Preparative Separation

The SARA column is dry-packed under vibration with the various packings, purged with nitrogen, and equilibrated with n-pentane. Pentane is recycled and replenished when necessary. This process removes air bubbles and any traces of entrained solvent impurities remaining from the preparation of the adsorbent. After an hour, recycling is stopped and the eluate discarded while the top reservoir is kept two-thirds full. A sample of 20-500 mg is placed into a tared cellulose thimble* and immediately immersed in the n-pentane in the top reservoir. Samples with high wax content may on occasion be extracted with cyclohexane. (However, note that most asphaltenes are soluble in the latter solvent.) The maltenes present in the sample are dissolved by the n-pentane, leached out of the thimble, and carried into the column. Recycling of the solvent is continued, and care is taken that there is always liquid in the top reservoir. The oils collect in the

*Cellulose thimbles should be double-walled and stored in a constant humidity chamber.

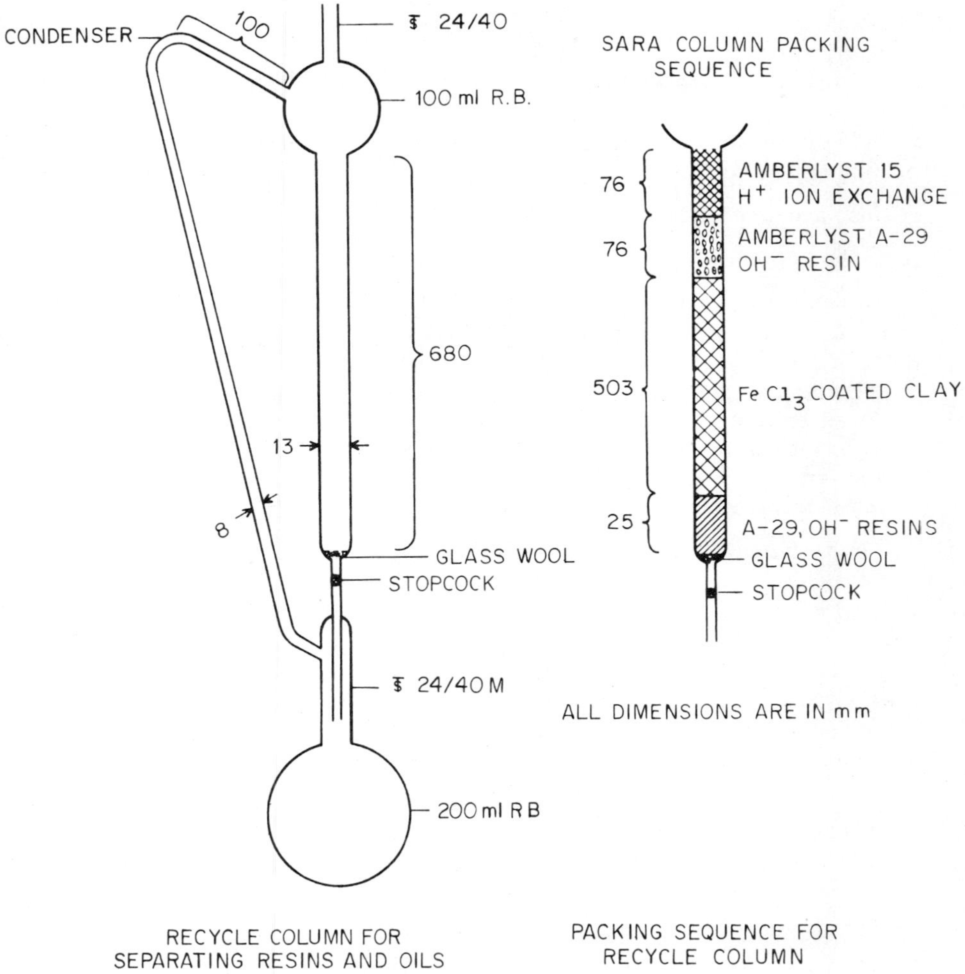

FIG. 4. Preparative SARA column [21]. (Courtesy ACS, I&EC Fund.)

bottom flask, the resins remain on the column, and the asphaltenes remain in the thimble.

The leaching action is continued by recycling until the eluate is colorless. All operations are carried out under nitrogen. The extraction time varies with the amount and nature of the sample and especially with the amount of asphaltenes and other insolubles in it. Petroleum residuals are commonly extracted overnight, while samples of tar sands, shale oils, and coal liquids may require 3 to 4 days. Coal liquids are particularly difficult to extract exhaustively because they contain benzene-insoluble components.

This delay could in principle be avoided by deasphalting the sample in a separate step. Often this is not worthwhile for two reasons: (a) while the result is obtained faster, more operator time is required; (b) those samples that take the longest to extract are also the ones most sensitive to any treatment. Even if benzene and n-pentane insolubles had been removed before placing them into the SARA apparatus, more insolubles will have formed during the operation, complicating and lengthening the dissolution process. Also, it is desirable with these sensitive materials to standardize the method of analysis in order to avoid differences caused by artifacts.

When the extraction is finished, the oils are recovered by evaporation of the n-pentane under nitrogen at 70°C. The thimble is removed, oven-dried at 100°C, and placed in a desiccator at constant humidity and room temperature until an equilibrium weight is attained, thus providing the weight of the asphaltenes. An alternate procedure is to extract the thimble with benzene for direct recovery of the asphaltenes. When studying coal liquids, this latter procedure must be followed in order to distinguish asphaltenes from the intractable benzene-insolubles.

In analytical SARA separations, resins are usually determined by difference, although they may be semiquantitatively recovered (approximately 90%) by exhaustive elution of the packing with chloroform. When the SARA column packings are used in individual columns for stepwise separation of the resins, as described in Chap. 11, recoveries approach 100%. However, when eluting a single SARA column as the one shown in Figure 4, certain strong bases and acids react with the ferric chloride to form new complexes which are insoluble in chloroform and, therefore, are not recovered.

The critical and time-consuming step in analyzing residuals and coal liquids is the exhaustive extraction of maltenes from the samples. If asphaltenes are absent, as in heavy distillates, this step is appreciably shortened.

Saturates and aromatics are obtained by the liquid chromatographic separation of the "oils" on either 0% water silica gel or 0% water basic alumina. The latter offers the advantage of being able to separate saturates, monoaromatics, di- + triaromatics, and polynuclear aromatics in one run, if this information is desired [21].

2. Analytical Separation

When a sample of heavy distillate or residual is free of asphaltenes, the SARA separation can be performed analytically in a medium- or high-pressure liquid chromatographic apparatus. A common arrangement of columns and detectors is shown in Figure 5, which depicts a closed system of two columns and three on-line monitors. Column 12 consists of the same four zones (in the same ratios) described earlier for the preparative SARA

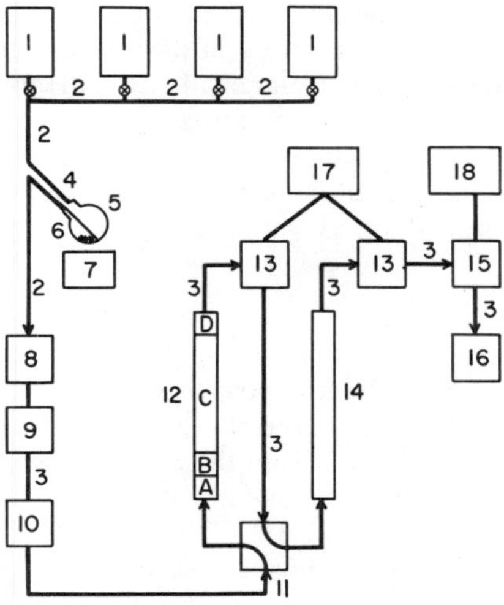

1. SOLVENT RESERVOIRS, 1ℓ.
2. TEFLON TUBING, 1/8" O.D.
3. TEFLON TUBING, 1/16" O.D.
4. ADAPTER WITH TWO 1/8" HOLES AND ₮ 24/40
5. 100 ml ROUND BOTTOM FLASK
6. 1/2" MAGNETIC BAR
7. MAGNETIC STIRRER
8. MILTON ROY MINIPUMP
9. FLOW-THROUGH PRESSURE GAUGE
10. INJECTION VALVE-SAMPLE LOOP
11. BYPASS VALVE
12. SARA COLUMN, 15" x 1/2"
    A. H^+ ION-EXCHANGE RESIN, 2"
    B. OH^- ION-EXCHANGE RESIN, 2"
    C. $FeCl_3$ -KAOLIN 10"
    D. OH^- ION-EXCHANGE RESIN, 1"
13. U.V. DETECTORS, 254 nm
14. SILICA-GEL COLUMN, 25" x 1/2"
15. DIFFERENTIAL R.I. DETECTOR
16. FRACTION COLLECTOR
17. U.V. DUAL CHANNEL RECORDER
18. R.I. RECORDER

FIG. 5. Schematic of integrated SARA apparatus [21]. (Courtesy ACS, I&EC Fund.)

column (Figure 4). Column 14 contains either fully activated (0% H_2O) silica gel (Davison Grade 923) or fully activated basic alumina (Alcoa F-20). The effluents from both columns are monitored at 254 nm with a dual-channel ultraviolet (uv) analyzer. Column 14 is also monitored with a differential refractometer. The maltene sample, dissolved in n-hexane, is injected into Column 12 by means of a calibrated valve, 10. As explained earlier, with the preparative SARA column, "resins" are completely retained on Column 12 and the uv response of this effluent is exclusively due to the aromatics fraction of the "oils." The SARA column proper does not separate saturate or aromatic hydrocarbon types except that the most highly condensed polyaromatics elute last because of weak interactions with the polystyrene matrix of the ion exchangers and somewhat stronger π-π interactions with the $FeCl_3$ [33]. The differential refractometer will show baseline separation of saturates from aromatics in the effluent from Column 14. When all the "oils" have entered Column 14, as indicated by return to baseline of the uv response, Column 12 is bypassed and the aromatics are desorbed from Column 14 with a chloroform ⟶ methanol gradient. The amount of "resins" is calculated by difference after gravimetric recovery of saturates and aromatics from Column 14.

3. Hydrocarbon Separation

Figure 6 shows typical detector responses from a medium pressure (<100 psig) integrated SARA experiment. The bottom curve (uv signal) depicts the skewed elution of "oils" from the SARA column. Slight tailing of the SARA fractions occurs for reasons given earlier. Its degree depends on the amount of sample charged. This "tail" is completely aromatic, as demonstrated in a study of many model systems using both differential refractometer and ultraviolet absorption detectors. The tailing of aromatics from the SARA column in no way inhibits the separation of saturates from aromatics on Column 14 as shown in Figure 6. The top curves, which are the responses of the uv and ri monitors to the effluents from Column 14, display well-separated peaks.

The saturates are always eluted in a highly symmetrical band which is completely resolved from the first aromatic components, provided the

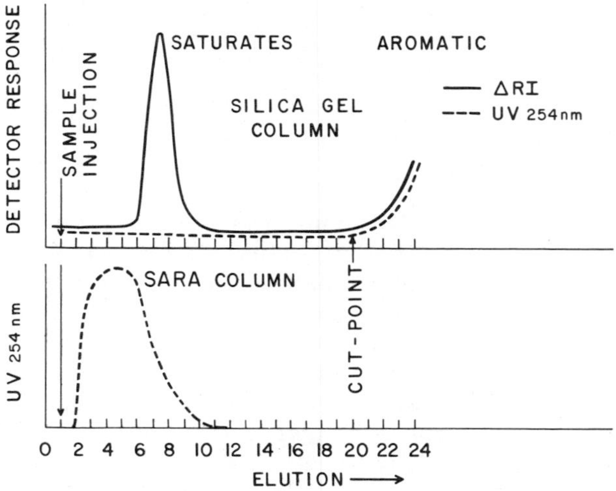

FIG. 6. Detector responses with SARA method [21]. (Courtesy ACS, I&EC Fund.)

silica or alumina gel had been completely activated (0% H_2O). The best way to remove final traces of water adsorbed from the air during packing operations and to ensure highest activity is to purge the column with dry N_2 while heating it to 400°C. The column is kept warm (100°C) and under N_2 until used. Columns and fittings should be stainless steel for this in situ activation.

If saturates and aromatic subtypes are to be separated on alumina in an integrated SARA experiment, Column 14 is eluted with a cyclohexane-chloroform-methanol gradient. Feeding the eluate through both channels of the uv analyzer, operated at different wavelengths, one can readily monitor the separation, as shown in Figure 7. For example, using the wavelengths 254 nm and 313 nm, one observes that monoaromatics (one benzene ring per molecule) give a response only at 254 nm. Condensed diaromatics respond at both 254 nm and 313 nm. Monoaromatics containing up to three cycloparaffin rings elute with a hexane-cyclohexane gradient on 0%-water alumina; naphthalenes will not elute with these solvents. If traces of water are left on the alumina, this separation cannot be achieved. Changes in (solvent) gradient are made only when the monitors indicate that a desired separation has been achieved. Di- + triaromatics are monitored at 313 nm

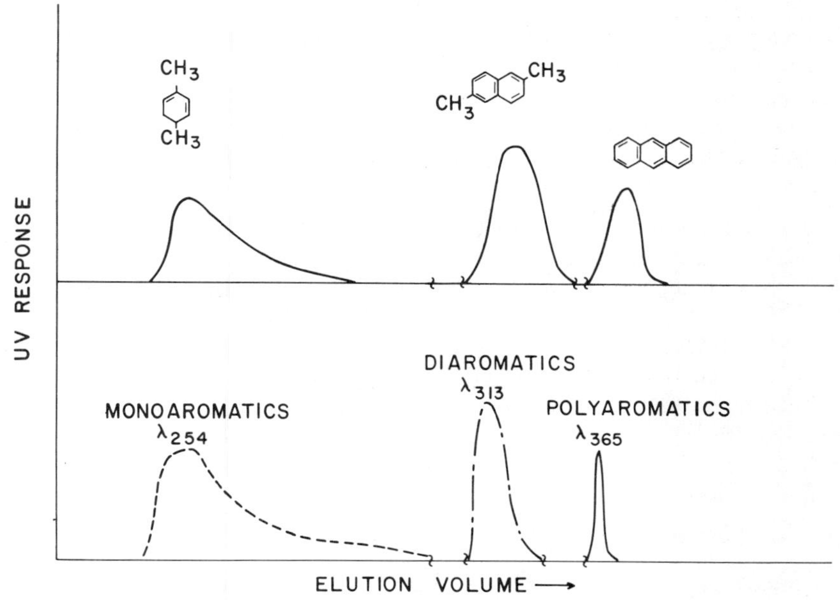

FIG. 7. Separation of aromatics on 0% water alumina [33]. (Courtesy ACS, Anal. Chem.)

and 365 nm. When the signal returns to baseline, a methanol gradient is started to elute the polyaromatics.

IV. OTHER SEPARATION SCHEMES (K. H. Altgelt)

As of now (1978), the SARA and USBM-API methods are probably the most universal separation schemes for heavy petroleum fractions and other fossil fuels. However, several other schemes are being used and offer certain advantages in special applications.

A. Analytical High Performance Liquid Chromatography

Suatoni's et al. [46, 47] rapid hydrocarbon group-type analysis by HPLC is a good example. It determines in minutes the amounts of saturates, aromatics, polar compounds (resins), and n-hexane insolubles (asphaltenes, etc.) in any medium- or heavy-fossil fuel sample. Such rapid analysis is possible because extremely small amounts of sample are charged (0.3 to 0.4 mg) and because the aromatics are eluted by backflushing.

The results are not equivalent to those obtained with the SARA method since only the "strong" acids and bases are retained as resins, while the weaker ones are eluted together with the aromatics [22]. The error introduced by this distribution becomes greater with petroleum residuals and synthetic fuels. The method works well with light- and middle-petroleum distillates. Suatoni and Swab showed with model compounds that polynuclear aromatics, even as large as coronene and p-diphenyl benzene, are eluted and thus separated from the polar compounds.

Quantitation is based on proper calibration. It turned out [46] that, within a given boiling range, the corresponding fractions (saturates, aromatics) of different crude oils usually had very close calibration factors, except for some highly naphthenic and highly paraffinic samples. Suatoni and Swab provide calibration factors plotted against boiling point for saturates and aromatics from crude oil fractions. They give only single calibration factors for topped tar sand bitumens and coal liquids. Results obtained on various rather diverse petroleum distillates and residuals closely matched those from ASTM D 2007. More detail is given in Chapter 6. Other examples of HPLC are found in Chapters 10, 14, and 17.

B. Preparative Separations

An alternative to the sequence in SARA is the chromatographic preseparation of a deasphaltened sample on deactivated silica or alumina with pentane or hexane into neutral and polar fractions (resins). These cuts can now be further chromatographed into narrower fractions without their mutual interference.

Snyder [45] developed a separation scheme on this basis. In its entirety it is rather complex, as Figure 8 shows. A petroleum distillate is separated on silica into a nonpolar fraction, S_0, and a polar one, S_{14}. S_0 is further chromatographed on deactivated alumina into saturates, mono-, di-, and polyaromatics, and a hetero-compound fraction. S_{14} is divided into cuts of increasing acidity by ion exchange chromatography. The first one, CE_0, contains all basic compounds. These are chromatographed on deactivated alumina into six cuts of increasing polarity. Complexation with Hg^{2+} removes sulfides. Chromatography of the six cuts on silica further separates them into different compound groups containing O, N, or both. This procedure takes advantage of the fact that silica separates polar compounds in somewhat different ways than alumina (see Chap. 10). Fractionation on charcoal isolates aliphatic hetero compounds from aromatic ones.

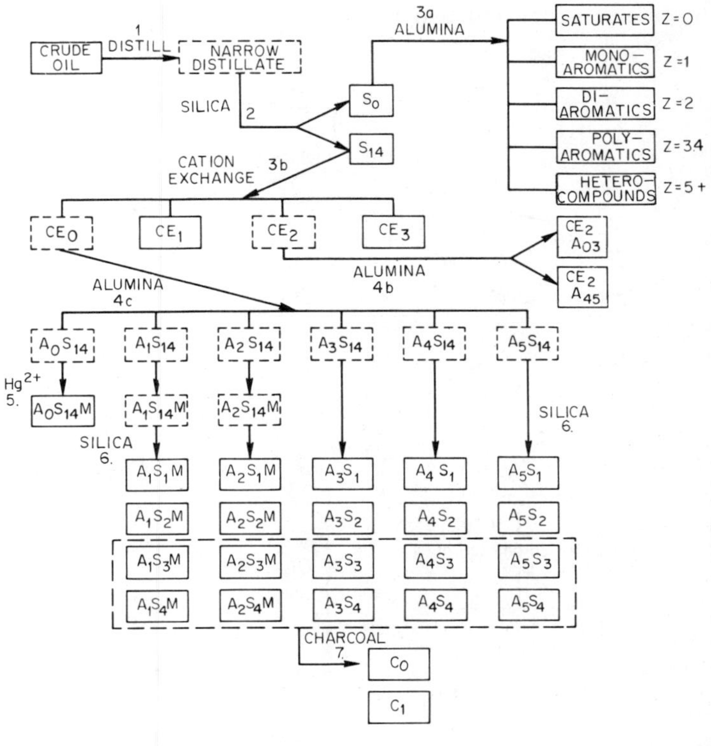

FIG. 8. Snyder's separation scheme [45]. (Courtesy Acc. Chem. Res.)

This separation scheme can easily be adapted to simpler group-type separations or to the isolation of only a few narrowly defined chemical species, for instance, phenols and pyridine derivatives. In either case, certain operations of the total scheme can be omitted, which makes the separation simpler and faster. One advantage of this scheme is the high degree of standardization which allows conclusions about the structure of a sample just from its distribution among the different fractions. Snyder achieved this standardization by use of adsorbents with carefully controlled degree of activation (water content) and of specific solvents and solvent mixtures. Details of both aspects are described in Chap. 10. Another advantage of Snyder's scheme is the avoidance of reactive adsorbents ($FeCl_3$, fully activated silica or alumina), which can change some constituents of the samples to be separated.

Oudin [34] used the same principle of preseparation of a petroleum sample into a neutral and a polar cut with subsequent separation into narrower fractions. However, having a different objective from Snyder, he took a different approach as shown in Figure 9.

Seifert and co-workers [38-40] developed a separation scheme that allows the determination and analysis of naphthenic acids in petroleum and

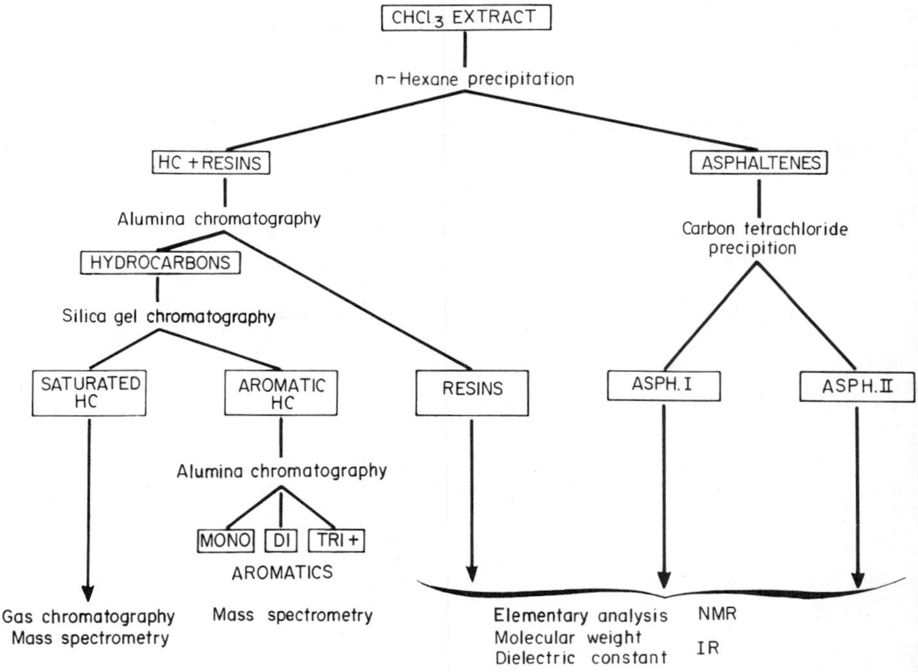

FIG. 9. Oudin's separation scheme [34]. (Courtesy Inst. Francais Petrole Rev., Editions Technip.)

the reliable identification and quantitation of the structural moieties [39], which bear the carbonyl groups. The complete scheme is shown in Figure 10. After extraction with alcoholic NaOH, the naphthenic acids are preseparated on a weakly basic ion exchange resin into weak and strong acids. Both cuts are again subdivided by ion exchange chromatography into a weaker and a stronger acid fraction [38]. The pure carboxylic acid fraction is reduced with $LiAlH_4$ to convert the acid groups via alcohols and tosylates [49] into the corresponding hydrocarbons. The resulting materials are then chromatographed on neutral alumina into saturates; mono-, di-, and triaromatics; nitrogen compounds; and an unidentified very polar fraction [40].

In more recent work, Seifert [41] chromatographed hexane-deasphalted crude oils and sediment extracts on silica columns. Elution with n-hexane gave baseline separation between saturates and aromatics and semiquantitative separation between mono- and polyaromatics. Benzothiophenes eluted with the polyaromatics. With benzene as the next solvent, phenols and carbazoles are obtained quantitatively together with some higher polynuclear aromatic hydrocarbons and other sulfur compounds. Amounts of 0.5-g material were chromatographed by this method in one day. Mass spectrometric analysis of the fractions demonstrated no interference by the polar constituents (7%) and the compounds in the benzene eluate (21%) with the separation of the hydrocarbons, even though they had not been preremoved but were obtained after the latter ones.

Al-Kashab and Neumann [2] also prefer the direct fractionation of hydrocarbon and heterocompounds of deasphalted residuals on a dual alumina-silica column and subsequent treatment of the polar fraction with cat- and anion-exchange resins into basic, acidic, and neutral materials. Their total separation scheme, shown in Figure 11, includes chromatography of the asphaltenes. Only in case of the highly polar asphaltene samples do Al-Kashab and Neumann first remove the basic and acidic compounds (true asphaltenes) with ion exchange resins and then chromatograph the remainder (occluded or peptized non-asphaltenes) on alumina-silica into saturates, aromatics, and heterocompounds. Separation of the saturates into n-alkanes and isoalkanes plus cycloalkanes by urea and thiourea addition completes the scheme.

Note the fact that the asphaltene fraction contained some nonasphaltenes. Observations like this, again, raise the question about the nature of asphaltenes and lead to closer investigations [28] of the concept of asphaltenes as a distinct fraction of petroleum and other fossil fuels. See also the discussion on page 211.

A common problem with routine fractionation of residuals is poor reproducibility. Drushel [11] overcomes this by deasphalting his samples in two steps and by using clay-silica as suggested by Aczel [1], instead of alumina in his chromatographic columns. Refluxing with n-heptane precipitates a major part of the asphaltenes. Evaporation of the filtrate and refluxing with n-pentane yields another, usually smaller, portion of asphaltenes.

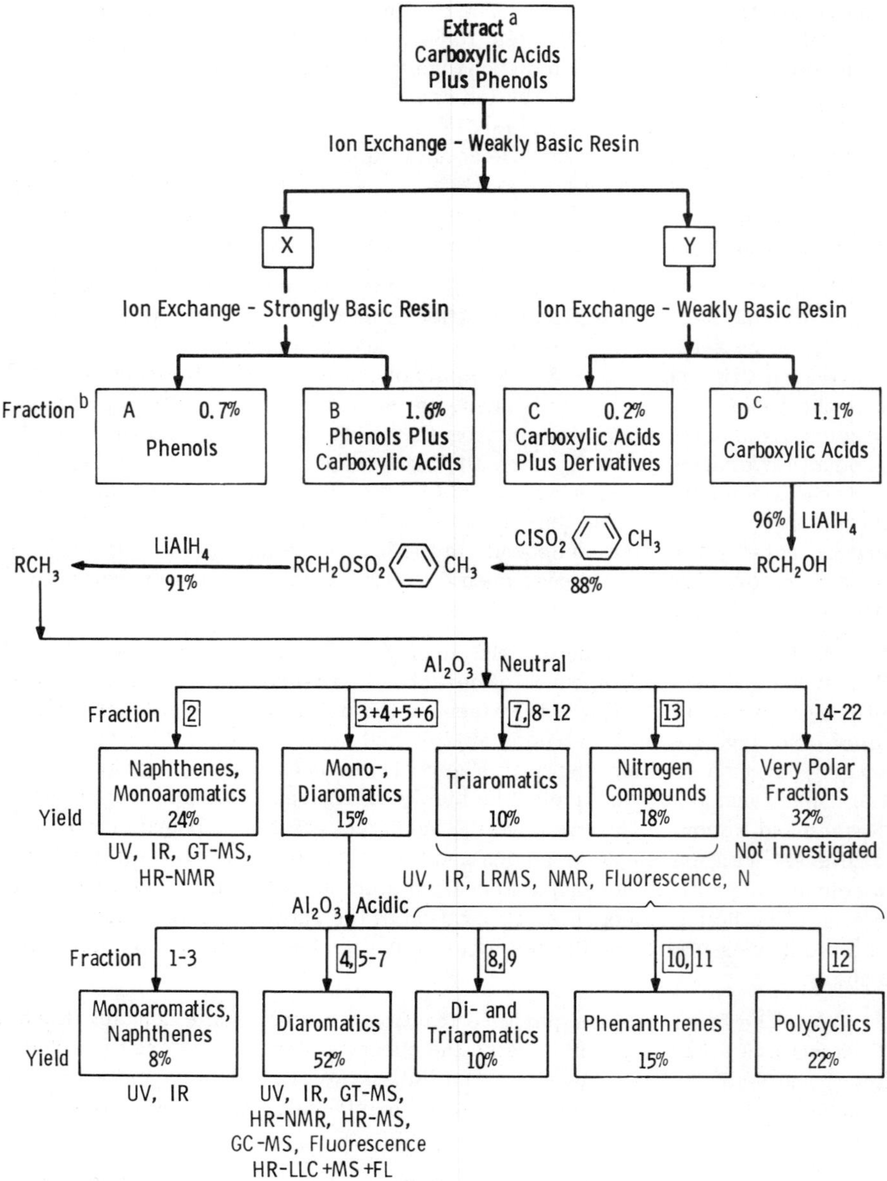

FIG. 10. Seifert's scheme of isolating and separating naphthenic acids [40]. (Courtesy Anal. Chem.)

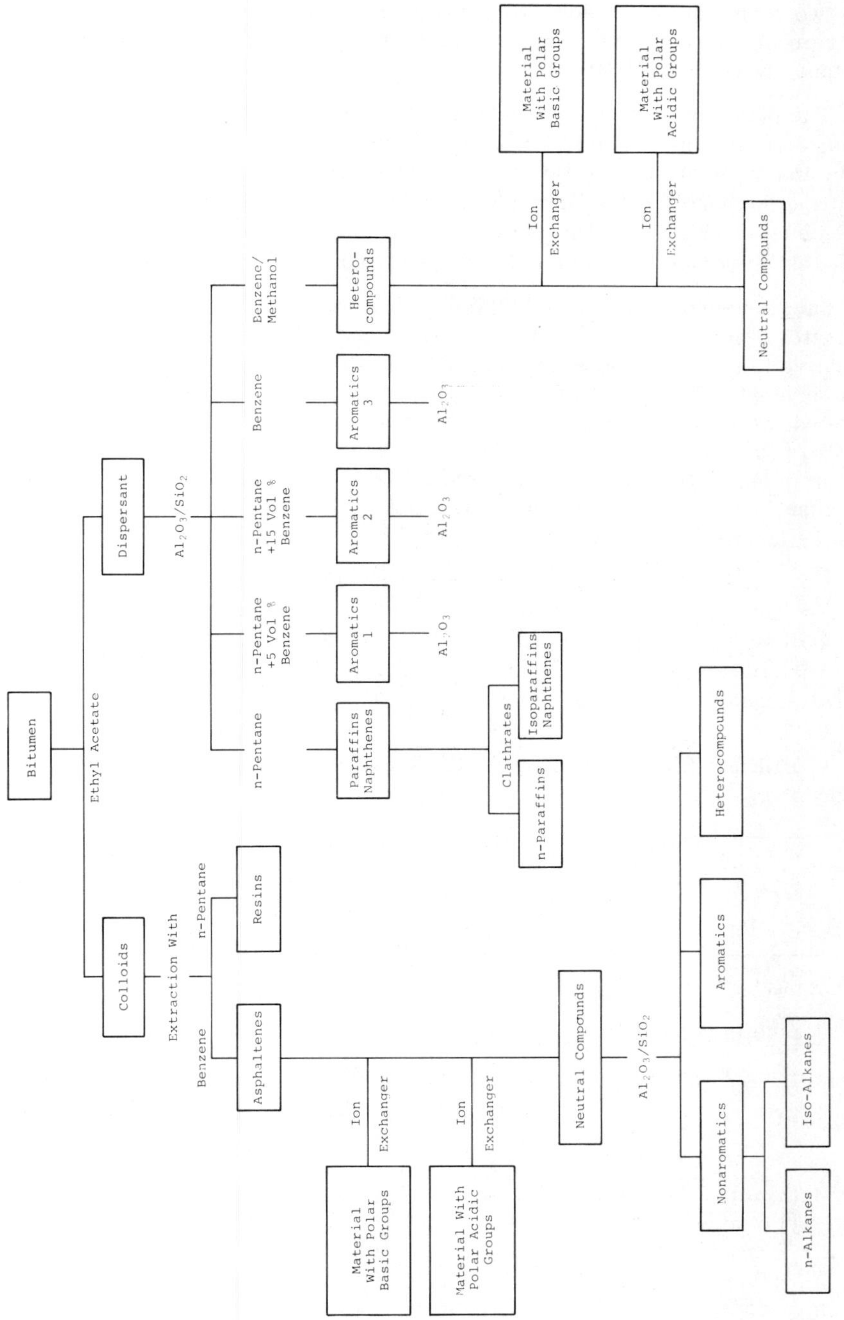

FIG. 11. Al-Kashab and Neumann's separation scheme [2]. (Courtesy Strassen and Tiefbau.)

The two-step treatment seems to avoid, or at least mitigate, occlusion and other problems that often reduce reproducibility in regular asphaltene-maltene separations.

The maltenes are chromatographed on clay-silica into only three fractions. Saturates are eluted with n-hexane or n-heptane, aromatics with benzene, and polar aromatics (heterocompounds) with an 80/20 mixture of benzene and methanol. If the limited degree of resolution obtained with this scheme is sufficient for the purpose, this is an excellent method of preparing sufficient material for elemental and spectroscopic analysis and other tests.

Sawatzky and co-workers [37] described two fast methods for separating saturates, mono-, di-, and polyaromatics, and polar compounds as an alternative to the much slower USBM-API method. Both are based on adsorption chromatography of the pentane-deasphalted sample on activated silica-alumina (weight ratio 37/47) with the solvents listed in Table 3. The preparative procedure (1.5-g sample) is about 10 times, the analytical one (150-mg sample) 100 times, faster than the USBM-API method. The authors speculate that their fraction eluted with hot pyridine might correspond to—or even be identical with—the basic fraction of the USBM-API method.

Combination of GPC with adsorption chromatography or any other method that separates by chemical functionality allows separating a sample in two dimensions, by molecular size and by chemical composition. Altgelt made use of this possibility in his study of asphalts and asphaltenes [3-5]. He fractionated asphalt samples first by GPC, precipitated the asphaltenes

TABLE 3. Eluant Sequence for Sawatzky et al. [37]. Two Separation Methods

Element	Volume (ml) System 1	Volume (ml) System 2	Fraction
n-Pentane	250	25	Saturates
5% Benzene in pentane	300	30	Monoaromatics
15% Benzene in pentane	300	30	Diaromatics
Benzene	100	10	Polyaromatics Polar compounds
60% Methanol, 20% benzene, 20% ethyl ether	50	12	Polar compounds (polyaromatics)
Methanol	100	12	
Pyridine (100°C)	100	10	Basic polar compounds
Benzene	100	10	

in each fraction with n-pentane, and chromatographed each maltene portion on deactivated alumina into an oil and three resin fractions. Coleman et al. [8] and Hayes et al. [13] revised this order by chromatographing their deasphalted samples first on silica or alumina and then separating those fractions further by GPC. The different merits of the two sequences will be discussed in Chapter 12.

McKay et al. [23] used GPC in extension of the USBM-API scheme to further separate their acid concentrate into four fractions, one of which consisted mainly of carboxylic acids in the form of dimers. Another fraction contained nitrogen compounds and phenols, and the last one consisted mainly of aromatic hydrocarbons. This latter concentrate could be separated even further by collecting narrower GPC fractions and by thin layer chromatography of the GPC cuts [24].

During the last few years, GPC has frequently been used in combination with other separation methods, mostly as an analytical tool rather than as part of a regular separation scheme. This aspect will be further discussed in Chap. 12. More separation schemes for heavy fossil fuel fractions [6] and for tar sand bitumens in particular [7a] have been reviewed recently.

Working with coal liquefaction products, Farcasiu [12, 48] felt none of the established separation methods was satisfactory. She had observed irreversible adsorption of the more polar constituents of coal liquids on activated alumina and loss or chemical changes of pyrrol-type compounds on strong ion exchange resins of the type used with the SARA method. Furthermore, she was interested not only in the composition of pentane-soluble materials (she calls them oils), but also in that of the pentane-insolubles (asphaltenes), and even in that of the benzene-insolubles (preasphaltenes or carbenes). The latter two fractions usually constitute more than 80% of coal liquids. Farcasiu developed a scheme designed to overcome the problems and to allow the fractionation and recovery of essentially the whole coal sample.

To minimize losses, she chose, as an adsorbent, silica, which in the beginning of a run contains 4% water and is later further deactivated by ethanol and methanol. Nine solvents, selected for increasing polarity and proton acceptor ability, elute the fractions listed in Table 4. Only about 5% or less of the solvent refined coal fractions remains on the column. Sample to adsorbent ratio is 1/25.

The structural features of material eluting in the 9 fractions were determined by analysis as well as with model compounds. Generally, both the number of hetero atoms and the molecular weight increase with increasing fraction number.

Farcasiu [12a] has extended her method to span the range from micro to macro separations (5 mg to several grams) and also to many more diverse liquefied coal samples. Several other groups are now using this technique also, but no published comment on it has come to our attention yet (1978).

TABLE 4. Farcasiu's Separation Scheme [12]. Adsorbent: Silica Gel Containing 4% Water

Classical description	Fraction	Elution solvent	Major compounds
Oils (1-3)	1	Hexane	Saturates
	2	Hexane/15% benzene	Aromatics
	3	Chloroform	Polar aromatics; non-basic N, O, S-heterocyclics
Asphaltenes (3-5)	4	Chloroform/10% Et_2O	Simple phenols
	5	Et_2O/3% EtOH	Basic nitrogen heterocyclics
Multifunctional compounds (5 through 9)	6	MeOH	Highly functional molecules (10 wt % heteroatoms)
	7	$CHCl_3$/3% EtOH	Polyphenols
	8	THF	Increasing O content and increasing basicity of nitrogen
	9	Pyridine	
	10		Noneluted, unknown materials

V. THOUGHTS ON THE USBM-API/SARA METHODS AND SHORTCUTS (M. L. Selucky and K. H. Altgelt)

The USBM-API and SARA separations are at present the most comprehensive and most widely applied methods for the study of heavy petroleum fractions and coal liquids. Despite their shortcomings of tedium, length of time, and possible loss or conversion of very polar and highly condensed polynuclear aromatic materials, they can be viewed as the yardstick by which other, faster methods should be measured. Of course, such a statement must be taken with a grain of salt. Different separations may complement the USBM-API/SARA methods; others may improve on them. Yet, for now at least, we have a standard that has been applied to many diverse samples and given a rather detailed description of each.

The original USBM-API method uses distillation to cut the crude sample into relatively narrow boiling point fractions. More recently, it has been demonstrated that such materials as Utah [32] or Athabasca [42] tar sands can be separated into classes of compounds without prior distillation, even without previous removal of asphaltenes [32]. The latter finding is at least indirectly supported by the separation of asphaltenes on ion exchangers [28]. If distillation is omitted, more caution is advised for making the cuts between fractions from either silica or alumina as suggested by Snyder's data of various classes of compounds. The separation of mono-, di-, and polyaromatics is not complete, even that of saturates and monoaromatics is sometimes only marginal. Fortunately, mass spectrometric methods for aromatic hydrocarbons, e.g., the method of Robinson and Cook [35], allow one to check the degree of overlap. Recently, HPLC has proven to be a good means of checking the elution of the different compound classes to ascertain proper cutting points for the standardized USBM-API method [42].

For routine separations, the USBM-API method is sometimes impractical for the reasons already stated, and short cuts are called for. These short cuts, however, should not be made indiscriminately, i.e., without knowing how the omission of certain separation steps will affect the results. A good analysis should answer questions raised with a sufficient degree of accuracy—no more, no less. Thus, the choice of a procedure will primarily depend on what we need to know about our sample. Selucky et al. [43] have studied in some detail how the omission of the ion exchange and complexation steps affect the separation of Athabasca bitumen. They treated their samples by two schemes: (a) the regular USBM-API method and (b) a scheme for the detailed analysis of the saturated and aromatic fractions which is shown in Figure 12. In both cases, the bitumen was deasphaltened (50 v of n-pentane) before the separation. The first separation step in the latter scheme was on activated silica and yielded fractions loosely termed hydrocarbons, polar-I, polar-II, and polar-III compounds. The hydrocarbon fraction contained saturated and aromatic hydrocarbons with single- and

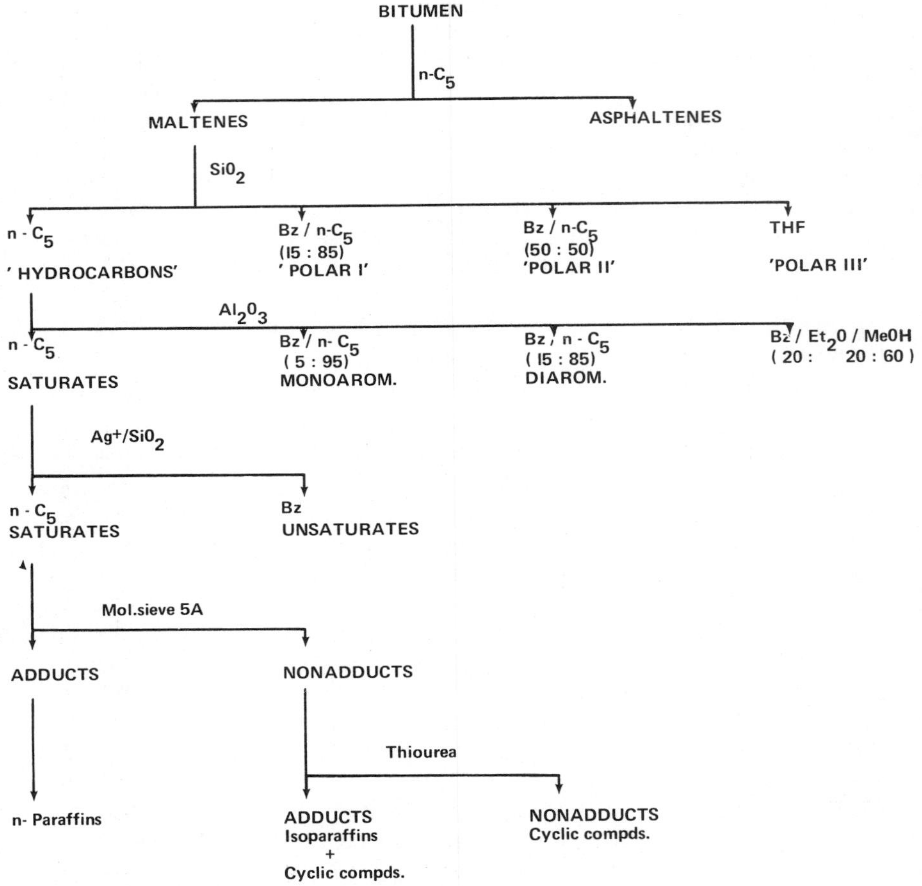

FIG. 12. Selucky's et al. separation scheme for hydrocarbons [42]. (Courtesy Fuel.)

multiple-ring systems up to about coronene. This was shown by HPLC on a μ-Porasil column and by comparison with the retention times of standards. The hydrocarbon fraction was further separated on alumina into saturated, mono-, di-, and polyaromatics.

The polar-I fraction was essentially free from mono- and diaromatic hydrocarbons as demonstrated by its separation on alumina [43]. The fact that elution with n-pentane + 15% benzene yielded only about 1% of material indicated a very low degree of overlap between this fraction and the mono- and diaromatics. The intermediate fraction of higher aromatics, eluted from alumina with benzene (see Sawatzky and George et al. [37]), was contaminated only slightly more—2.6%. Thus, the omission of the ion-exchange and complexation steps had not led to an appreciable overlap of the three hydrocarbon fractions with the polar constituents. Also, a comparison of the yields of the various classes from this separation and from the USBM-API procedure suggested that polar, heterocyclic material did not substantially interfere with the short-cut separation—at least not with that of the Athabasca bitumen used in this study. More evidence was obtained by rechromatography of acid, base, and neutral Lewis-type compounds isolated from the USBM-API procedure on both silica and alumina. On silica only 0.29% material was carried over to the "hydrocarbon fraction." On alumina only 0.1% material was carried over into the 15% benzene/n-pentane fraction and 0.24% into the Ba/Et_2O/MeOH fraction. The latter fraction represents the "polyaromatic-nonspecified polar material" of the USBM-API procedure. It should be pointed out that these results which were obtained with only one type of bitumen cannot be generalized to extend to materials that are high in nitrogen or possibly sulfides.

This kind of checking a separation might also be desirable for other types of short-cut procedures, e.g., the HPLC methods of Suatoni et al. [46, 47]. In spite of some doubts raised [22], the foregoing evidence shows that short cuts can be used in certain cases, e.g., with Athabasca bitumen. However, while the USBM-API and SARA methods seem to be generally applicable for heavy materials, the suitability of a short-cut procedure should be doubted unless proved otherwise.

Where the distribution of the hydrocarbons in nondistilled samples is of interest, an extension of the USBM-API and SARA methods is needed. Mass spectrometry is not sufficient for the quantitative analysis of saturates in residuals because its sensitivity varies with the C-number within a class of compounds. Selucky and Chu et al., therefore, treated the saturates fraction further [42], as shown in Figure 12. They passed it through an Ag^+-silica or aluminosilicate [29] column to remove residual unsaturates, adducted it with Molecular Sieve 5A to separate n-alkanes, and clathrated it with urea and thiourea to concentrate iso-alkanes and cyclic hydrocarbons. With most heavy bitumens (n- and iso-alkanes), this method yields sufficient material for reasonable quantification by gas chromatography. However, with biodegraded oils (e.g., Athabasca), which contain only negligible amounts of open-chain compounds, the adductions did not substantially improve the extent of separation. Here they only increased the number of fractions, thus, making quantitative evaluations more difficult. In these cases, the relative proportions of mono-, di-, tri-, and higher cyclic saturates cannot be determined well.

The SARA method in its more sophisticated form assembles the USBM-API columns into a single chromatographic train with concurrent asphaltene precipitation. The use of a pump and detectors make it much faster than the USBM-API procedure and allows to select proper cutting points with greater ease. The results of the two methods, however, are not always identical, especially if the sample contains asphaltenes. Asphaltenes can give rise to a number of problems. It is known that their yield varies with the nonsolvent used for their precipitation and with other factors. For instance with Athabasca bitumen, n-hexane gives 80% of the yield obtained with n-pentane, n-heptane only 67% [46]. It also makes a difference whether the whole bitumen is extracted from the tar sand and the asphaltenes are precipitated from it the regular way or only the maltenes are extracted from the tar sand and the asphaltenes are determined by difference or by subsequent extraction.

Asphaltenes from different sources may have different solubilities in a given solvent. For instance, cyclohexane, often claimed to be a solvent for asphaltenes, produced 0.7% precipitate from Athabasca bitumen, which amounted to 4% of total asphaltenes [30]. Further, in the separation of Athabasca asphaltenes on ion exchangers, about 13% of sample could not be eluted from the anion exchanger with MeOH/CO_2 because of its insolubility in this solvent [44]. Benzene saturated with CO_2 eluted the residual sample with ease. Anomalies like these make comparisons between different separation schemes difficult, even between such similar procedures as the USBM-API and the SARA methods.

To summarize, while departures from the generally applicable USBM-API and SARA procedures are not being advocated, in those cases where they are necessary for practical reasons, the influence of short cuts should be examined before a method is adopted for routine use with different types of material. With nondistilled samples additional caution should be exercised, as with increasing boiling point of the sample separation into compound classes becomes more difficult.

REFERENCES

1. T. Aczel, Preprints, Div. Petrol. Chem., ACS 17(4):F66 (1972).
2. M. Al-Kashab and H. J. Neumann, Strassen and Tiefbau, 30(3):44 (1976).
3. K. H. Altgelt, Separ. Sci 5:855 (1970).
4. K. H. Altgelt, Bitumen, Teers, Asphalte, Peche 21:475 (1970).
5. K. H. Altgelt, Strassen and Tiefbau 29:26 (1975).
6. K. H. Altgelt and T. H. Gouw, Advances in Anal. Chem. 13:71 (1975).
7. J. W. Bunger, Preprints, Div. Fuel Chem., ACS 19(2), Los Angeles, California, April 1974, p. 231.

7a. J. W. Bunger, Preprints, Div. Petrol. Chem., ACS 22(2):716 (1977).

8. H. J. Coleman, D. E. Hirsch, and J. E. Dooley, Anal. Chem. 41:800 (1969).
9. J. J. Cummins, R. E. Poulson, and W. E. Robinson, Preprints, Div. Fuel Chem., ACS 20(2), Philadelphia, Pennsylvania, April 1975, p. 154.
10. J. E. Dooley, D. E. Hirsch, C. J. Thompson, and C. C. Ward, Hydrocarbon Processing 53(11):187 (1974).

11. H. V. Drushel, Preprints, Div. of Petrol. Chem., ACS 21(1):146 (1976).
12. M. Farcasiu, Fuel 56:9 (1977).
12a. M. Farcasiu, personal communication (1978).
13. M. H. B. Hayes, M. Stacey, and J. Standley, Fuel 51:27 (1972).
14. W. E. Haines and C. J. Thompson, ERDA Rept. of Invt., LERC/RI-75/5, BERC/RI-75/2, July 1975, 30 pages.
15. G. K. Hartung and D. M. Jewell, Anal. Chem. Acta 26:514 (1962).
16. D. E. Hirsch, R. L. Hopkins, H. J. Coleman, F. O. Cotton, and C. J. Thompson, Anal. Chem. 44:915 (1972).
17. D. M. Jewell and G. K. Hartung, J. Chem. and Eng. Data 9(2):297 (1964).
18. D. M. Jewell and R. E. Snyder, J. Chromatog. 38:351 (1968).
19. D. M. Jewell, J. H. Weber, J. W. Bunger, H. Plancher, and D. R. Latham, Anal. Chem. 44:1391 (1972).
20. D. M. Jewell, R. G. Ruberto, and B. E. Davis, Anal. Chem. 44:2318 1972).
21. D. M. Jewell, E. W. Albaugh, B. E. Davis, and R. G. Ruberto, I&EC Fundamentals 13:278 (1974).
22. D. M. Jewell, personal communication.
23. J. F. McKay, D. M. Jewell, and D. R. Latham, Anal. Chem. 43:645 (1972).
24. J. F. McKay and D. R. Latham, Anal. Chem. 45:1050 (1973).
25. J. F. McKay, T. E. Cogswell, J. H. Weber, and D. R. Latham, Fuel 54:50 (1975).
26. J. F. McKay, J. H. Weber, and K. R. Latham, Anal. Chem. 48:891 (1976).
27. J. F. McKay, T. E. Cogswell, P. J. Amend, P. M. Harnsberger, and D. R. Latham, Preprints, Div. Fuel Chem., ACS 21(7):52 (1976).
28. J. F. McKay, P. J. Amend, T. E. Cogswell, P. M. Harnsberger, R. B. Erickson, and D. R. Latham, Preprints, Div. Petrol. Chem., ACS 22(2):708 (1977).
29. S. Lam and E. Grushka, J. Chromatog. Sci. 15:234 (1977).
30. D. L. Mitchell and J. G. Speight, Fuel 52:149 (1973).
31. J. R. Morandi and R. E. Poulson, Preprints, Div. Fuel Chem., ACS 20(2), Philadelphia, April 1975, p. 162.
32. A. G. Oblad, J. D. Seader, J. D. Miller, and J. W. Bunger, Oil Shale and Tar Sands, AIChE Symposium Series 155, 72 (J. W. Smith and M. T. Atwood, eds.), New York, 1976, pp. 73-74.
33. H. H. Oelert and A. D. Giehr, J. Chromatog. 13:465 (1975).
34. J. L. Oudin, Inst. Francais Petrole Rev. 25:470 (1970).
35. C. J. Tobinson and G. L. Cook, Anal. Chem. 41:1548 (1969).
36. T. C. S. Ruo, M. L. Selucky, and O. P. Strausz, Report to Joint CIC/ACS Meeting, Montreal, June 1977.

37. H. Sawatzky, A. E. George, G. P. Smiley, and D. D. Montgomery, Fuel 55:16 (1976).
38. W. K. Seifert and W. G. Howells, Anal. Chem. 41:554 (1969).
39. W. K. Seifert, R. M. Teeter, W. G. Howells, and M. J. R. Cantow, Anal. Chem. 41:1638 (1969).
40. W. K. Seifert and R. M. Teeter, Anal. Chem. 42:750 (1970).
41. W. K. Seifert, Advances in Organic Geochemistry, Madrid, 1975.
42. M. L. Selucky, Y. Chu, T. C. S. Ruo, and O. P. Strausz, Fuel 56:369 (1977).
43. M. L. Selucky, T. C. S. Ruo, Y. Chu, and O. P. Strausz, Preprints, Div. Petrol. Chem., ACS 22(2):695, New Orleans, March 1977.
44. M. L. Selucky, T. C. S. Ruo, S. S. Kim, and O. P. Strausz, Report to the Fossil Fuel Conference, Saratoga, Wyoming, July 1977.
45. L. R. Snyder, Acc. Chem. Res. 3:290 (1970).
46. J. C. Suatoni and R. E. Swab, J. Chromatog. Sci. 13:361 (1975).
47. J. C. Suatoni, H. R. Garber, and B. E. Davis, J. Chromatog. Sci. 13:367 (1975).
48. D. D. Whitehurst, M. Farcasiu, and T. O. Mitchell, Natl. Techn. Inf. Service PB-257569, U.S. Dept. of Commerce, Springfield, Va., 1976.
49. Reaction product of the intermediate alcohol with p-toluene solfonyl chloride.

Chapter 10

ADSORPTION AND PARTITION CHROMATOGRAPHY

Lloyd R. Snyder
Technicon Instruments Corp.
Tarrytown, New York

Dennis L. Saunders
Research Department
Union Oil Company of California
Brea, California

I. INTRODUCTION

David Day [103] first introduced the possibility of adsorption chromatography for petroleum analysis at the turn of the century. At that time he demonstrated the separation of petroleum into fractions with different properties by passing it through columns of Fuller's earth. Day expressed the hope that ". . . we will be able to obtain complete resolutions [in this fashion] . . . ," and further noted the possibility of ". . . characterizing a lubricant oil by the correct percentage of the component hydrocarbons instead of by densities, inflammation points and viscosities. . . ." Almost 80 years later his hope concerning "complete resolution" remains unfulfilled. Nevertheless, adsorption chromatography has made possible the rapid and accurate analysis of the major compound types present in a variety of petroleum products.

Adsorption chromatography was much used by the API Research Project 6 group under Rossini in the 1940s and later [60], in the pioneering characterization of the individual hydrocarbons in petroleum. By the midfifties several of the large oil companies were heavily involved in the analysis of petroleum fractions by means of preliminary adsorption separations, followed by more detailed spectral analysis [40]. A decade later this separation technique had been developed to the point where it could be predictably applied to several aspects of petroleum analysis: routine assay of certain compound types in heavy distillates [65], total assay of the major hydrocarbon and sulfur classes in any petroleum fraction [66], and the qualitative analysis of the less prevalent nitrogen and oxygen-containing compounds [67]. In the late sixties high performance liquid chromatography (HPLC) began to evolve from the independent development of liquid chromatography (LC) technology, application of GC theory, and the development of efficient stationary phases and detectors. However, applications of HPLC to petroleum analysis have begun to appear only recently. Today adsorption chromatography is an indispensable tool in the petroleum analytical laboratory for use in both quality control and research. In the near future we should expect to see an increasing number of applications of HPLC to petroleum analysis. Partition chromatography was introduced by Martin and Synge in the early 1940s [48, 49], but never achieved the same widespread application and utility enjoyed by adsorption chromatography in petroleum analysis. However, the adaptation of this technique to HPLC has opened up new possibilities, such as the separation of polynuclear aromatic hydrocarbons by reversed-phase partition LC, and analysis for specific petroleum additives.

In this chapter we shall discuss some of the major applications of adsorption and partition chromatography to petroleum analysis. In order to fully appreciate the concepts and accomplishments of those who have developed these applications, it seems worthwhile to briefly review the fundamentals of the adsorption and partition processes. We shall first review the basic principles of the separation mechanisms, the various techniques of separation, and the unique characteristics of stationary phases and solvents. The balance of the chapter is devoted to various applications classified according to the molecular types separated.

A. Principles of Adsorption Chromatography

We shall outline the fundamental principles of adsorption. More detailed discussions of the theory and practice of adsorption chromatography are available in several recent monographs [27, 34, 37, 54, 68, 91].

The stationary phase in adsorption chromatography, the adsorbent, is a porous material with high surface area. For petroleum separations polar adsorbents such as silica (SiO_2) or alumina (Al_2O_3) are most useful, although nonpolar adsorbents such as charcoal have been used. Components of a sample mixture are attracted to the surface by reversible intermolecular forces: (a) dispersion, (b) dipole, (c) hydrogen bonds, and (d) weak covalent bonds. The strength of adsorption depends upon the adsorbent and the functional groups present in the sample molecule, and the position of the equilibrium depends on the relative strengths of adsorption of sample and solvent. The wide range of strengths of intermolecular forces for different functional groups gives rise to an extremely wide range of adsorption strengths for different molecular types. For example, alkyl groups are only weakly adsorbed because they are attracted to the surface by dispersion forces alone [56, 68]. On the other hand, alcohols are attracted by much stronger dipole and hydrogen bonding forces. Because alkyl groups are so weakly adsorbed, in comparison to other functional groups, homologs of a given molecular class (for example, alkyl benzenes) are adsorbed to nearly the same extent.

The weak adsorption of alkyl groups has been used effectively to separate petroleum samples into a series of compound types or classes. For example, it is possible to separate the saturated hydrocarbons, alkyl monoaromatic hydrocarbons, and alkyl diaromatic hydrocarbons from most petroleum samples as three distinct bands. This unique ability of adsorption chromatography to separate by compound type is a distinct advantage in petroleum analysis because

1. In light distillates it is possible to simplify the further analysis of individual components by isolating primary compound classes (e.g., saturates and aromatics).
2. In heavier, more complex fractions it is extremely difficult, or impossible, to analyze for individual compound types by any other technique.
3. Many properties of petroleum products are a function of the relative amount of different compound types rather than the specific amount of individual compounds.

Another characteristic of adsorption chromatography is its unique selectivity toward the separation of geometrical isomers. This is a consequence of the geometry of the surface and the different ways individual isomers can align themselves on the surface. Thus, adsorption chromatography allows the separations of certain isomer-pairs which are virtually unattainable by partition chromatography or any other technique. The wealth of information about relative adsorption strengths gives rise to yet another advantage of adsorption chromatography. With the data and theory developed over the past 15 years it is possible to predict with reasonable accuracy the retention of specific compound classes. Conversely, it is possible to gain qualitative information about unknown mixtures from their chromatographic profile. Thus, the spectral analysis of chromatographic cuts can frequently be interpreted in terms of more than one compound type. By means of relative chromatographic retention it is often possible to eliminate one or more of these types. For example, a mixture of saturated hydrocarbons and alkyl aromatics cannot be distinguished by ir from alkyl-substituted aromatics alone, but adsorption chromatography allows the separation of saturates from aromatics so that the first possibility can be eliminated.

There are limitations to adsorption chromatography. It does not lend itself to the separation of homologs, and its ability to resolve very polar (especially ionic) molecular types is relatively poor. Finally, because there is some unavoidable overlap in the separation of certain compound types, it is impossible to obtain clean separations in some cases, particularly with wide-boiling-range or high-boiling samples.

B. Principles of Partition Chromatography

In liquid-liquid partition chromatography, the stationary phase is a liquid dispersed over the surface of a finely divided inert support, and the mobile phase is another liquid which is saturated with the stationary phase. The sample distributes between the stationary phase and mobile phase liquids in a manner analogous to liquid extraction. When the stationary phase liquid is the more polar of the pair, the mechanism is called normal partition. When the stationary phase is the less polar of the pair, the mechanism is called reversed-phase partition. Partition chromatography can be performed in columns, on thin layers, or on paper. Paper chromatography is a special case of partition chromatography wherein the paper acts as a support.

Recently column packings have become available in which the stationary liquid phase is chemically bonded to the support. This type of stationary phase may be better described as a surface modified adsorbent than as a true liquid phase, but its retention resembles that of unbonded liquid phases—in equivalent systems—enough to warrant classification of the mechanism as partition [25].

An important difference between partition and adsorption chromatography is the selectivity for different members of a homologous series. Thus, it is often possible using partition chromatography to separate a series of molecules differing only in their degree or type of alkyl substitution [45]. There are a wide variety of stationary liquid phases available which give the chromatographer a fairly broad range of selectivities for different compound types. On the other hand, partition chromatography is not well suited for the separation of compound classes. There are several disadvantages of partition chromatography on mechanically held liquid phases.

1. The mobile phase must be saturated with stationary phase to prevent washing it off the support.
2. The temperature must be held constant to avoid changing the mutual solubilities of the phases.
3. Gradient elution (solvent programming) cannot be used.
4. Mobile phases are limited to solvents which are immiscible with the stationary phase liquid.

Chemically bonded phases circumvent these problems, but the range of bonded phases available is more limited and their sample capacity is lower than that of mechanically loaded phases. Another disadvantage of partition chromatography is that it is not yet possible in general to predict the elution order a priori (there are a few notable exceptions [30]). The selection of phases for a particular mixture therefore remains a largely empirical problem.

II. TECHNIQUES OF SEPARATION

Liquid chromatographic separations have been accomplished with three basic techniques: frontal, displacement, and elution chromatography. The frontal technique, in which the sample itself serves as the mobile phase, has found little application to petroleum analysis, while the displacement technique and especially the elution technique are widely used.

A. Displacement Chromatography

In this technique, relatively large amounts of sample are introduced to a column of adsorbent, followed by a polar, strongly adsorbing solvent. As the solvent moves through the column, it pushes or displaces the sample ahead of it, so that during the entire separation the sample forms a discrete slug (no admixture of solvent). The primary application of displacement chromatography in petroleum analysis is the well known FIA (Fluorescent Indicator Adsorption) analysis of gasoline-range samples [6, 38]. Here three dyes are first added to the sample, the sample separates into saturates, olefins, and aromatics during the analysis, and each of these three fractions is permeated by one of the three dyes (which have adsorption characteristics similar to saturates, olefins, and aromatics, respectively). The separation is stopped just before sample leaves the column and is examined under ultraviolet light, which reveals the separated fractions via fluorescence of the contained dye. The analysis is then completed by volumetric assay of the three separated fractions using the length of the zones as a measure of their volumes. The use of FIA analysis has been extended to kerosene range samples [55], but for heavier petroleum fractions [38, 92] it is complicated by two factors: increasing sample viscosity and increasing sample complexity. As sample viscosity increases, the separation process becomes slower and less efficient. The greater structural diversity of higher boiling fractions also leads to increase in the number of compounds whose retention is similar, but whose compound type is different, thereby resulting in overlapping of compound types into the various fractions. At the same time, displacement chromatography is an insufficiently powerful technique to yield the desired resolution of high-boiling samples into more than about three compound types.

B. Elution Chromatography

This general procedure, whereby relatively small amounts of sample are introduced to a column of adsorbent, then washed from the column with relatively large volumes of solvent, can be usefully subdivided into several subtechniques, which differ in both their intended application and experimental technique. We will discuss in turn the following four elution procedures: low-pressure column chromatography, linear elution adsorption chromatography (LEAC), thin-layer chromtography (TLC), and high-performance liquid chromatography (HPLC).

1. Low-Pressure Column Chromatography

This is the classical form of adsorption chromatography as applied to all kinds of petroleum analysis. Glass columns, coarse particle sizes, and pressures of 1 to 10 psi are typical; for some examples, including experimental details, see Refs. 40, 85, and 92. Typically the separations are run in an overloaded condition with sample/adsorbent weight ratios greater than 0.01. The retention of different compounds and compound classes therefore varies with sample size (nonlinear isotherm separation; Sec. 3.A).

When this technique is used to separate one or more compound-class fractions from a sample (for whatever purpose), it is customary either to collect numerous small fractions for subsequent analysis and pooling of similar fractions, or to monitor the column effluent during the separation. This approach serves two purposes. First, it allows selection of the optimum cutpoint between adjacent compound-class fractions, which yields the minimum cross-contamination of the two fractions. Second, it can provide information on the relative purity of each composite fraction after recovery and pooling. In some cases, however, predetermined fractions can be collected without intermediate testing and pooling. We will illustrate these points with some examples.

In one routine procedure for the separation and analysis of the major hydrocarbon classes in high boiling samples [66], the column effluent is monitored at two wavelengths in the ultraviolet during elution from the column of monoaromatics, diaromatics, and triaromatics. These measurements allow selection of the best cutpoints between these three compound-class fractions. Extrapolation of the elution curve for monoaromatics also permits an estimate of the overlap of monoaromatics into the diaromatics fraction—a quantity which is not easily obtained by spectral analysis of the total diaromatics fraction. This extrapolation procedure is greatly simplified by the now well documented [66] observation that the concentration of a given compound class in the column effluent declines exponentially with column effluent volume. Therefore, plots of the logarithm of effluent concentration (or detector signal) versus effluent volume are generally linear.

Another example of low-pressure column chromatography is in the routine assay for total saturated hydrocarbons in heavy distillates. Although numerous such procedures have been reported (see review of Ref. 92), each of these is based on one of two approaches. Several fractions of initial column effluent can be collected and assayed for total weight of sample. Typically, it is found that the sample concentration in each such fraction decreases rapidly, as the balance of saturates leaves the column, then begins to rise as monoaromatics begin to elute. A cutpoint is arbitrarily selected, the various fractions prior to the cutpoint are pooled into a single saturates fraction, and total sample saturates are then calculated from the weight of this composite fraction. Alternatively, it has been found [92] in elution from alumina that the effluent volume required to elute 99.5%+ of the sample saturates can be predicted from the sample boiling point. In this case a predetermined effluent volume is collected, which contains all the sample saturates. Any contamination of this fraction by monoaromatics is easily determined by uv analysis of the fraction, with subsequent correction of the saturates analysis.

A final example of low-pressure column chromatography is in the qualitative and quantitative analysis of individual oxygen and nitrogen compound classes [85]. Here the complexity of this sample fraction is such that no single procedure can achieve complete separation of individual compound classes. Therefore, it is both suitable and convenient to collect several arbitrary fractions by stepwise elution with a series of solvents of increasing strength. The individual fractions are then further analyzed by spectral or chemical techniques.

2. Linear Elution Adsorption Chromatography (LEAC)

This technique refers to a general procedure that was developed in the early 1960s [70, 71, 76]. It is carried out quite similarly to low-pressure column chromatography as above, but with the following important differences:

1. Sample size is much reduced so that all separations are in the linear region of the adsorption isotherm, and the retention is independent of sample concentration. Standardized adsorbents are used, which means there will be predictable, repeatable separations.
2. Predetermined effluent volumes are collected for each compound class of interest so that individual subfractions need not be collected, and the column effluent need not be monitored during the separation.
3. The scale of the separation is kept small so that each analysis relies on a small column (e.g., 400 by 8 mm); this means the space required for carrying out the separation is small, the cost of adsorbent and solvents is low, adsorbent can be discarded after a single use, and the separation is completed rapidly.
4. Several columns are run at the same time by an eluent manifold; this allows a large number of assays of a given type (e.g., alkyl thiophenes [71]) to be run at the same time by a single operator, making the cost per analysis low. Figure 1 shows a typical LEAC apparatus arrangement.
5. Since the amount of sample analyzed is small (e.g., 10 to 100 mg), uv or colorimetric analysis of the collected fractions is used for analysis of the separated fractions; multiwavelength analysis permits determination of as many as three compound classes per fractions (e.g., indoles, carbazoles, and benzcarbazoles [86]).

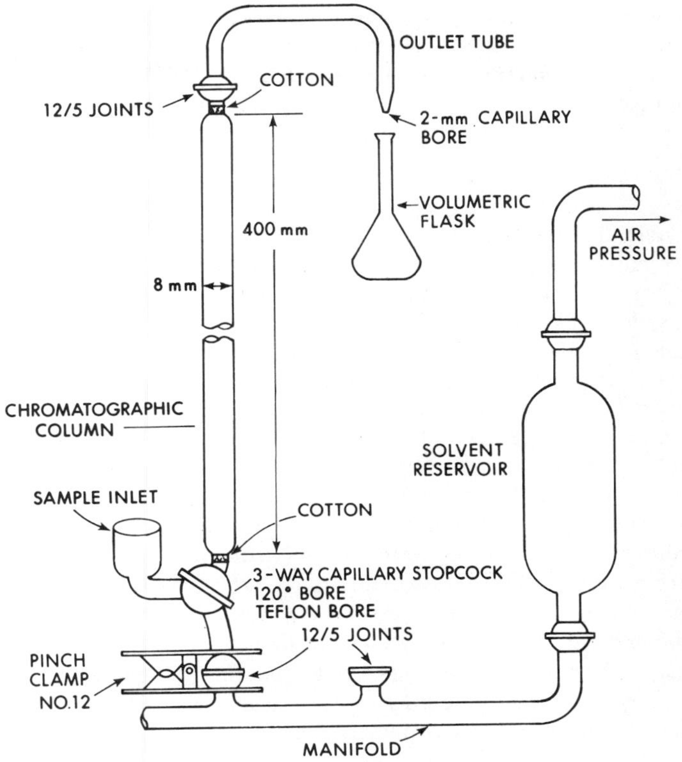

FIG. 1. Apparatus for LEAC analysis.

The technique of LEAC has now been applied to a wide variety of compound-class assays for various hydrocarbon, sulfur, and nitrogen compound classes [65, 71]. It is particularly useful for the analysis of trace constituents that are not readily analyzable by, for example, direct mass spectrometry of the total sample. However, it has also been used for the total analysis of various aromatic subtypes in cracked gas oils [72]. A single operator can easily carry out 8 to 16 assays per day for as many as five different components per assay. The equipment required is quite simple, apart from the need for a spectrophotometer to analyze collected fractions. LEAC assays have been reported for the analysis of olefins, polyaromatics, sulfides, indoles, and several other compound classes. A complete listing is given in Secs. VI and VII. In addition, the LEAC procedure can be applied to the analysis of additives in petroleum samples. The technique of LEAC has also been suggested [73] for the qualitative characterization of narrow fractions obtained from low-pressure column chromatography. Because the reproducible and predictable nature of LEAC permits the assignment of different possible petroleum compound classes to different LEAC fractions, the measurement of the LEAC retention volume range of a given fraction defines its possible compositional range. This technique has, however, not been much used.

3. Thin-Layer Chromatography (TLC)

This technique was introduced as a practical technique by Kirchner [36], but remained little used until the late 1950s when Stahl popularized it [94]. The unique aspects of TLC are: (a) the stationary phase is coated onto a flat glass, plastic, or aluminum foil sheet rather than being confined to a column; (b) solvent flow occurs by capillary action, rather than by force of pressure or gravity; (c) development is stopped before the mobile phase reaches the end of the stationary phase bed; and (d) strongly retained components are easily recognized at the origin, in contrast to column development in which they are eluted very late or never. Once a TLC separation is completed, separated zones can be visualized by their color or fluorescence or by various sprayed-on reagents which react with the sample to produce colored or fluorescing products. Qualitative information can be obtained from the relative migration distance (rf values), or by comparison with known compounds run simultaneously on the same plate. Quantitative evaluation can be accomplished by comparing the color density of the separated zones. Alternatively, the adsorbent containing the desired component can be removed, followed by elution of the sample and determination by spectrophotometry, spectrofluorometry, and so on.

TLC has been found useful for the analysis of additives in petroleum products. In one example, the antioxidant additive 2,4 dimethyl, 6-tert-butyl phenol is determined in turbine fuel [4]. First, an aliquot of the fuel containing the additive is concentrated by low-pressure column chromatography. Second, an aliquot of the concentrate is separated on a thin layer of silica. The antioxidant is visualized by spraying the plate with phosphomolybdate reagent and quantified by comparing the darkness of the appropriate colored zone with standards, using a densitometer.

4. High-Performance Liquid Chromatography (HPLC)

This technique is characterized by long, narrow (reusable) columns, small-diameter particles of column packing, high-pressure flow, and continuous detection of the eluted components. Apparatuses for HPLC include accurate-delivery high-pressure pumps, gradient elution facilities, small-volume sampling valves, and detectors capable of measuring samples in the low-nanogram range.

Applications of HPLC to the analysis and preparative separation of chemical mixtures now number in the thousands, yet few applications to petroleum analysis have so far appeared. Two excellent examples of the possible application of HPLC to petroleum analysis have recently appeared [97, 98], one of which allows the rapid analysis of virtually any petroleum sample for saturates, aromatics, polar compounds, and asphaltenes. Samples are analyzed by injecting a few microliters of hexane-solubles solution into a column packed with 10-μ silica gel. Once the saturates have eluted, the aromatics are backflushed in order to obtain a sharp, narrow peak. The eluted groups are detected by a differential refractometer and quantitated by electronic integration.

III. STATIONARY PHASES

In order to understand and appreciate the applications of adsorption and partition chromatography to petroleum, it is important to be aware of the properties and peculiarities of column packings. Only the fundamentals are reviewed here to serve as an introduction to the more detailed discussions available in the literature [25, 27, 34, 37, 54, 68, 91].

A. Adsorbents

The dependence of the adsorption equilibria (and therefore the separation) upon sample concentration can best be described by the adsorption isotherm: a plot of the equilibrium concentration of adsorbed solute versus the concentration of nonadsorbed solute at constant temperature. The desirable case is a linear isotherm, wherein the adsorbed concentration is directly proportional to the nonadsorbed concentration; retention is then independent of sample size, which simplifies the interpretation of the separation and yields improved resolution. Nonlinear isotherms, which give separations in which retention is difficult or impossible to predict, are a consequence of the heterogeneity of adsorption sites on the adsorbent surface [68]. The distribution of adsorption sites of different strength determines the adsorbent activity and therefore degree of sample retention. Adsorbents with high activity give high retention volumes, however, the isotherm remains linear only at very low concentrations. At higher concentrations the retention volume becomes a function of the sample concentration. The adsorbent linear capacity ($\Theta_{.1}$) is defined as the maximum amount of sample which can be separated per gram of adsorbent while maintaining retention time within 90% of the value for very small samples. The linear capacity of an adsorbent can be increased by eliminating the strongest adsorption sites (which are usually present in low concentration). The elimination of strong sites is accomplished by deactivating the adsorbent with water or some other very strongly adsorbed deactivating agent, such as methanol or ethylene glycol. The water is added in a carefully controlled amount so that it is just sufficient to completely cover the strongest sites.

Figure 2 illustrates how adsorbent deactivation affects the separation. Figure 2A illustrates the shape of the isotherm of an adsorbent before (curve a) and after deactivation (curve b). While the undeactivated adsorbent has a greater affinity for the sample at all concentrations as indicated by the greater slope, it begins to curve at much lower concentrations than the deactivated adsorbent. The consequences of this are shown in Figures B and C. Figure 2B shows that at low concentration the undeactivated adsorbent exhibits greater retention, but the retention time is not constant over much of the sample size range, thus the sample capacity ($\Theta_{.1}$) is much higher for the deactivated adsorbent. Figure 2C illustrates the enhancement of sample capacity with adsorbent deactivation.

Experimentally, adsorbent deactivation is very easily accomplished. First, the adsorbent is oven dried to remove any water remaining on the surface; second, a predetermined quantity of liquid water is added (usually 0.5 to 10% by weight); third, the water-adsorbent mixture is tumbled until any lumps disappear; and fourth, the adsorbent is allowed to equilibrate in a closed container for 8 or more hr. If highly reproducible separations are required, the activity of the adsorbent can be exactly adjusted to a standardized level. This is done by first measuring the activity by determining the retention of a test compound. Then, by adding portions of dry adsorbent or additional increments of water and reequilibrating, the retention of the test compound can be finely adjusted to a predetermined value. Large batches of standardized adsorbent can be prepared, bottled in small containers, and stored for a year or more without change in activity. A detailed procedure for preparing standardized alumina is given in Appendix A.

There are a number of general characteristics which can be used to classify adsorbents. Adsorbents may be either polar or nonpolar. Polar adsorbents such as silica, alumina, or magnesia preferentially retain polar solutes. On polar adsorbents retention increases approximately in the sequence

saturated hydrocarbons < alkylhalides, aromatics < ethers

< esters, ketones, aldehydes < alcohols, amines < carboxylic acids

It is possible to calculate relative compound-type retention on alumina for most petroleum-related compounds [87]. Examples of relative elution order of different petroleum compound classes are also given in Secs. VI and VII. Table 1 lists the relative adsorbtivity (ϵ_1) on alumina of petroleum hydrocarbon and sulfur compound classes for both the parent compound and the alkyl substituted compounds found in 600 to 1000° F fractions. Figure 3 shows the range of relative adsorption of oxygen and nitrogen compound classes in 600 to 1000° F petroleum fractions. Nonpolar adsorbents such as charcoal, preferentially adsorb more polarizable compounds such as

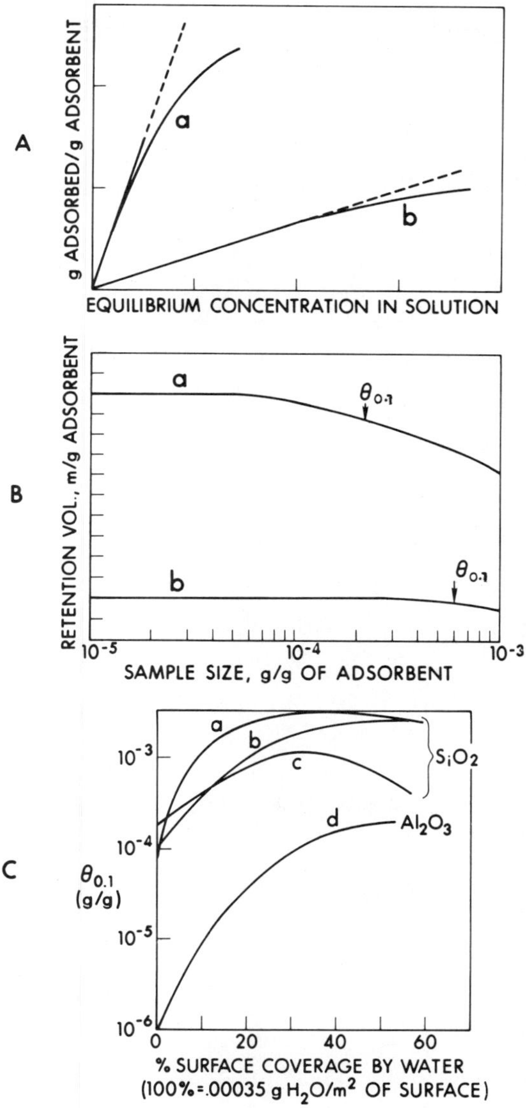

FIG. 2. The effect of adsorbent deactivation. A. Adsorption isotherms for undeactivated (a) and deactivated (b) adsorbents. B. Retention versus sample size for activated (a) and deactivated (b) adsorbents. C. The variation of sample capacity with the degree of deactivation.

TABLE 1. Relative Adsorptivity of Petroleum Compound Classes on Alumina [87] (partial listing)

Compound type	Parent compound	Relative adsorptivity (ϵ_1) Parent compound	Petroleum fractions (600-1000° F)
Monoaromatic hydrocarbons	Benzene		-.06 to -.09
Diaromatic hydrocarbons	Naphthalene	0.04	.06 to .07
	Biphenyl	0.06	.07 to .08
	ϕ-CH_2-ϕ	0.06	.07 to .08
	ϕ-$(CH_2)_{10}$-ϕ	0.04	.04 to .05
Triaromatic hydrocarbons	Fluorene	0.15	0.16 to 0.17
	Phenanthrene	0.17	0.17 to 0.18
	Anthracene	0.18	0.18 to 0.19
Tetraaromatics	3,4-Benzphenanthrene	0.17	0.17 to 0.19
	Pyrene	0.20	0.20 to 0.22
	Benzfluorene	0.25	0.25
	Triphenylene	0.24	0.22 to 0.24
	Benzanthracene, chrysene	0.26	0.24 to 0.26
	Naphthacene	0.39	0.35 to 0.39
Pentaaromatics	Benzpyrene	0.28	0.28 to 0.29
	Dibenzfluorene	0.31	0.31 to 0.32
	Dibenzanthracene	0.31	0.31 to 0.32
	Pentacene	0.53	0.48 to 0.53
Thiophenes	Thiophene	-0.17	-0.05 to -0.10
	Benzothiophene	0.04	0.06 to 0.07
	Dibenzothiophene	0.15	0.15 to 0.18
	Tribenzothiophene	0.22	0.20 to 0.22
Sulfides	Dimethyl sulfide	-0.03	0.00 to 0.01
	Methyl phenyl sulfide	0.04	0.04 to 0.05

TABLE 1. (continued)

Compound type	Parent compound	Relative adsorptivity (ϵ_1) Parent compound	Relative adsorptivity (ϵ_1) Petroleum fractions (600-1000° F)
Sulfides (cont.)	Methyl naphthyl sulfide	0.15	0.13 to 0.15
Disulfides	Dimethyl disulfide	-0.02	-0.01 to 0.01
	CH_3—S—CH_2-S—CH_3	0.20	0.12 to 0.15
Thiols	Methyl mercaptan	0.00	0.01 to 0.03
	Benzenethiol	0.98	0.54 to 0.65
Sulfoxides	Dimethyl sulfoxide	0.90	0.33 to 0.38
	ϕ—SO—CH_3	0.39	0.27 to 0.30

aromatics, chloro, bromo, or iodo derivatives, and sulfur compounds. Retention also increases with molecular weight.

Adsorbents may be acidic or basic. Acidic adsorbents such as silica preferentially adsorb amines, while basic adsorbents such as alumina preferentially retain acids, phenols, and pyrroles. Adsorbents of a given type may vary widely in physical characteristics such as surface area, pore size, shape, and particle diameter. High surface areas are generally desirable because they offer high sample capacity, however, the very highest surface area adsorbents tend to have greater heterogeneity and have small pore diameters which tend to exclude the larger sample molecules from reaching the internal surface. Particles may have irregular, spherical, or pellicular structure. Irregular particles are usually the lowest cost for a given particle size. Pellicular adsorbents (solid spherical core coated with a thin shell of porous adsorbent) offer the best efficiency for a given particle size, but they are the most expensive and have the lowest sample capacity.

Silica is the most widely used adsorbent. It offers higher efficiency and sample capacity than other adsorbents and it is available in a greater variety of forms. Because the silica surface is mildly acidic, it preferentially retains basic compound types. Silicas having a surface area of 300 to 500 $m^2\ g^{-1}$ are generally preferred. Silica is activated by heating to 150 to

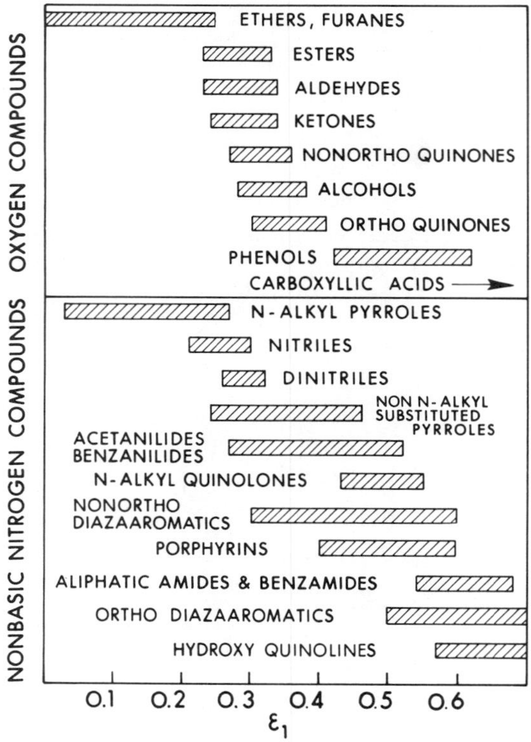

FIG. 3. Grouping of different oxygen and nitrogen compound types by separation on 3.8% H_2O-Al_2O_3 [87].

200°C for 8 to 16 hr. Temperatures above 200°C should be avoided, because surface activity is irreversibly destroyed.

Alumina is commonly used in the analysis of petroleum samples. Its popularity is partly related to its superior ability to separate a given compound class by aromatic ring number. Also, untreated alumina is basic, and therefore preferentially retains acidic compounds. Aluminas with surface areas of 100 to 200 $m^2\ g^{-1}$ are generally preferred. Activation of alumina is carried out by heating to 200 to 400°C for 8 to 16 hr.

Charcoal is a nonpolar adsorbent and is commonly used to separate ionic and/or aliphatic compounds (which are not adsorbed) from aromatics. Charcoal has also been used to separate compound classes isolated by elution from a polar adsorbent into subfractions by aromatic ring number; for example, the separation of the basic nitrogen fraction into pyridines, quinolines, and phenanthradines [89].

Many other adsorbents have been used over the years, but none appear to have significant advantages over silica, alumina, and charcoal for petroleum separations. Numerous examples of modified adsorbents have been described. Notable examples of these modifications are (a) addition of silver salts to enhance selectivity of olefins and aromatics [15, 33]; (b) addition of heavy metals to enhance the adsorption of sulfur compounds [13, 58]; and (c) modification with bisulfite in order to selectively adsorb aldehydes [1].

Table 2 lists the properties of a few of the adsorbents currently available. A more complete description of column packings for HPLC is available in the literature.

B. Stationary Phases for Partition Chromatography

In partition chromatography the stationary phase is a liquid which is either mechanically held or chemically bonded to a porous support. In this section we briefly describe supports, liquid phases, and bonded phases.

Supports for partition chromatography should be inexpensive, rigid, inert materials of moderately high surface area (10 to 50 $m^2\ g^{-1}$). In the past, diatomaceous materials such as celite or chromosorb have been popular [91]; however, low surface area silicas have become popular recently. Silica has the advantage of being less fragile than diatomaceous materials and it is available in a greater variety of particle sizes and forms [46].

Liquid phases are coated onto the support to form the stationary phase. The stationary liquid may be polar, nonpolar, nonviscous, viscous, or even semisolid, but it must be immiscible with the mobile phase liquid [91]. Examples of stationary phase liquids for normal partition are: oxydipropionitrile, carbowax, and triethylene glycol. Stationary liquids for reversed-phase partition are usually saturated hydrocarbons such as squalane, octadecane, or hydrocarbon polymers. The two phases of a ternary mixture such as ethanol-water-isooctane have also been used together as the mobile and stationary phase pair [30]. For nonvolatile stationary liquids the support may be coated by slurrying with an appropriate volume of 0.5 to 10% solution of liquid phase in a volatile solvent. Slow evaporation of the solvent then leaves an evenly coated layer of liquid phase on the support surface. Alternatively, the uncoated support is packed into the column and is coated in situ by passing small portions of stationary liquid or large volumes of mobile phase (saturated with stationary liquid) through it. Final liquid phase loadings average about 10% by weight for porous supports and about 1% for pellicular supports.

Bonded phases chemically combine the support with the stationary liquid. As already described, bonded phases have several advantages over conventional partition phases. Their major disadvantages are that sample capacity is low, laboratory procedures for their preparation are not generally

TABLE 2. Some Commercial Adsorbent Properties

Type	Trade name	Manufacturer[b]	Form[a]	Particle size (μ)	Surface area ($m^2 g^{-1}$)
Coarse porous silica	Code 62 silica	Davison	I	74-250 μ	340
	Code 12 silica	Davison	I	74-840	740
	Porasil A	Waters	S	37-75	400
	EM Type 40	EM Laboratories	I	63-200	650
	EM Type 60	EM Laboratories	I	40-63, 63-200	500
	EM Type 100	EM Laboratories	I	63-200	400
	Bio-Sil A	Bio-Rad Laboratories	I	74-150	400
Pellicular silica	Corasil II	Waters	S	37-50	14
	Pellosil HC	Whatmann	S	37-44	8
	Perisorb A	EM Laboratories	S	30-40	14
	Vydac	Separations group	S	30-44	12

Micro-particle porous silica	Spherosil	Rhone-Progil (France)	S	5, 10, 20	600
	Li Chrosorb Si-60	EM Laboratories	I	5, 10, 20	500
	μ Porasil	Waters	I	8-12	400
Porous alumina	Alcoa F-20	Alcoa	I	Coarse	100-200
	Aluminum Oxide 150	EM Laboratories	I	63-200	70
	Chromasep	Tracor	I	5, 10	70-90
	Li Chrosorb Alox T	EM Laboratories	I	5, 10, 20	70-90
	Spherisorb A1	Rhone-Progil (France)	S	5, 10, 20	95
Pellicular alumina	Pellumina HC	Whatmann	S	37-44	8
Pellicular charcoal	Pellicarb	Whatmann	S	37-44	—

[a] I = irregular, S = spherical.
[b] Sources: Davison Chemical, Baltimore, MD; Waters Associates, Milford, MA; E. M. Laboratories, Elmsford, NY; Bio-Rad Laboratories, Richmond, CA; Whatman, Inc., Clifton, NJ; Separations Group, Hesperia, CA; Alco, Pittsburgh, PA; Tracor, Austin, TX.

available, and commercially available products are quite expensive. Few applications of bonded phases to petroleum analyses have been reported, but owing to their widespread and growing popularity, a number of applications are to be expected in the near future. At present, chemically bonded phases are available with the following functional groups: octadecyl, p-alkylphenyl, phenyl, alkylnitrile, nitro, fluoroether, alkylamine. Majors' recent review [46] of stationary phases for HPLC includes an excellent section on bonded phases.

IV. SOLVENTS

A. Physical Properties

Of the hundreds of solvents that can be used as the mobile phase in liquid chromatography, fewer than 2 dozen have found widespread use (Table 3). The list of preferred LC solvents is small because useful solvents require a combination of physical and chromatographic properties. Solvents of low viscosity are preferred in order to minimize the separation time. For example, cyclohexane and pentane have similar chromatographic properties, but the viscosity of cyclohexane is 4 times higher and therefore an equivalent separation with pentane would be completed in 1/4 the time. Petroleum applications often require preparative separations. In order to recover the separated sample, the eluent is removed by distillation or evaporation. In either case it is important to use low boiling point solvents to minimize sample vaporization or thermal degradation. Also, solvents which may react with the adsorbent, the sample, or air should be avoided. For example, acetone should not be used for separation on alumina because it reacts to form diacetone alcohol. The purity of a solvent must also be considered. Even reagent-grade solvents may have small but significant quantities of nonvolatile residues. Some solvents such as tetrahydrofuran contain small amounts of antioxidant additives which become concentrated in the recovered fractions. Ethers may contain explosive peroxides and should be used with caution in preparative applications. A recent monograph on solvents [59] gives detailed procedures for solvent purification.

B. Solvent Strength

Solvent strength is the chromatographic property of the solvent which controls the migration of a sample. Thus, weak solvents result in longer retention times for a given solute or compound class, while stronger solvents result in shorter retention times. The solvent strength depends on the chromatographic system. For example, in adsorption chromatography pentane is a very weak solvent; however, in reversed-phase partition chromatography it is a very strong solvent.

In adsorption chromatography solvent strength is indicated by the eluent strength parameter $\epsilon°$. Table 3 lists $\epsilon°$ values for the most commonly used LC solvents. Eluents of intermediate strength can be prepared by mixing two or more solvents; however, the resulting solvent strength ($\epsilon°_{AB}$) is a nonlinear function of the composition as shown in Figure 4. The solvent strength of any mixture can be calculated (see Ref. 68). Tables 4 and 5 list the solvent strength for a variety of mixed solvents on alumina and silica respectively. The ratio of retentions for a given solute eluted with two solvents differing in solvent strength by an amount Δ ° is approximately $10^{8\Delta\epsilon°}$. Thus, with the available range of solvent strengths (-0.25 to 0.95) it is possible to vary the retention of a given compound by a factor of about 10^{10}. In practical terms this means that small changes in solvent composition can result in rather large changes in retention. In the separation of petroleum samples it is often necessary to make one or more changes in solvent composition in a given analysis in order to elute the various compound classes under optimum conditions. In one example [51], pentane is used to elute a saturate fraction from dry alumina, then the solvent is changed to ethyl ether in order to elute the aromatics, and finally, the solvent is changed to a mixture of benzene and methanol in order to displace the strongly adsorbed oxygen and nitrogen compounds. Without the solvent changes, the later fractions would never be eluted.

Solvent strength in partition is less well defined than in adsorption systems. For normal partition chromatography, the more polar solvents are stronger solvents, whereas the opposite is true for reversed phase partition. The Hildebrand solubility parameter δ (listed in Table 3) has been proposed as a useful solvent strength scale for partition chromatography. We should expect to see more work relating to the quantitative prediction of solvent strength in the future. Recently [83], another solvent strength scale P' based upon the Rohrschneider constants has been proposed; these values are also listed in Table 3. Solvent polarity increases with P' or δ.

C. Selectivity

Selectivity refers to the difference in retention between two components or compound classes. In adsorption and partition chromatography, selectivity can be controlled somewhat by changing the composition of the mobile phase while maintaining solvent strength constant. Normally there will exist several solvents of suitable strength for a given separation. However, the relative retention of different compounds or compound types can vary greatly for different solvents—even those of equal $\epsilon°$ value. This suggests the use of different solvents to achieve specific separations of interest, particularly after an initial rough separation with a series of solvents of different strength. Table 6 is an example of how solvent composition can affect selectivity. While retention (k') remains fairly constant for the two compounds shown in

TABLE 3. Properties of Common LC Solvents

Solvent	$\epsilon°$ [a]	δ [b]	η [c]	RI[d]	UV cutoff[e]	BP[f]	TLV[g]	ρ [l]	P'	Selectivity parameters[m] I_{ap}	χ_e	χ_d	χ_n
Fluoralkanes	-.25	~5.5		1.25									
Fluoroethers[h]	-.13	~5.5	0.44[i]		210	50.	—	—	—	—	—	—	—
n-Pentane	0.00	7.1	0.24	1.358	210	36.1	500	.626	—	—	—	—	—
n-Hexane	0.00	7.3	0.31	1.375	210	68.7	500	.659	0.0	0.00	—	—	—
Isooctane	0.01	7.0	0.50	1.391	210	99.2	—	.692	-0.4	-0.10	—	—	—
Carbon disulfide	0.15	10.0	0.37	1.628	380	46.2	20	1.270	1.0	0.11	—	—	—
Isopropyl ether	0.28	7.3	0.37	1.368	220	68.3	250[k]	.724	2.2	-0.15	0.54	0.11	0.35
Isopropyl chloride	0.29	8.3	0.33	1.378	225	34.8	—	.862	—	—	—	—	—
Toluene	0.29	8.9	0.59	1.497	285	110.6	100	.867	2.3	0.00	0.32	0.24	0.44
Benzene	0.32	9.2	0.65	1.501	280	80.1	25	.879	3.0	0.04	0.29	0.28	0.43
Ethyl ether	0.38	7.4	0.23	1.352	220	34.6	400	.713	2.9	-0.22	0.55	0.11	0.34
Chloroform	0.40	9.1	0.57	1.446	245	61.2	25	1.480	4.4	-0.10	0.28	0.39	0.33
Methylene chloride	0.42	9.6	0.44	1.424	245	39.8	500	1.335	3.4	0.20	0.34	0.17	0.49

Tetrahydrofuran	0.45	9.1	0.55	1.408	220[j]	66.	200	.889	4.2	-0.15	0.41	0.19	0.40
Acetone	0.56	9.4	0.32	1.359	330	56.2	1000	.790	5.4	-0.10	0.36	0.24	0.40
Acetonitrile	0.65	11.8	0.37	1.344	210	81.6	40	.782	6.2	0.04	0.33	0.26	0.41
Isopropanol	0.82	10.2	2.3	1.378	210	82.4	400	.785	4.3	-0.30	0.54	0.20	0.26
Ethanol	0.88	11.2	1.20	1.361	210	78.5	1000	.789	5.2	-0.23	0.51	0.21	0.28
Methanol	0.95	12.9	0.60	1.329	210	64.7	200	.791	6.6	-0.40	0.49	0.15	0.36
Water	large	21.	1.00	1.333	210	100.0	—	.998	(9)	—	0.40	0.34	0.26
N,N-dimethyl formamide	large	11.5	0.92	1.431	270	153.0	—	.949	6.4	0.00	0.41	0.21	0.38

[a] Solvent strength on alumina.
[b] Solubility parameter.
[c] Viscosity in cP at 20°C.
[d] Refractive index at 20°C.
[e] Wavelength for 1 A absorbance in a 1-cm cell.
[f] Boiling point °C, 760 torr.
[g] Threshold limit value (ppm) American Conference of Government Industrial Hygienists (1971).
[h] Data from 3M Co., FC-78.
[i] cP, 25°C.
[j] Usually contains uv absorbing antioxidant but can be purified by distillation.
[k] Forms peroxide quickly.
[l] Density, g ml^{-1} at 20°C.
[m] See Refs. 83 and 91.

TABLE 4. Binary Solvent Strength on Alumina [68]

$\epsilon°$	Solvent mixture	Alternate mixture
0.00	100% P	—
0.05	8% I-P	1.5% M-P
0.10	8% B-P	4.0% M-P
0.15	16% B-P	3.5% MA-P
0.20	28% B-P	5.0% MA-P
0.25	49% B-P	8.0% MA-P
0.30	83% B-P	2.5% D-P
0.35	54% M-P	2.0% D-B
0.40	84% M-P	7.0% D-B
0.45	24% MA-B	14.0% D-B
0.50	42% MA-B	6.0% A-B
0.55	66% MA-B	4.0% IPA-B
0.60	100% MA	7.0% IPA-B
0.65	100% A	10.0% IPA-E
0.70	21% IPA-B	20.0% IPA-E
0.75	40% IPA-B	3.0% MeOH-E
0.80	75% IPA-B	7.0% MeOH-E
0.85	17% MeOH-MA	17.0% MeOH-E
0.90	42% MeOH-MA	40.0% MeOH-E
0.95	100% MeOH	40.0% MeOH-E

Solvent mixture refers to % v/v of indicated pairs: P, pentane; I, isopropyl chloride; B, benzene; M, methylene chloride; MA, methyl acetate; D, diethyl amine; A, acetonitrile; IPA, isopropyl alcohol; E, ethyl ether; MeOH, methyl alcohol.

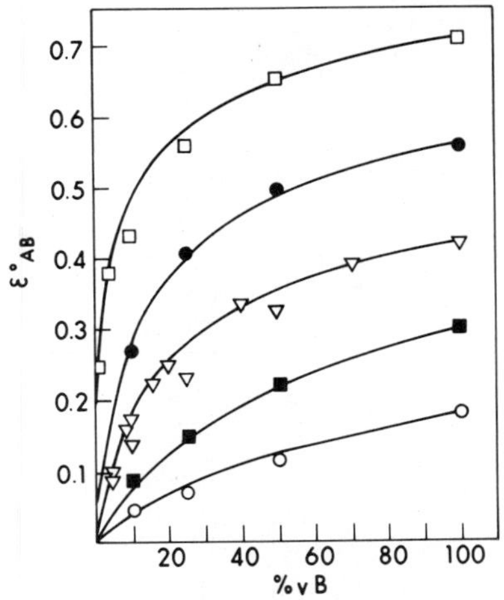

FIG. 4. The variation of mixed solvent strength ϵ°_{AB} with solvent composition. Adsorbent: alumina with 3.9 wt % H_2O. Solvent B: pentane. Solvent B: ○, CCl_4; ■, n-propyl chloride; ▽, CH_2Cl_2; ●, acetone; □, pyridine.

TABLE 5. Binary Solvent Strength on Silica [68]

ϵ°	Solvent mixture	Alternate mixture
0.00	100% P	—
0.05	4% B-P	3% M-P
0.10	11% B-P	7% M-P
0.15	26% B-P	14% M-P
0.20	52% B-P	4% E-P
0.25	100% B	11% E-P
0.30	22% E-B	4% A-P
0.35	62% E-B	10% A-P
0.40	25% A-B	11% AN-P
0.45	65% A-B	31% AN-P
0.50	100% AN	3% MeOH-M
0.55	6% MeOH-M	4% MeOH-E
0.60	13% MeOH-M	8% MeOH-E
0.65	26% MeOH-M	18% MeOH-E
0.70	56% MeOH-M	40% MeOH-E
0.75	100% MeOH	—

Solvent mixtures refer to % v/v of indicated pairs: P, pentane, B, benzene; E, ethyl ether; A, acetone; M, methylene chloride; AN, acetonitrile; MeOH, methanol.

Table 6, different solvent mixtures result in large changes in relative retention (α). A few examples may clarify this area (however, for a general discussion, see Refs. 68, 84, and 91). On alumina, polycyclic aromatic hydrocarbon classes are separated by aromatic carbon number with pentane-pyridine solutions as solvent. Thus, anthracenes and phenanthrenes are eluted together, benzanthracenes and chrysenes separate together, and so on. However, ethyl ether or methylene chloride solutions in pentane can provide significant resolution of these isomeric hydrocarbon classes, separating, for example, phenanthrenes from anthracenes. Consequently, an initial separation with pentane-pyridine as solvent can be used to isolate phenanthrenes plus anthracenes, while a subsequent separation of the latter group of compounds with methylene chloride-pentane as solvent can separate the anthracenes from the phenanthrenes.

Similarly, it is found that basic solvents (e.g., diethyl amine-pentane or pyridine-pentane solutions) preferentially retain N-H or O-H-containing compound types on alumina. Consequently an initial separation with a nonbasic solvent, followed by reseparation with a basic solvent, should provide resolution of compound types with N-H and/or O-H groups from compounds without this functionality.

Finally, compound types containing two (or more) weakly adsorbing functional groups often overlap onto compound types containing a single stronger-adsorbing group. Fractions containing such mixtures can be reseparated with certain solvents that preferencially retain the difunctional

TABLE 6. Solvent Selectivity on Alumina [91]

Solvent (pentane solution) (v/v%)		k' 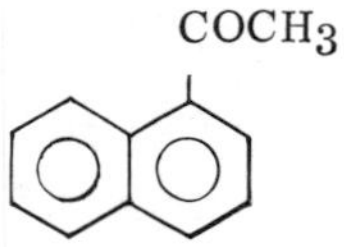	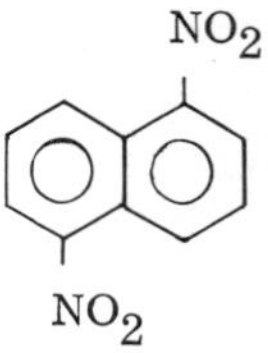	α
50	Benzene	5.1	2.5	0.5
25	Diethyl ether	2.5	2.9	1.16
23	CH_2Cl_2	5.5	5.8	1.05
4	Ethyl acetate	2.9	5.4	1.8
5	Pyridine	2.3	5.4	2.4
0.05	CH_3SOCH_3	1.0	3.5	3.5

compound types; e.g., very dilute solutions in pentane of very polar (large $\epsilon°$) solvents, such as 1% v/v acetonitrile/pentane.

Selectivity in partition chromatography is controlled in a similar manner. If components of a mixture are inadequately resolved with one solvent mixture, a new mixture is chosen with a similar solvent strength but a different chemical composition. The choice of a new solvent mixture is guided by considering the types of solvent-solute intermolecular interactions which can occur. The dominant interactions are dispersion, dipole orientation, and hydrogen bonding. Recently, the degree with which various solvents can undergo these types of interactions has been quantitated [83, 90, 91]. Table 3 lists the following selectivity parameters: I_{ap} (dispersion), χ_e (proton donor), χ_d (proton acceptor), and χ_n (dipole orientation). By choosing a new solvent mixture having the same strength (δ or P') but differing in the magnitude of one or more of these parameters, the relative retention of components of the mixture will be changed, provided the individual components of the mixture differ in the way they interact with the solvent. For example, if we substitute ethanol (P' = 5.2, $\chi_e = 0.51$) for acetone, (P' = 5.4, $\chi_e = 0.36$) as our solvent, we would expect to improve the selectivity between samples which can donate protons relative to those which cannot.

V. CLASSIFICATION OF APPLICATIONS

A recent review [2] includes an excellent summary of past applications of adsorption chromatography to petroleum analysis. These far outnumber those of partition. Although there are several examples of unique separations by partition, few have reached the status of routine procedures. In the following sections the applications of adsorption and partition chromatography to petroleum analysis will be categorized by the compound class analyzed. This is a somewhat arbitrary division since a number of procedures have been developed to analyze a wide variety of compound types in a single sample. A detailed description of these integrated procedures is covered in Chapter 9 on separation schemes; see also Refs. 67, 69, 75, 85, 89.

As indicated previously, many different chromatographic techniques can be applied to the analysis of petroleum. Before looking at specific examples of these applications it is useful to classify the various immediate goals of the separation, since the specific goal usually determines which technique will be favored for a particular application.

A. Goals of the Separation

Routine analysis by compound type is one of the more important applications of adsorption chromatography. In some cases the compound types of interest are sufficiently resolved to allow analysis of the sample by gravimetric or volumetric means alone (following separation). More often partially separated fractions must be examined further to correct for overlap of one or more compound types into adjacent fractions. Alternatively, subsequent analytical measurements may be required to distinguish different subtypes within a given fraction. Whatever the approach, routine compound-type analysis procedures normally emphasize speed and convenience. Perhaps the most widely used routine procedures are those adopted by the American Society for Testing and Materials [6-11].

Preparation of compound-type fractions for qualitative analysis by other techniques has been widely used in establishing the individual compound-types that are present in petroleum and related samples (refinery streams, shale oil, hydrogenated coal, and so on). Ideally, the separation yields "pure" fractions containing only the compound type to be characterized. Commonly a number of small, arbitrary fractions are collected by successive elution with one or more solvents, and these individual fractions are each examined further. In some cases the elution of a compound type of interest can be followed by monitoring the column eluent (e.g., by ultraviolet absorption), which allows its immediate collection as a single fraction.

Preliminary separations and sample clean-up, i.e., isolation of compound-type fractions for further separation into pure compounds, has been used extensively for the analysis of gasoline and naphtha-range samples. Most often a separation by adsorption into saturates, olefins, and aromatic subclasses is followed by gas chromatographic (GC) separation of a particular compound-type fraction. Also, preliminary separation by GPC into narrow molecular size cuts which are then analyzed in detail by GC, or reversed phase partition chromatography has been used effectively. Such preliminary separations both simplify each fraction for the final separation into individual compounds and facilitate the identification of pure compounds.

Preparation of compound-type fractions to calibrate other analytical methods represents another important use of adsorption chromatography—particularly in the development of direct mass spectrometric methods for the analysis of petroleum distillates. Normally the mass spectral characteristics of a given compound type vary with molecular weight, variations in the distribution of subtypes within the fraction provided by adsorption, and even the source of the sample (e.g., straight run, catalytically cracked, and so on). However, the necessary calibration data can usually be obtained if well-separated compound-type fractions are available from a range of samples varying in both boiling point and source. Following development of a mass-spectral method, these same calibration fractions can be reblended to give new samples of known composition, and the mass-spectral assay can then be tested by application to these synthetic samples.

Characterization of unknown compound-types via separation per se is used less commonly, but can be a valuable adjunct to follow-up characterization procedures. The relative retention of a compound-type of interest with respect to known compound-types, or its inclusion in certain fractions from a standard separation scheme, can be used to confirm or rule out possible compound types which cannot be fully characterized by available methods. One example of this is provided by the identification of benzothiophenodibenzofurans (BDF) in a heavy California distillate [69]. High-resolution mass spectrometry confirmed the presence of compounds with the formula $C_nH_{2n-26}OS$, beginning with the parent compound $C_{18}H_{10}OS$. The presence of these compounds in certain fractions from a standard separation scheme (alumina and silica separations) established that only the BDFs were compatible with the molecular formula and separation characteristics of this class of compounds. However, the use of separation sequence to identify specific compound classes must be used with caution, as illustrated in one study [73] that showed several incorrect assignments of this type.

A discontinuity in the column eluent or appearance of a distinct elution band is usually assumed to indicate the elution of a new compound type. This information in combination with other analytical data on fractions collected around this discontinuity is often used to infer conclusions with respect to sample structure and/or the relative separation of one compound-type fraction from another (see Figure 5 and Sec. VII.A).

Isolation of pure compounds directly by adsorption chromatography was used originally in the characterization of low-boiling distillates by API Project 6. Today this application is of historical interest only, as it has

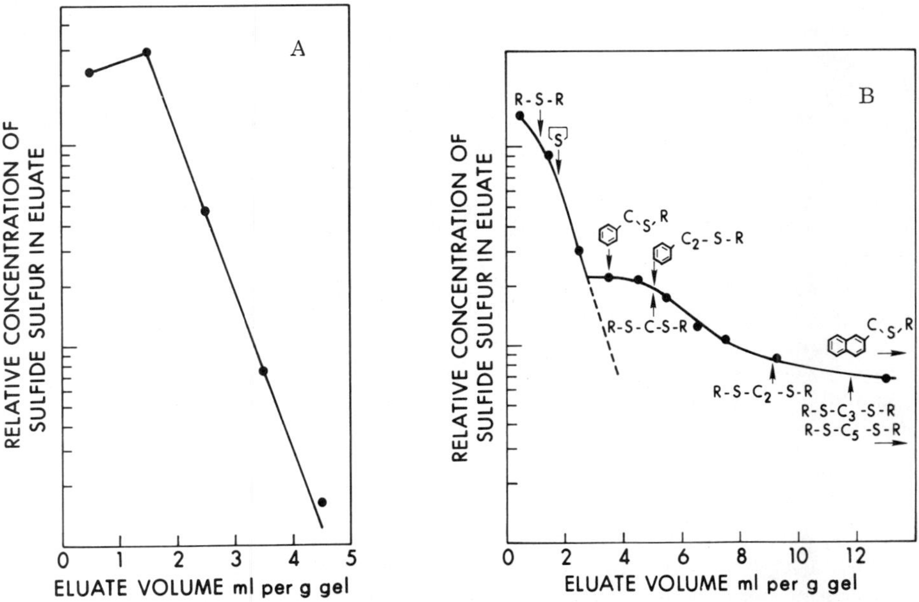

FIG. 5. Elution of light gas oil sulfides from alumina by pentane [71]. A. Low-sulfur sample; B. High-sulfur sample.

been superceded by GC. However, the determination of individual petroleum additives is in some cases done by adsorption chromatography, especially by TLC. The increased separation power of HPLC appears to be reviving interest in the use of LC for the isolation of individual compounds. For example, high-performance adsorption and reversed phase partition chromatography were used for the determination of benzo(a)pyrene and other PNAs in used engine oils and other petroleum samples [31, 32, 53, 57, 95, 110].

B. Chemical Reaction Plus Adsorption Chromatography

Many petroleum separation schemes have been reported in which chemical conversion of certain compound classes was used to facilitate their separation. In general, such schemes should be avoided. Usually the conditions required for near-quantitative conversion of one compound class lead to undesired side reactions of these and other compounds. The usual result is an increase in sample complexity, rather than a simplification, plus considerable doubt as to whether final fractions are contaminated by artifacts from the chemical reaction. A possible exception is given in Sec. VII.A.

In separations by adsorption chromatography, unintentional reaction or loss of the sample is a frequent concern. Under the best of conditions most adsorbents are potentially reactive toward certain compounds, particularly under unfavorable experimental circumstances. Therefore it seems wise in general to use adsorbents and conditions that have been shown not to result in alteration of the sample (e.g., discussion of Ref. 85). Particular suspicion should be directed at reactive chemical compounds or adsorbents, such as acidic clays, ferric chloride, florisil, and heavy metal-impregnated adsorbents. In our experience, few separation problems are so difficult as to require the use of these potentially reactive adsorbents.

VI. APPLICATION TO PETROLEUM HYDROCARBONS

Adsorption chromatography has played a major role during the past 35 years in both the qualitative and routine quantitative analysis of the different hydrocarbon classes in petroleum. Here we will divide our discussion into specific areas: saturated hydrocarbons, olefins, mono- through tetra-aromatics, and higher polycyclics.

A. Saturated Hydrocarbons

The isolation of this compound-class from the remainder of the sample is undoubtedly the single most common application of adsorption chromatography in routine petroleum analysis (see Ref. 92 for a review). Normally a saturated hydrocarbon solvent such as pentane or hexane is used with columns of either silica or alumina, in a low-pressure column mode. Under these conditions the overlap of saturates and monoaromatics is usually quite low (contamination of each fraction less than 1%). The separation of saturates from monoaromatics is favored by the greater retention of monoaromatics on higher surface area, thermally activated adsorbents; e.g., an 800 m^2 g^{-1} silica preactivated at 150 to 200°C. Several routine procedures for the determination of saturates in a variety of petroleum products are available [7, 8, 47, 92]. The separation of saturates from olefins is more difficult. For gasoline-range samples reasonable separations can be achieved by either FIA (displacement) analysis [6, 38] or by high-performance liquid chromatography [97]. The latter separation requires small-particle columns (10 μ) and a very nonpolar fluorocarbon solvent. Presumably, this separation could be enhanced by incorporating 1 to 5% silver nitrate into the adsorbent as a surface coating. Good separations of olefins from saturates in higher boiling samples have not yet been reported.

The use of cyclohexane as solvent has been recommended [62] for the separation of samples containing waxes, asphaltenes, or other pentane insolubles. Apparently this solvent can dissolve such samples completely without affecting the normal saturates-monoaromatics separation. Separation of the total saturate fraction into saturate subtypes by adsorption chromatography is not used. One study [92] showed some tendency of polycyclic saturates to be retained more strongly than paraffins, but the degree of separation was insufficient to be analytically useful. The use of perfluoroalkane solvents does increase overall saturates retention to the point where individual compounds can be separated [74, 97], but no useful compound-class separations result. Mair et al. [45] were able to separate certain branched and cyclo paraffins by partition and Sato et al. [61] reported the precipitation of waxes from the saturated hydrocarbon fraction of an asphalt.

B. Aliphatic Olefins

Olefins are normally present in significant amounts only in cracked samples. For these samples adsorption chromatography was used earlier to determine total olefins directly by the FIA method [6, 38], and more recently by HPLC [97]. Figure 6 shows the HPLC separation [97] of saturates, olefins, and aromatics in a gasoline. The peak height of the aromatics was enhanced by backflushing. In these latter procedures it is likely that cyclic diolefins partially overlap into the aromatics fraction. LEAC analysis has been used to determine total olefins in higher boiling samples and in straight run samples as well [75] by separating olefins (plus saturates) from aromatics, then determining olefins in that fraction by iodine complexation and uv analysis. If it were desired to improve on these past methods for separating olefins from both saturates and aromatics (to give a pure olefin fraction), it seems clear that the following approach would be preferred:

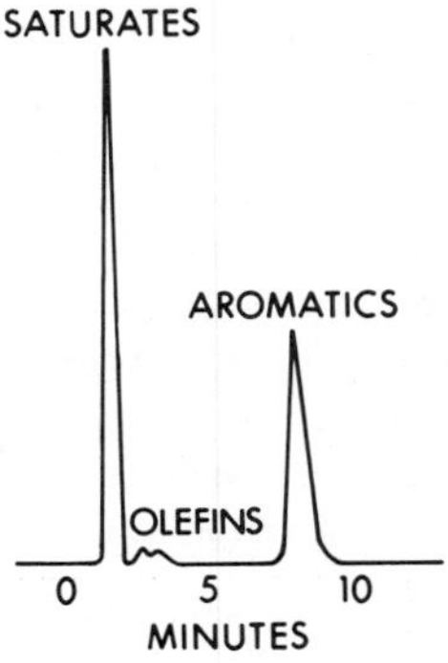

FIG. 6. Chromatogram of a motor gasoline with aromatics backflushed. Column = μ-porasil, 30 cm by 4 mm; mobile phase FC-78; flow rate: 3.5 ml min^{-1}.

1. Use of a high surface area silica in the dry state
2. Variation in the solvent strength by adding fluorocarbon (as in Refs. 74 and 97) to either pentane or hexane.

Finally, if one is willing to go to some lengths, it should be possible to isolate a pure aliphatic olefin fraction from any sample in the following way. First, carry out the separation just described on dry silica with pentane-fluorocarbon as solvent, and collect several fractions during the elution of olefins from the column. Each of the latter fractions can then be reseparated on silver nitrate-impregnated silica with the same solvent. Under these conditions any saturates or aromatic contaminants in the olefins fraction should be separated during the second separation. The various final olefins fractions can then be composited for further analysis.

C. Mono- Through Tetra-aromatics

It should first be stated what is included in these various compound classes, and this is summarized in Table 7. In general, these four compound classes

TABLE 7. Composition of Different Compound-Class Fractions Obtainable by Separation on Alumina

Fraction	Hydrocarbon types	Nonhydrocarbons
Monoaromatic	Benzenes	Thiophenes
Diaromatics	Naphthalenes	Benzothiophenes
	Biphenyls	Alkyl sulfides
	Diphenyl alkanes	Benzofurans
Triaromatics	Fluorenes	Dibenzothiophenes
	Phenanthrenes	Dibenzofurans
	Anthracenes	Other compounds
	Pyrenes	
Tetraaromatics	Benzfluorenes	
	Chrysenes, etc.[a]	
	Other compounds	

[a]And other isomers.

can be resolved from each other by elution from water-deactivated alumina by solvents of varying strength (beginning with pentane and ending with a solvent such as 10% ethyl ether/pentane; see Tables 3-5 and 9; some examples are given in Refs. 66 and 72. The overlap between adjacent fractions ranges from minor for the mono-, di-aromatics separation, to significant for the tri-, tetraaromatics separation. As in the case of all such separations, higher boiling samples show poorer resolution of separated compound classes. Nevertheless, adequate separations of these various compound classes are usually possible, especially for lower-boiling and/or catalytically cracked samples. The variation of separation with sample boiling point is discussed in Ref. 76.

Separations of these aromatic subtypes on silica is usually much poorer than on alumina (see discussion of Ref. 76). However, silica does show preferential retention of polyphenyl alkanes relative to fused aromatics of similar unsaturation [77]. Thus, the diaromatic fraction from an alumina separation should be separable on silica into fused and poly-phenyl alkane fractions.

D. Higher Polycyclic Aromatics

These refer to pentaaromatics and higher derivatives. Normally these are present in low concentration in distillate samples and cannot be adequately resolved in residual samples. There are some exceptions of interest, however. The lightly alkylated polycyclics such as benzpyrene and coronene occur in the bottoms fractions from various reforming and hydrocracking processes, and these compounds are of special interest in some applications. When the nitrogen and oxygen contents of the sample are quite low (as is usually the case for such hydro-processed samples), the polycyclics of interest can be well resolved by chromatography over water-deactivated alumina. Individual compounds can be distinguished if high-pressure LC is used (especially with small-particle, high-efficiency columns). Usually gradient elution or stepwise elution will be required to cover the wide range of retentions involved. For the case of samples containing significant amounts of nitrogen compounds, the latter will normally completely obscure the presence of these trace polyaromatics. What is needed then is a preliminary separation of the sample over silica to yield a predominantly hydrocarbon fraction, free of nitrogen and oxygen compounds [85]. The latter can then be reseparated over alumina to yield resolution of the desired polycyclics. Appendix B is a detailed LEAC routine procedure for the analysis of coronenes in hydrotreated samples.

For the high-resolution separation of individual polycyclic hydrocarbons such as the benzpyrenes, the following procedure is recommended. First, prepare a hydrocarbon concentrate free of both nitrogen-oxygen compounds and the bulk of lower hydrocarbons. This is most easily accomplished on silica. The resulting polycyclic concentrate can next be subjected to high-pressure LC by reversed phase partition or on a deactivated alumina. If resolution is inadequate for some parts of the latter separation, due to sample complexity, these subfractions can be isolated and reseparated on an activated alumina with solvents such as methylene chloride-pentane (see discussion of Ref. 78). By changing solvent and adsorbent activity, it is possible to change separation factors on alumina for individual polycyclic isomers by as much as a factor of 30. Alternatively, initial separation of a polycyclic concentrate via silica plus alumina separation can be followed by high-resolution mass spectral analysis of the concentrate.

Finally, several procedures have been reported for the simultaneous separation and total analysis of heavy distillates in terms of aromatic hydrocarbon types; e.g., one procedure [66] yields saturates, monoaromatics, diaromatics, polyaromatics, and hetero compounds containing N or O. Table 8 lists a variety of applications of adsorption and partition chromatography to the analysis of hydrocarbon components.

Sample	Compound class isolated	Comments	References
Distillates < 400° F	Saturates, olefins, aromatics	Routine FIA	6, 38
Distillates 400-1000° F	Saturates	Rapid, routine low-pressure column separation method	92
Distillates < 400° F	Saturates, olefins, aromatics	HPLC alternative to FIA	97
Distillates 400-1000° F	Saturates	ASTM procedure	10
Distillates 400-1000° F	Saturates	Improved ASTM method	47
Distillates < 1000° F	Olefins	Determination by uv of I_2-Olefin Complex	75, 82
Crude oils, distillates, residues	Saturate, aromatics, polar compounds, asphaltenes	Rapid HPLC method	98
Cracked gas oils 400-1000° F	Mono-, di-, tri-, tetra-aromatics	Routine LEAC method	72
Solvents 100-450° F	Aromatics	Benzene, toluene, ethylbenzene and total aromatics determined by GC	22
Crude oils and tars	Benzo(a)pyrene	Adsorption and partition HPLC	57
Wax	Polycyclic aromatics	Partition TLC	102
Used engine oils	Polycyclic aromatics	Reversed phase partition HPLC	101
Distillate 650-1000° F	Saturates, monoaromatics, diaromatics, polyaromatic and polar compounds	Dual packed silica alumina column	28
Distillates 400-1000° F	Saturates, monoaromatics, diaromatics, polyaromatics, and polar compounds	0%, 4% H_2O-Al_2O_3 coupled-columns	66

VII. APPLICATION TO SULFUR, OXYGEN, AND NITROGEN COMPOUNDS

Traditionally adsorption has been used for the partial separation of these hetero compound classes rather than near-complete separation, as in the case of hydrocarbon classes. When successful, partial separations can yield fractions that allow the subsequent analysis of individual compound classes by other analytical methods; e.g., mass spectrometry, uv, and colorimetry. We should keep in mind certain goals of such hetercompound separations:

1. Concentration of compound classes of interest to a level at which they can be determined in a given fraction, and elimination of interferences from major hydrocarbon classes.
2. Separation of the benzologs of the same general class (e.g., thiophenes, benzothiophenes, and dibenzothiophenes), since these usually cannot be readily distinguished by high-resolution mass spectrometry from benzologs of differing cycloalkyl substitution.
3. Separation of isomeric compound classes (e.g., cycloalkyl indoles from quinolines) that cannot be readily distinguished by mass spectral analysis.
4. Achievement of such separations conveniently, reproducibly, and without significant loss or alteration of compounds of interest.

The following discussion will concentrate on heterocompounds known to be present in petroleum; for example, those found in the total heterocompound analysis of one California petroleum [67]. Additional information on the adsorption of other possible compound types is reviewed in Ref. 87 for alumina as adsorbent.

A. Sulfur Compounds

The principal sulfur types present in petroleum distillates consist of free sulfur, mercaptans, sulfides, polysulfides, and thiophenes, along with their various benzologs and cycloalkyl derivatives. Mercaptans and disulfides are found in the gasoline fraction of some crudes, but their concentrations in higher boiling distillates are normally negligible. Our discussion of the separation of these compounds will therefore be brief. Compounds containing both a sulfur atom and one or more additional hetero atoms (S, O, N) in the same molecule are found in significant amounts in distillates boiling above 1000°F, but these compounds are usually present in minor amounts, if at all, in lower boiling fractions. The only exception is that of nonvicinal disulfides discussed below. Adsorption chromatography is quite limited in its ability to separate these polyfunctional compounds in high boiling fractions, because of the complexity of such samples, and because the sulfur atom has little effect on molecular adsorption. The following discussion will therefore emphasize the separation of monosulfides and thiophene derivatives.

1. Mercaptans and Vicinal Disulfides (R-S-S-R')

These compounds normally separate with the diaromatic hydrocarbons on alumina, with minor overlap into the monoaromatic fraction [79, 87]. On silica these compounds tend to be eluted after the diaromatics. An exception is the thiophenols, which on alumina (but not silica) are strongly retained and elute after the pentaaromatics.

Because of the strong complexation of thiols with mercuric ion it can be anticipated that mercuric ion-impregnated adsorbents will preferentially retain thiols.

2. Thiophene Derivatives

These compounds and their higher benzologs generally overlap the elution of hydrocarbons of similar aromatic ring number, as first observed by Lumpkin and Johnson [40]. Thus, with either alumina or silica as adsorbent, thiophenes elute with the monoaromatics, benzothiophenes elute with diaromatics, and so on. As in the case of the hydrocarbons, better separation of the thiophene benzologs is obtained on alumina versus silica. There is a slight tendency for earlier elution of each thiophene class relative to the hydrocarbon class of similar aromatic ring number [85, 87]. The separation of thiophene-benzolog classes (thiophenes, benzothiophenes, among others) by elution from alumina is sufficient for the subsequent assay of each benzolog in its fraction by high-resolution mass spectrometry. The alkyl plus cycloalkyl thiophenes can be determined as a class by LEAC [71]. Saturates, monoaromatic hydrocarbons, and thiophenes are eluted from alumina as a single fraction, with thiophenes determined in that fraction by uv absorption at two wavelengths (245 and 270 nm). FCC samples cannot be analyzed by this method [65] because total thiophene contents of these samples are low and aromatic olefins are present which interfere with the final assay.

3. Sulfide Derivatives

Monosulfides overlap the elution of diaromatic hydrocarbons from alumina [79, 87] but are retained more strongly on silica. Therefore, the diaromatic/sulfide fraction from an initial alumina separation can be further separated on silica to give separate diaromatics and alkyl sulfides fractions [71]. Benzothiophenes are retained with the diaromatics fraction.

So-called nonvicinal disulfides (e.g., I) and aromatic sulfides (e.g., II) are eluted just after the monosulfides, and largely separate with triaromatic hydrocarbons in alumina chromatography [71, 87]. The earliest

$$\mathrm{R{-}S{-}CH_2{-}S{-}R'}\ \text{(I)} \qquad \phi\mathrm{{-}CH_2{-}S{-}R}\ \text{(II)}$$

evidence for their presence in petroleum was afforded by the shape of the elution curve during chromatographic separation of gas oil samples on alumina [71] (see Figure 5). In Figure 5a a low-sulfur light gas oil was charged to a column of 4% water-alumina, the column was eluted with pentane, small fractions of effluent were collected, and sulfides were measured in each fraction by the uv absorption of the iodine complex at 308 nm. The sample of Figure 5a contained only monosulfides, and a normal monomodal elution curve is observed with absorbance falling off linearly in this log-linear plot, as expected (see Sec. II.B). In Figure 5b, a high-sulfur light gas oil was separated similarly. Here the later elution of a second sulfide class is evident and comparison of this chromatogram with retention volumes for various nonvicinal sulfides (arrows in Figure 5b) suggests that the second fraction in Figure 5b is due to compounds of types I and II. Nonvicinal difunctional sulfides (i.e., I and II) are predicted to be retained even more strongly on silica, just as in the case of monosulfides. Therefore, further separation of initial triaromatics-nonvicinal sulfides fractions from alumina chromatography using silica as adsorbent in the second separation should yield final fractions of triaromatics separated from nonvicinal sulfides. Some contamination of the nonvicinal sulfides fraction by nitrogen and oxygen compounds would be expected, however. The alkyl monosulfides and nonvicinal disulfides can each be determined by means of LEAC [71]. Elution of the sample from alumina separates these two compound types for subsequent assay as the iodine complex, and also eliminates basic nitrogen compounds that interfere in the direct assay of the total sample for sulfides.

Partition chromatography using mercuric acetate plus acetic acid on deactivated silica as stationary phase was found useful for isolating sulfides from other compound types [52]. The sulfides were retarded while thiols were irreversibly adsorbed.

4. Difunctional Compounds Containing Sulfur

Little is known about the possible structures of compounds containing S—N, S—O, and S_2. Inferential evidence suggests that the sulfur atom occurs mainly in thiophene rings that are part of large polyaromatic ring systems (see Sec. V.A). If this is the case, such compounds will be separated with corresponding compounds that lack one sulfur atom; i.e., where a benzene ring replaces the sulfur-containing thiophene ring.

5. Sulfoxides

These compounds are not believed to be present in petroleum, but can be formed readily as a result of air oxidation of aliphatic sulfides present in the sample. They elute after the pentaaromatic hydrocarbons from alumina and are held quite strongly on silica (see further discussion of Ref. 69).

6. Separation after Conversion to Sulfones

A possible exception to the undesirable aspects of reaction of the sample before separation is afforded by the separation scheme of Drushel and Sommers [20]. These workers used peroxide oxidation to convert various thiophene and sulfide compounds in petroleum to the sulfones. Their approach was a preliminary separation of a thiophene-sulfide concentrate by chromatography on silica, oxidation of this fraction to yield sulfones of all sulfur compounds, rechromatography of the reaction product to isolate the more polar sulfones from the hydrocarbons present in the original fraction, and reduction of the sulfone fraction with $LiAlH_4$ to yield the original thiophenes and sulfides.

The overall merit of this scheme is that (a) comparatively mild and reversible reactions are employed, and (b) there is no other way in which thiophene derivatives can be separated from hydrocarbons of similar aromatic ring number. This approach could be further improved by a more detailed initial separation of the sulfur compounds into thiophene, benzothiophene, sulfide, and so on fractions using successive separations on alumina and silica as above. The oxidation of each of the resulting fractions could then be optimized for the specific compound class present and reaction of the sulfides could be altogether avoided. The final fractions would then correspond to pure compound classes.

7. Free Sulfur

Little is known about the chromatography of free sulfur, but one study [50] indicates that it separates with the monoaromatics when eluted from alumina with heptane.

8. Sulfonates

Routine ASTM procedures are available for the analysis of petroleum sulfonate sodium salts [9] and barium and calcium sulfonates [11], by elution from silica.

B. Oxygen Compounds

The major oxygen compounds in petroleum distillates boiling below 1000° F include the following classes: carboxylic acids, phenols, furan derivatives, and aliphatic carbonyl compounds (ketones, esters) [67]. In addition, lesser amounts of aromatic ketones (e.g., fluorenones) and dihydrobenzofurans are present in some samples. Compounds containing two or more oxygen groups per molecule are present in small amounts in higher boiling distillates, and are probably important constituents of samples boiling above 1000° F, as well as of air-blown asphalts. Amides, although oxygen-containing compounds, are discussed in Sec. VII.C on nitrogen compounds.

Figure 3 indicates the relative retention of various oxygen and nitrogen compound types on deactivated alumina. These compounds are ranked by the solvent strength ($\epsilon^\circ = \epsilon_1$) required to give a retention of 1 ml g^{-1} on 3.8% water-alumina. This figure clearly indicates the impossibility of cleanly separating some compound classes with a single adsorbent. This problem is lessened by the use of two or more adsorbents.

The separation characteristics of oxygen and nitrogen compounds are conveniently discussed in terms of a standard separation scheme first developed some years back [85]. Successive separation on alumina and silica is used with a series of standard solvents (see Table 9). 30 standard fractions result as shown in Table 9. Thus, each of the initial five fractions from alumina is further separated on silica into five final fractions. These fractions are designated by the solvent used in each separation, e.g., A_1S_2 is the third silica fraction from the second alumina fraction of a particular sample. Higher fraction numbers for alumina (e.g., A_4, A_5) refer to compounds that are relatively strongly held on alumina, while higher silica fraction numbers (e.g., S_3, S_4) refer to compounds that are strongly retained on silica. This classification scheme will be further illustrated in the following examples.

1. Carboxylic Acids

These compounds are strongly retained on silica (S_4 fraction) and irreversibly held on basic alumina. Adsorption is not usually employed in the initial isolation of these compounds from petroleum. Rather, alcoholic NaOH extraction or anion exchange chromatography is preferred for reasonably clean-cut separations of this class of compounds (e.g., Refs. 42 and 64). However, further separation of an initial acidic fraction (from basic extraction) on silica has been used to purify the carboxylic acids from contaminating phenols [63, 64], which concentrate into the preceding S_2 and S_3 fractions.

TABLE 9. Standard Solvent Series for Separation of Nitrogen and Oxygen Compounds on Alumina and Silica [85]

Fraction[b]	Solvent	Solvent strength $\epsilon°$
	Alumina[a]	
A_0	Pentane	0.00
A_1	9% v/v Benzene/pentane	0.12
A_2	41% v/v Benzene/pentane	0.24
A_3	70% v/v Benzene/pentane	0.30
A_4	7% v/v Acetonitrile/benzene	0.48
A_5	50% v/v Methanol/benzene	0.48^+
	Silica[c]	
S_0	Pentane	0.00
S_1	4% Benzene/pentane	0.04
S_2	19% v/v Benzene/pentane	0.12
S_3	52% v/v Benzene/pentane	0.20
S_4	50% v/v Methanol/benzene	0.20^+

[a]4% water-alumina (Alcoa F-20, a 200 $m^2\ g^{-1}$ basic alumina).
[b]Designation of eluate fraction eluted by 5 ml of indicated solvent/g adsorbent.
[c]4% water-silica (Davison Code 62, a 400 $m^2\ g^{-1}$ silica).

2. Phenols

These compounds are concentrated into the $A_5S_{2,3}$ fractions. Thus, they are strongly held on alumina, but have an intermediate retention on silica. This is a consequence of the preferential retention of acidic compounds on basic alumina (but not on silica). The relative separation of the phenols in the scheme of Table 9 is further illustrated in Figure 7.

Strongly retained nonhydrocarbons, such as the phenols, exhibit little tendency to be separated into benzologous classes (e.g., phenols, naphthols, etc.) on either alumina or silica. The adsorption of the hydroxyl group in these compounds is sensitive to the nature of adjacent substituent groups in the molecule, and this effect completely overshadows the effect of additional aromatic rings on adsorption. Possibly separation on charcoal (see pyridines below) might yield the desired separation of the phenol benzologs, but this remains to be demonstrated. In one light gas oil [81] 2-hydroxybiphenyls were not resolved from other phenols by adsorption, but were separable by cation exchange chromatography.

3. Furans

The furans, like the thiophenes, separate with the hydrocarbons of similar aromatic ring number. Thus, benzofurans are found in the diaromatics fraction, dibenzofurans separate largely with triaromatics, etc. As in the case of the thiophenes, there is a tendency toward earlier elution of the furans relative to hydrocarbons of similar aromatic ring number, so that considerable amounts of dibenzofurans are found in the diaromatics fraction, and significant overlap of triaromatics and naphthobenzofurans occurs.

4. Aliphatic Carbonyl Compounds

The major monofunctional compounds of this class are believed to be aliphatic ketones and esters, respectively [69]. These are concentrated into the $A_{1,2}S_{2-4}$ fractions of Table 9. Similar difunctional compounds also occur in significant amounts in higher boiling distillates and are concentrated into fractions $A_{3-5}S_4$. The separation of these compounds from adjacent oxygen and/or nitrogen classes (see Figure 7) is by no means sharp,

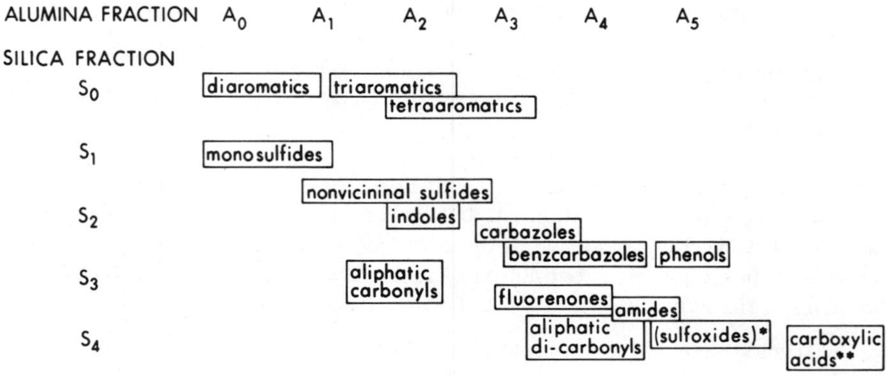

FIG. 7. Separation of various petroleum compound classes in standard scheme of Table 9 [67, 87]. *, Artifacts not present in original petroleum; **, not eluted from alumina even with 50% v/v methanol/benzene.

and there is normally substantial contamination by aromatic compounds, such as carbazoles and amides. Fortunately the aliphatic oxygen compounds can be cleanly resolved from these interfering aromatics by charcoal chromatography [85]. Aliphatics are not retained on charcoal under the conditions used in Ref. 85, while aromatics (particularly those with two or more aromatic rings) are strongly retained.

5. Aromatic Carbonyl Compounds

Fluorenones were not found in the California crude studied in Ref. 67 but were reported in another crude oil [39]. They are predicted to separate in the A_3S_3 and adjacent fractions of Figure 7. They can also be separated from petroleum by extraction with strong base [39].

6. Dihydrobenzofurans

These were found in minor amounts in a California crude [67]. They concentrate into the A_1S_{1-3} fractions.

7. Aromatic O-O and O—S Compounds in Heavy Distillates

Little is known of these structures, although inferential evidence suggests [69] that difurano and thiophene-furano structures predominate. If so, these compounds are expected to separate with hydrocarbons of similar aromatic ring number, e.g., benzothiopheneo-dibenzofuran with the penta-aromatics.

C. Nitrogen Compounds

The principal nitrogen compounds in petroleum fractions boiling below 1000°F include the following: pyridine derivatives, indole derivatives, and aromatic amides. In addition, minor amounts of di-nitrogen (N—N), $N—O_2$, and N—S compounds are found in some crudes. Aliphatic nitrogen compounds and primary or secondary aromatic amines appear to be absent from most petroleums, but may be present in FCC, hydrotreated or coker fractions.

1. Pyridine Derivatives

Pyridines, quinolines, phenanthridines, and higher benzologs are found in distillates boiling to 1000°F. The nitrogen atom in these compounds is quite sensitive to substitutional effects—like the phenols discussed in Sec. VII.B—and as a result these compounds have a wide range of retention in adsorption. The pyridine derivatives are distributed across the A_{1-4} fractions of Table 9 when separated on alumina. On silica, because bases are more strongly retained, they are concentrated into later fractions, S_{2-4}. Because of these characteristics, adsorption is a poor means of isolating the pyridines from other compound types. Fortunately, clean separations of this compound class are possible by means of cation exchange [85], and this is a preferred first step in the separation of these nitrogen bases.

Like the phenols, the pyridine benzologs cannot be effectively separated from each other by elution from alumina or silica. However, charcoal chromatography is reasonably effective in resolving the various benzologous classes from a cation-exchange concentrate of these bases. Pyridines, quinolines, and phenanthridines have been separated in this fashion [89].

A detailed analysis of pyridine derivatives in petroleum was obtained by Drushel and Sommers [21], who employed a complex separation scheme involving adsorption, extraction, and gas chromatography.

2. Indole Derivatives

The predominant compounds of this class are the N—H (rather than N—R) derivatives of indole, carbazole, and 1,2-benzcarbazole, with lesser amounts of 3,4-benzcarbazoles. These are distributed over alumina fractions A_{2-4} with some tendency for the indoles to concentrate into early-eluting fractions, and the benzcarbazoles to be found in later fractions. On silica these compounds elute somewhat earlier, being concentrated into the $S_{2,3}$ fractions (because the acidic N—H group is less strongly retained on silica versus alumina). However, the separation of these benzologs on alumina or silica is relatively poor. Better separations are afforded by cation exchange which separates most of the more basic indoles from the less basic higher benzologs. The lower molecular weight homologs also tend to predominate, allowing partial recognition of each benzologous series from the mass spectral parent peak distribution for the unseparated mixture of benzologs. However, it is anticipated that better separations of the individual benzologs will be afforded by charcoal chromatography, which has so far not been attempted.

The 3,4-benzcarbazoles are eluted from alumina well before the 1,2-isomers, permitting partial separation of these two subtypes [86]. The 2,3-benzcarbazoles have not yet been observed in petroleum, but they too can be separated from the other isomers. On water-deactivated alumina the 2,3-benzcarbazoles are eluted between the other two isomer groups. However, with more active aluminas, e.g., 1-2% water-deactivated, the 2,3-benzcarbazoles are more strongly retained than either of the other two isomeric groups [78].

The N-alkyl derivatives of the indoles, carbazoles, and benzcarbazoles are believed to be minor constituents of some crudes, although the evidence is somewhat inferential at present [67, 89]. The nonacidic compounds are less strongly adsorbed on alumina, concentrating into the A_{1-2} fractions, but are held in the same silica fractions ($S_{2,3}$) as the N-H derivatives. Alumina chromatography is apparently the method of choice for the separation of N—H from N—R indole derivatives.

LEAC methods have been reported for the assay of indoles, carbazoles, and benzcarbazoles in both FCC [86] and the crude oil [88] distillates. A preliminary cation exchange separation to remove interfering pyridine derivatives is followed by a separation of the indole benzologs on alumina, and a three-wavelength uv analysis of the total fraction (at 275, 292, and 305 nm). These procedures work well for the FCC samples because pyridine and indole derivatives constitute almost all of the polar compounds in these samples. Their accuracy for straight run samples is somewhat poorer, however, with recent data [80] suggesting errors of 25 to 40% in some cases. Further work must be done before LEAC assays for indole derivatives in petroleum distillates can be accepted without reservation.

The separation scheme of Drushel and Sommers [21] resulted in a fairly detailed analysis of indoles and carbazoles in petroleum.

3. Aromatic Amides

The detailed molecular structure of these compounds has not yet been fully determined [67, 69, 89], although 2-quinolones are an important subtype in most crudes. The amides are strongly adsorbed on both alumina and silica (concentrating into the $A_{4,5}S_{3,4}$ fractions), which facilitates their separation from other polar compounds. They are also separable by means of cation-exchange chromatography by virtue of their weak basicity. Since they are aromatic compounds, they can be separated from aliphatic di-carbonyls by means of charcoal chromatography.

Despite the combined use (as above) of various adsorption and ion-exchange separations, only two distinct amide types have so far been isolated from petroleum: the quinolones and another uncharacterized amide class described in Ref. 81. The infrared spectra of amide fractions suggest that several additional amide compound types remain to be separated and identified.

4. Aromatic N—N and N—O—O Compounds in Heavy Distillates

The di-azaaromatics (N—N) have been observed (e.g., Ref. 89) in the strong-base fraction from cation-exchange chromatography as minor constituents with the normal pyridine derivatives. They are therefore probably indolo- or quinolino-pyridines, quinolines, etc. They would be expected to concentrate into fractions $A_{3-5}S_{3-4}$. The N—O—O compounds observed in one crude distillate [89] were concentrated into the $A_{2,3}$ fractions of a weak-base concentrate from cation exchange. It is suspected that they are amide derivatives, but nothing more of their structure or separation characteristics is known.

Table 10 lists a variety of applications of adsorption chromatography to the analysis of sulfur, oxygen, and nitrogen compounds.

VIII. APPLICATION TO PETROLEUM ADDITIVES

Although liquid chromatography has been successfully applied to the analysis of compound types in petroleum fractions for many years, relatively few publications have dealt with the analysis of additives in petroleum products. Modern lubricating oils contain a large number of additives such as antioxidants, detergents, antiwear agents, corrosion inhibitors, viscosity index improvers, ashless dispersants, and pour point depressants. Similarly, fuels contain antioxidants, oil, detergents, dyes, metal deactivators, dispersants, and corrosion inhibitors. Many additive types are fairly polar and this simplifies the separation problem, since the products to which they have been added are generally refined to reduce or remove naturally occurring polar materials. The technique most commonly applied to the analysis of petroleum products is TLC, probably because it is low cost, fast, because specific reagents can be used to identify the additive in the presence of interfering components, and because of the widespread popularity of the technique over the past 2 decades. Undoubtedly, column chromatographic techniques could be adapted to many of the TLC procedures described in the literature. It is likely that many applications of HPLC to petroleum additives will soon appear. In the following section we divide the discussion between the analysis of additives in lube oils, fuels, and miscellaneous additives and analyses not previously described.

A. Lubricating Oil Additives

It is usually necessary to concentrate the additives in a lube oil in order to facilitate detection and quantitation in the final analytical procedure. Preseparation schemes include extraction with a polar solvent (such as methanol), dialysis through a rubber membrane, or column chromatography. It is often desirable to follow the depletion of antioxidant additive in lube oils. Commonly used antioxidants include the hindered phenol type (2,4-dimethyl-6-tert butyl phenol, 2,6-ditert butyl-4-methyl phenol, etc.), and nitrogen types

TABLE 10. Selected Separations of Sulfur, Oxygen, and Nitrogen Compounds

Sample	Compound classes isolated	Comments	References
Distillates 400–1000°F	Alkyl sulfides, nonvicinal aromatic and/or polysulfides, alkyl thiophenes	Routine LEAC method	71
Vacuum gas oils	Sulfides, thiophenes	Separation after oxidation	20
200–250°C Distillate	Benzothiophenes	Separation by adsorption and several other techniques	29
Asphaltenes	Porphyrins	Several schemes	99
Crude oil	Acids, phenols, carbazoles, amides	Several separation steps including adsorption	43
	Pyridine homologs	Selective adsorption on Ag_2O	100
Crude oil extract	Acids, phenols	Extraction, ion exchange and adsorption	64
Crude oil	Carbazoles and indoles		
Cracked and straight-run distillates	Indole, carbazoles, benzcarbazoles	Rapid LEAC method	86, 88

(naphthyl amines, diphenylamines, phenothiazines). Hindered phenol types behave as relatively nonpolar compounds when separated by adsorption on polar adsorbents, because the bulky alkyl groups adjacent to the oxygen sterically prevent strong adsorption interactions with the surface. As a result, the heavier aromatic components of the oil tend to interfere. This problem is avoided in TLC procedures by making use of visualization reagents which specifically interact with the phenol (see Chap. 14 and Table 3). Quantitation is accomplished by comparing the spot color density with those of standards using a densitometer or by eluting the spot from the plate, developing the color in a test tube, and determining the concentration by spectrophotometry. Nitrogen-containing antioxidants can be separated from other components of a lube oil concentrate on silica or alumina plates using moderately polar solvents. Table 11 illustrates some of the TLC data for the separation of antioxidants in petroleum products.

Antiwear additives include metal dialkyl or diaryl dithiophosphates and trialkyl or triaryl phosphates. These additives are quite polar and can be determined in lube oils without preseparation by TLC using a polar solvent. In one example [35], a small (0.5-ml) sample of oil is diluted with toluene and chromatographed on a silica plate with toluene to move the base oil components to the top of the plate. The plate is then dried and rechromatographed with a mixed solvent containing toluene, methyl ethyl ketone, acetic acid, and pyridine, and then the plate is sprayed with dithizone reagent which forms a colored product with the zinc. In this way it is possible to follow both the disappearance of the additive and the formation of degradation products. It should be possible to adapt the TLC procedure to a preparative scale column method in order to collect sufficient quantities of the degradation products for structural analysis.

In another example [16], sulfonate and phenate based detergents were determined by first concentrating the additives by dialysis followed by TLC of the concentrate on silica using a solvent mixture containing chloroform, ammonia, and isopropanol. The developed plate is then visualized with pinacryptol yellow and examined under ultraviolet light. Figure 8 shows typical TLC chromatograms of certain lubricant additives.

B. Fuel Additives

Applications of adsorption and partition chromatography to fuel additives are somewhat limited. Amos [4] reported a simple, rapid method for the estimation of 2,4-dimethyl-6-tert butyl phenol in aviation turbine fuel by TLC. In this example the additive is concentrated by evaporation of most of the fuel. The concentrate is then chromatographed on silica using carbon tetrachloride and the plate is visualized by spraying with phosphomolybdic acid. The spots are compared with a densitometer to standards run on the

TABLE 11. Phenolic Antioxidants [16]

Phenol	Approx. rf × 100			Color reaction			
	A	B	C	D	E	F	G
2,4-dimethyl-6-tert butyl phenol	35	43	30	O-YG	M	OY	R
2,6-ditert butyl-4-methyl phenol	74	88	66	O-Y	M	OY → W	OBr → W
2,6-ditert butyl phenol	73	87	66	Br-B	P	Y	R
4,4'-methylene bis (2,6-ditert butyl phenol)	46	62	36	BrO	P	O	O
2,6-ditert butyl dimethyl amino-4-methyl phenol	0	0	0	P	P	O	R
4,4-bis (2,6-ditert butyl phenol)	51	68	39	YG	YBr	Y	Y

Nitrogen-containing Antioxidants

Nitrogen compounds	Approx. rf × 100		Color reaction			
	A	B	D	G	F	H
α-Naphthylamine	12	7	BG	Br	Br	Pk
Phenyl-alpha-naphthylamine	53	47	O	R	P	O
Octylated phenyl-alpha-naphthylamine	66	59	O	R	P	OR
beta-naphthylamine	7	6	Br	RPk	Pk	Br
Phenyl-beta-naphthylamine	38	37	P	P	P	R
N,N'-disec. butyl-p-phenylene diamine	6	0	R	Pk	Pk	RBr
Phenothiazine	24	27	MB	W	G-Gr	M → B

(continued)

TABLE 11 (continued)

Nitrogen compounds	Approx. rf × 100		Color reaction			
	A	B	D	G	F	H
Dioctylphenothiazine	47	49	G–Gr	G	G	G–Br
Diphenylamine	50	42	GB	O	O	Pk
Dioctyldiphenylamine	78	62	B → RBr	O	Y	Pk

Solvents: A = 3% ethyl acetate in isooctane
B = carbon tetrachloride
C = 3% di-isopropyl ether in isooctane
} unlined tank, 20° C

Sprays: D = DCLQ after 20 min
E = DCLQ after 20 min, resprayed with N/2 ethanolic potassium hydroxide
F = p-nitrophenyl diazonium fluoroborate
G = tetra-azotized o-dianisidine
H = chloramine T

Key to Color Code

Y	Yellow	B	Blue	G	Green	Pk	Pink
O	Orange	P	Purple	Gr	Grey	M	Mauve
R	Red	Br	Brown	W	White		

Examples of uses of symbols

YG	yellow-green
O-Y	orange spot with yellow center
OY → W	orange-yellow turning to white on standing

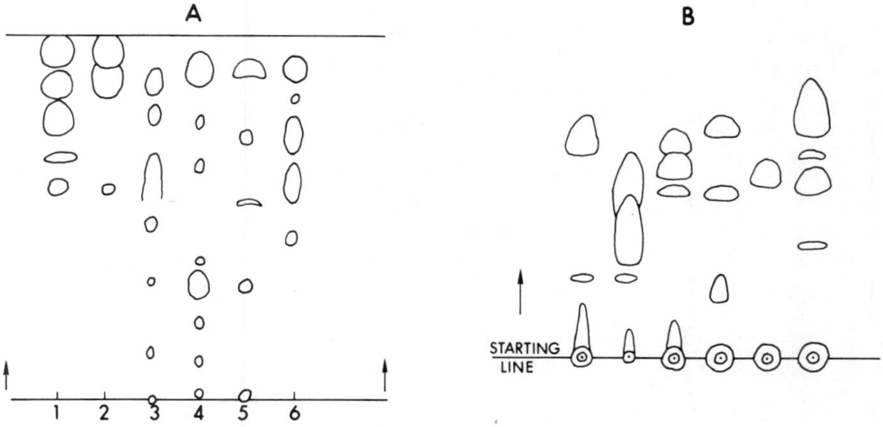

FIG. 8. TLC of lubricating oil additives [35].

same plate. With this method the author was able to simultaneously run 10 samples within 3 hr. Similarly the analysis of nitrogen containing antioxidants in gasolines has also been reported [17]. Gasoline dyes have also been separated by TLC [26, 44]. Various contaminants in fuels have been analyzed by paper chromatography and TLC. For example, one procedure was given for the estimation of heavy oils in gasoline [24] and others [12, 14] for the contamination of gasoline with kerosene. No successful determination by adsorption chromatography of polyimide or succinimide based detergent additives has yet been reported, probably because these polyfunctional compounds are chemisorbed.

C. Miscellaneous Additives

Lynes [41] described the analysis of alkanol amines in cutting oils using TLC. Soutar [93] followed gas oils and kerosenes marked with furfuraldehyde by TLC on silica gel using aniline acetate as a spray reagent. Dobies [19] measured hydroxybenzophenone light absorbers in wax. Antioxidants were measured in mineral oil [23] and transformer oils [96] using, respectively, silica plates and alumina columns, and C_1-C_{12} alkyl phenols were measured in oil by TLC on polyamide layers [18]. Table 12 lists some representative examples of the application of adsorption and partition chromatography to additives in petroleum products.

TABLE 12. Selected Applications to Petroleum Products

Product	Additive analyzed	Comments	References
Lubricants	C_1-C_{12} alkyl phenols	Polyamide layers	18
Aviation turbine fuels	2,4 dimethyl-6-tert butyl phenol	Rapid quantitative method	4
Aviation lubricants	Nitrogen antioxidants	TLC, silica gel	17
Mineral oil	2,6 ditert butyl-4-methyl phenol		23
Transformer oils	2,6 ditert butyl-4-methyl phenol	Column chromatography on alumina + polarography	96
Gasoline	Antioxidants, phosphoric acid esters		35
Lubricants	Detergent inhibitors, zinc dialkyl dithiophosphates	TLC	35
Lubricants	Antioxidants, dithiophosphates, organic phosphates, chlorinated compounds, corrosion inhibitors	TLC	16
Cutting oils	Alkanol amines	TLC, silica gel	41
Gas oils and kerosenes	Furfuraldehyde	Furfuraldehyde used as a marker	93
Gasoline	Kerosene, heavy oil	Contaminants in gasoline	12, 14, 24
Wax	Hydroxybenzophenone uv absorbers	TLC on extract	19
Lubricants	Amine antioxidants, synthetic esters, tricresyl phosphate, zinc dialkyl dithiophosphate, calcium phenate, calcium sulfonate	Comparison of TLC and HPLC methods	5
Lubricants	Antioxidants, phenates, sulfonates, zinc dialkyl dithiophosphates, polymers	Quantitative TLC	3

APPENDIX A. Preparation of Standardized 4.0% H_2O-Al_2O_3

The following procedure is meant to serve as an example of the ease in which standardized adsorbents can be prepared. This procedure results in an adsorbent with exactly adjusted activity which is used in the analysis of coronenes (Appendix B). Adsorbents of other activity are prepared in an identical way, but different nominal water content and different retention times for the test compound (naphthalene in this example) are used.

First, calcine Alcoa F-20 alumina at 400°C for at least 8 hr, then cool to room temperature in a closed 1-gallon can. Cautiously add 4 ml of distilled water per 100 g of alumina and mix thoroughly. Vent the container at intervals to release pressure. Store for 16 hr and test the activity as indicated below.

Weigh an empty 400-mm by 8-mm column, pack it with adsorbent to be tested, and reweigh to obtain the net adsorbent weight W. Mount the column in the apparatus (see Figure 1) and charge 50 μl of 0.1 wt. % naphthalene in isooctane to the column using vacuum at the column outlet. Elute the sample with pentane and collect seven 5-ml fractions. Obtain the absorbance of cuts 2 through 7 in 1-cm cells at 275 nm to give A_2, A_3, A_4, etc., and calculate the following factors:

$$F_2 = (A_2/(A_2 + A_3 + A_4 + A_5 + A_6 + A_7) = A_2/S$$

$$F_3 = (A_2 + A_3)/S$$

$$F_4 = (A_2 + A_3 + A_4)/S$$

$$F_5 = (A_2 + A_3 + A_4 + A_5)/S$$

$$F_6 = (A_2 + A_3 + A_4 + A_5 + A_6)/S$$

$$F_7 = (A_2 + A_3 + A_4 + A_5 + A_6 + A_7)/S$$

Plot the values of F_2, F_3, and so on on probability paper versus the eluate volume which corresponds to each factor (F_2:10, F_3:15, F_4:20; etc.). Interpolate the value of V_R for F = 0.500 (50%) and calculate $R = V_R/W$. Depending on the value of R, add either water or calcined alumina to the nominal 4% H_2O-Al_2O_3 to give standardized 4.0% H_2O-Al_2O_3.

R	Add
1.80-1.70	0.5 ml H_2O per 100 g of adsorbent
1.70-1.63	0.4 ml
1.63-1.55	0.3 ml

R	Add
1.55-1.48	0.2 ml
1.48-1.42	0.1 ml
1.42-1.36	Add nothing, adsorbent is standardized
1.36-1.31	2.5 g of calcined Al_2O_3, per 100 g adsorbent
1.31-1.25	5.0 g
1.25-1.21	7.5 g
1.21-1.17	10.0 g
1.17-1.12	12.5 g

If either water or calcined alumina is required, shake the resulting mixture in a closed container and permit it to stand 16 hr before use.

APPENDIX B. Analysis of Alkyl Coronenes in Nitrogen-Free Hydrocarbons

The following procedure is meant to serve as an example of the ease with which routine LEAC procedures can be carried out.

Pack an 8-mm (OD) by 185-mm column with standardized 4.0% H_2O-Al_2O_3 after first placing a plug of cotton at the bottom of the column. Pour the adsorbent into the column while gently tapping the column to obtain maximum density. Continue tapping until no further settling of the adsorbent is observed and place a cotton plug at the top of the adsorbent bed. Place the column in the apparatus and charge 100 μl of sample into the column inlet using vacuum at the column outlet. Elute the sample with 12% v/v dichloromethane in pentane until 40 ml has been collected. The air pressure on the solvent reservoir should be adjusted to give a flow rate of 2 to 3 ml min^{-1}. Discard the initial fraction, change the eluent to 30% v/v dichloromethane-pentane, and collect a 25-ml coronene fraction. Shake the flask containing the coronene fraction and measure its absorbance in 5-cm cells at 301, 307, 313, and 385 nm (A_{301}, A_{307}, A_{313}, and A_{385}). Calculate the coronenes concentration as follows:

$$\text{coronenes (ppm)} = \frac{144}{d}\left(A_{307} + 0.3(A_{301} + A_{313}) - 2.8\, A_{385}\right)$$

where

d = density of sample

REFERENCES

1. S. Adachi, J. Chromatog. 17:295 (1965).
2. K. H. Altgelt and T. H. Gouw, "Advances in Chromatography," vol. 13 (J. C. Giddings, E. Grushka, R. A. Keller, J. Cazes, eds.), Marcel Dekker, New York, 1975.
3. R. Amos, Z. Anal. Chem. 236:350 (1968).
4. R. Amos, J. Inst. Petrol. 54:9 (1968).
5. R. Amos and S. G. Perry, J. Chromatog. 83:245 (1973).
6. ASTM D 1319, Hydrocarbon Types in Liquid Petroleum Products by Fluorescent Indicator Adsorption, 23:665 (1975).
7. ASTM D 2002, Isolation of Representative Saturates Fraction From Low-Olefinic Petroleum Naphthas, 24:141 (1975).
8. ASTM D 2003, Isolation of Representative Saturates Fraction From High-Olefinic Petroleum Naphthas, 24:144 (1975).
9. ASTM D 2548, Analysis of Oil Soluble Sodium Petroleum Sulfonates by Liquid Chromatography, 24:475 (1975).
10. ASTM D 2549, Separation of Representative Aromatics and Nonaromatics Fractions of High Boiling Oils by Elution Chromatography, 24:478 (1975).
11. ASTM D 2894, Analysis of Calcium and Barium Petroleum Sulfonates by Liquid Chromatography, 24:850 (1975).
12. M. Babitz and A. Rocker, Israel J. Technol. 4:271 (1966).
13. G. M. Badger, N. Kowanko, and W. H. F. Sasse, J. Chromatog. 13:234 (1964).
14. S. Bentaur, M. Babitz, and A. Rocker, Israel J. Technol. 3:220 (1965); cited also in 66:47892z (1967).
15. L. R. Chapman and D. F. Kuemmel, Anal. Chem. 37:1598 (1965).
16. J. P. Coates, J. Inst. Petrol. 57:209 (1971).
17. R. B. Delves, J. Chromatog. 26:296 (1967).
18. P. F. Diamond, J. Chromatog. 32:419 (1968).
19. R. S. Dobies, J. Chromatog. 35:370 (1968).
20. H. V. Drushel and A. L. Sommers, Anal. Chem. 39:1819 (1967).
21. H. V. Drushel and A. L. Sommers, Anal. Chem. 38:19 (1966).
22. E. R. Fett, D. J. Christoffersen, and L. R. Snyder, J. Gas Chromatog. 6:572 (1968).
23. H. Fischer, Schmiertechnik 13:154 (1966); abstracted in J. Inst. Petrol. 53:10A (1967).
24. H. G. Goebgen and J. Brockmann, Wasser und Luft Betrieb 13:204 (1969).
25. E. Grushka, ed., "Bonded Stationary Phases in Chromatography," Ann Arbor Science, Ann Arbor, Michigan, 1974.
26. H. Hausser, Arch. Kriminol 125:72 (1960).
27. E. Heftmann, ed., "Chromatography," 3rd ed., Van Nostrand Reinhold, New York, 1975.

28. D. E. Hirsch, R. L. Hopkins, H. J. Coleman, F. O. Cotton, and C. J. Thompson, Anal. Chem. 44:915 (1972).
29. R. L. Hopkins, M. J. Coleman, C. J. Thompson, and H. T. Rall, Anal. Chem. 41:2041 (1969).
30. J. F. K. Huber, C. A. M. Meijers, and J. A. R. J. Hulsman, Anal. Chem. 44:111 (1972).
31. R. E. Jentoft and T. H. Gouw, Anal. Chem. 40:923 (1968).
32. R. E. Jentoft and T. H. Gouw, Anal. Chem. 40:1787 (1968).
33. G. Jurriens, Riv. Ital. Sostanze Grasse 42:116 (1965).
34. B. L. Karger, L. R. Snyder, and C. Horvath, "An Introduction to Separation Science," Wiley, New York, 1973.
35. F. C. A. Killer and R. Amos, J. Inst. Petrol. 52:315 (1966).
36. J. G. Kirchner, J. M. Miller, and G. J. Keller, Anal. Chem. 23:420 (1951).
37. J. J. Kirkland, ed., "Modern Practice of Liquid Chromatography," Wiley-Interscience, New York, 1971.
38. H. S. Knight and S. Groennings, Anal. Chem. 28:1949 (1956).
39. D. R. Latham, C. R. Ferrin, and J. S. Ball, Anal. Chem. 34:311 (1962).
40. H. E. Lumpkin and B. H. Johnson, Anal. Chem. 26:1719 (1954).
41. A. Lynes, J. Chromatog. 23:316 (1966).
42. J. F. McKay, D. M. Jewell, and D. R. Latham, Separ. Sci. 7:361 (1972).
43. J. F. McKay, T. E. Cogswell, D. R. Latham, Preprints, Div. Petrol. Chem., ACS, 19:25 (1974).
44. G. Machata, Arch. Toxikol 127:1 (1961).
45. B. J. Mair, M. J. Montjar, and F. D. Rossini, Anal. Chem. 28:56 (1956).
46. R. E. Majors, American Laboratory 13 (October 1975).
47. K. Marron, The Determination of the Saturates Content of Lube Oils, in "Recent Analytical Developments in the Petroleum Industry" (D. R. Hodges, ed.), Wiley, New York, 1974.
48. A. J. P. Martin and R. L. M. Synge, Biochem J. 35:91 (1941).
49. A. J. P. Martin and R. L. M. Synge, Biochem J. 35:1358 (1941).
50. M. F. Millson, J. Chromatog. 50:155 (1970).
51. Unpublished work of this laboratory (Union Oil Co.).
52. W. L. Orr, Anal. Chem. 38:1558 (1966).
53. E. D. Pellzzari and C. M. Saparacino, Anal. Chem. 45:378 (1973).
54. S. G. Perry, R. Amos, and P. I. Brewer, "Practical Liquid Chromatography," Plenum, New York, 1972.
55. S. G. Perry, J. Inst. Petrol. 58:47 (1972).
56. M. Popl, J. Mostecky, V. Dolansky, and M. Kuras, Anal. Chem. 43:5181 (1971).
57. M. Popl, M. Stejskal, and J. Mostecky, Anal. Chem. 46:1581 (1974).
58. H. Prezler and R. Oehler, Chem. Tech. 15:31 (1963).

59. J. A. Riddick and W. B. Bunger, "Techniques of Chemistry," vol. II, "Organic Solvents" (A. Weissburger, ed.), 3rd ed., Wiley, New York, 1970.
60. F. D. Rossini, B. J. Mair, A. J. Streiff, "Hydrocarbons From Petroleum," Reinhold, New York, 1953.
61. S. Sato and Y. Imazumi, Sekiyu Gakkai Shi 8:432 (1965).
62. R. D. Schwartz and D. J. Brasseaux, Anal. Chem. 30:1999 (1958).
63. W. K. Seiffert, Anal. Chem. 41:562 (1969).
64. W. K. Seiffert and W. G. Howells, Anal. Chem. 41:554 (1969).
65. L. R. Snyder, in "Hydrocarbon Analysis," Special Technical Publication No. 389, Am. Soc. Test Mats., Philadelphia, 1965, p. 399.
66. L. R. Snyder, Anal. Chem. 37:713 (1965).
67. L. R. Snyder, Accts. Chem. Res. 3:290 (1970).
68. L. R. Snyder, "Principles of Adsorption Chromatography," Marcel Dekker, New York, 1968.
69. L. R. Snyder, Anal. Chem. 41:1084 (1969).
70. L. R. Snyder, Anal. Chem. 33:1535 (1961).
71. L. R. Snyder, Anal. Chem. 33:1538 (1961).
72. L. R. Snyder, Anal. Chem. 36:774 (1964).
73. L. R. Snyder, Anal. Chem. 38:1319 (1966).
74. L. R. Snyder, J. Chromatog. 36:476 (1968).
75. L. R. Snyder, Anal. Chem. 37:149 (1965).
76. L. R. Snyder, Anal. Chem. 33:1527 (1961).
77. L. R. Snyder, "Principles of Adsorption Chromatography," Marcel Dekker, New York, 1968, p. 281.
78. L. R. Snyder, J. Chromatog. 20:463 (1965).
79. L. R. Snyder, "Principles of Adsorption Chromatography," Marcel Dekker, New York, 1968, p. 264.
80. L. R. Snyder, Preprints, Div. Petrol. Chem., ACS, 15(C43) (1970).
81. L. R. Snyder, Anal. Chem. 41:314 (1969).
82. L. R. Snyder, Anal. Chem. 34:771 (1962).
83. L. R. Snyder, J. Chromatog. 92:223 (1974); J. Chromatog. Sci. 16:223 (1978).
84. L. R. Snyder, J. Chromatog. 63:15 (1971).
85. L. R. Snyder and B. E. Buell, Anal. Chem. 40:1295 (1968).
86. L. R. Snyder and B. E. Buell, Anal. Chem. 36:767 (1964).
87. L. R. Snyder and B. E. Buell, J. Chem. & Eng. Data 11:545 (1966).
88. L. R. Snyder and B. E. Buell, Anal. Chim. Acta 33:285 (1965).
89. L. R. Snyder, B. E. Buell, and H. E. Howard, Anal. Chem. 40:1303 (1968).
90. L. R. Snyder, B. L. Karger, and R. A. Keller, Use of the Solubility Parameter in Predicting Chromatographic Retention and Eluotropic Strength, in "Gas Chromatography 1970" (R. Stock, S. G. Perry, eds.), Elsevier, Amsterdam, 1971, p. 125.

91. L. R. Snyder and J. J. Kirkland, "Introduction to Modern Liquid Chromatography," 2nd ed., Wiley-Interscience, New York, 1979.
92. L. R. Snyder and W. F. Roth, Anal. Chem. 36:128 (1964).
93. N. M. Soutar, Analyst (London) 95:316 (1970).
94. E. Stahl, ed., "Thin-Layer Chromatography. A Laboratory Handbook," Springer-Verlag, New York, 1969.
95. W. Strubert, Chromatographia 6:205 (1973).
96. J. C. Suatoni, Anal. Chem. 38:1271 (1966).
97. J. C. Suatoni, H. R. Garber, and B. E. Davis, J. Chromatog. Sci. 13:367 (1975).
98. J. C. Suatoni and R. E. Swab, J. Chromatog. Sci. 13:361 (1975).
99. J. M. Sugihara, J. F. Branthaver, G. Y. Wu, and C. Weatherbee, Preprints, Div. Petrol. Chem., ACS, 15(C5) (1970).
100. S. Tabak and M. R. M. Verzola, J. Chromatog. 51:334 (1970).
101. C. G. Vaugh, B. B. Wheals, and M. J. Whitehouse, J. Chromatog. 78:203 (1973).
102. H. Woggon and D. Jehle, Plaste Kaut. 13:460 (1966).
103. L. Zechmeister, Historical Introduction, in "Chromatography" (E. Heftmann, ed.), 2nd ed., Reinhold, New York, 1967, Chap. 1.

Chapter 11

ION EXCHANGE AND COORDINATION CHROMATOGRAPHY

Douglas M. Jewell
Gulf Research and Development Company
Pittsburgh, Pennsylvania

I. INTRODUCTION

The use of ion exchange resins in the petroleum field is relatively new (approximately 10 years old) but it is gaining rapidly in popularity [7, 11]. Although ion exchange resins have been used for many years in other fields, applications were always limited to aqueous systems or mixtures of water and polar organic solvents. The problems in applying this technique to petroleum fractions were due primarily to (a) a partial solubility of the resins in pure organic solvents and (b) the lack of sufficient ionization in the hydrocarbon matrix of petroleum for classical ion exchanging to occur.

A. Properties of Ion Exchange Resins

With the use of macroreticular cross-linked polystyrene polymers at room temperature, limitation (a) was overcome. The incorporation of polar functional groups such as amines and sulfonic acids into the polystyrene framework resulted in the formation of a new family of compounds possessing many useful properties, including ion exchanging, thus overcoming limitation (b).

Macroreticular ion exchange resins have a number of advantages. They are more resistant to organic fouling than the gel-type resins. Their porous structure (100 to 1200 A) permits good permeation of organic solvents and affords reasonable surface areas. The large pores are especially important for application to large molecules, such as asphaltenes. The rigid structure of the polymer provides excellent physical strength and a minimum of volume change (swelling) with different polar solvents. The fixed nature of the ionic functional group provides an electrostatic surface on the polymer. Polar compounds in petroleum will migrate (from n-pentane or n-hexane solution) to the polymer surface and form new bonds which may vary from very weak hydrogen bonds to very strong ionic ones. The very weak bonds can be dissociated by moderately polar solvents, whereas strong ionic bonds must be broken by displacement with a stronger ion or ligand. In practice, one can find corresponding relationships between the ion strengths of compounds, as defined by nonaqueous titrimetry [2], and their isolation on anionic or cationic macroreticular ion exchange resins. Acids and bases, as isolated and defined by these techniques, are readily related to the classical Brønsted acid-base systems.

B. Coordination Chromatography

Petroleum contains polar nonhydrocarbons which are too weak to be detected as either a Brønsted acid or base and are, therefore, not removed by ion exchange resins alone. These remaining nonhydrocarbons contain at least one heteroatom (nitrogen, oxygen, or sulfur) which can react as a Lewis base with numerous transition metal salts to form stable coordination complexes [3]. Deposits of such metal salts as a thin film on an inert support (e.g., clay, firebrick, cellulose) can be used in coordination chromatography [4]. The term also applies if the metal is incorporated in a macroreticular ion exchange resin such that the primary reaction is one of metal-ligand coordination. Since the bulk of nonhydrocarbons in petroleum are nonbasic or nonacidic, coordination chromatography is an important addition to the tools for petroleum analysis, offering the potential to design selective isolation techniques based on established (solution) coordination chemistry. The reaction of ferric chloride with carbazoles in solution [5] and with numerous nitrogen ligands in petroleum is known [3]. Ferric chloride on clay is particularly effective for isolating neutral nitrogen compounds [4]. Nickel ions incorporated in ion exchange resins will isolate sulfur compounds (sulfides) from petroleum [10].

II. PREPARATION OF ADSORBENTS

A. Cation Exchange Resins

The most readily available macroreticular ion exchange resins today are manufactured and sold under the Amberlyst or Amberlite labels by Rohm & Haas Chemical Company (Philadelphia). Amberlyst 15 is a typical strong cationic exchange resin available as the sodium salt of sulfonic acid bonded to the polymer matrix. In order to exhibit maximum effectiveness for petroleum applications, the resin must be converted to the hydrogen form by stirring with 10% aqueous HCl in methanol. This is followed by exhaustive washing with methanol, benzene, and finally n-pentane in order to remove all entrained organic impurities and adsorbed water. The resin is air dried and stored in airtight plastic or glass containers.

B. Anion Exchange Resins

Amberlyst A-27, A-29, or Amberlite IRA-904 are typical strong anionic exchange resins. These are quaternary ammonium salts (chloride) possessing different degrees of cross-linking and, therefore, have different swelling properties. They are all suitable for petroleum applications, but the IRA-904 is the most available and effective. The preferred form for maximum effectiveness (highest capacity) is the hydroxide counter ion. This is prepared by stirring with 10% potassium hydroxide in methanol followed by exhaustive sequential washing with methanol, benzene, and n-pentane. Air drying, preferably in-vacuo at 30 to 40° C, and storing in plastic or glass containers, completes the preparation. A slightly weaker form of the anion resin can be prepared by converting the OH^- to HCO_3^- with CO_2-saturated methanol.

C. Ferric Chloride on Clay

Many transition metal salts possess moderate to high solubility in methanol. 5 to 10% solutions of the salt are stirred with commercially available clays (for example, Attapulgus, Floridin, and kaolin) in such a ratio that the solid is always covered with the liquid phase. The pure kaolin clays are preferred because of (a) their excellent adsorptivity and high capacity for many salts and (b) their white appearance, which permits visual observation of the coating process. Since kaolin is not readily available in the United States in the 100-mesh particle size (usually in colloidal form only), it must be blended or substituted with other minerals. The methanol slurry is filtered, washed several times with benzene and n-pentane, and air dried in partial vacuum by thin-film evaporation at 30°C. An alternative procedure uses a chloroform solution of anhydrous ferric chloride for direct slurrying with clay. The powder is stored in dark plastic containers. In the case of ferric chloride, the adsorbent will have a bright yellow color and usually contain 0.5 to 5.0% elemental iron. If the preparation steps have exposed the adsorbent to excess heat or moisture, hydration will cause the appearance to be reddish; this form is unsuitable for quantitative applications to petroleum.

III. CHROMATOGRAPHIC PROCEDURES

A. Adsorption Steps

Figure 2 in Chap. 9 shows an example of a chromatographic column with solvent recycle which is ideal for ion exchange resins and coated adsorbents. For the preparative isolation of acids, bases, and neutral nitrogen compounds, the recycle column should be water-jacketed since desorption and regeneration steps will involve exothermic reactions. The main function of this column design is to provide continuous percolation of the adsorbent for complete sample elution with a minimum amount of solvent and personal attention. A column is dry packed while vibrating it right under the top reservoir. Then it is wetted with n-pentane. Recycling for several hours by distillation removes residual air. Since the recovery-distillation flask may contain a trace of benzene which was entrained in the adsorbent, it is replaced before proceeding with a sample.

The sequence of the packings in a column is determined by the need for removing the acids and bases before the neutral nitrogen compounds. Whether to use the anion or the cation exchange resin first is not nearly as important as being consistent in the sequence from sample to sample. Since many polar molecules contain both acidic and basic groups, different quantities of acids and bases will be obtained if the sequence is reversed. If separate columns are used for removing acids, bases, and neutral hetero compounds, each step can be checked by titration [2]. If the combination, or SARA, column is used and if nonaqueous titrimetry is unavailable, infrared spectra of the effluent oils can be utilized. When all polar nonhydrocarbons are removed, ir will show the absence of any O—H, N—H, and C=O absorption bands.

Samples are weighed and placed in a tared cellulose thimble. Thimbles used for quantitative separations must be stored in a constant humidity chamber (approximately 42% relative humidity is satisfactory). Double-walled thimbles are preferred for large samples. If micro quantities (<100 mg) or sample properties do not permit direct transfer to a thimble, the material may be quantitatively transferred to a vial (using benzene or methylene chloride as solvent), which is small enough to fit inside the cellulose thimble. After removing the solvent under nitrogen and determining the sample weight, the vial is scratched in several points around its barrel before placing in the thimble. The vial is then carefully crushed with pliers, leaving the sample mostly dispersed on glass inside the thimble.

Thimble and sample are immediately immersed in the n-pentane in the reservoir on top of the preconditioned cation exchange column. The solvent is then recycled for 12 to 24 hr to completely extract (by leaching action) the maltenes through the thimble. The length of extraction time depends on the quantity and nature of sample; nonpetroleum samples (kerogens, coal liquids, etc.) often require more than 24 hr. The n-pentane effluent from the first ion exchange resin contains the maltenes less bases. This solution is reserved for subsequent percolation through the anion exchange resin. The thimble containing now only the asphaltenes is removed from the reservoir, dried first at room temperature, and then at 100°C for 1 hr. It is kept in a constant humidity chamber until it reaches a constant weight. The asphaltenes are determined by the weight increase. When the vial technique is used, or if the actual recovery of asphaltenes is desired for subsequent studies, the thimble is extracted in a soxhlet with benzene, and the asphaltenes are ultimately determined gravimetrically.

The maltene solution is subsequently percolated through a preconditioned column of the anion exchange resin. The effluent, which contains the maltenes less acids and bases, is percolated through a preconditioned column of ferric chloride-clay to remove the neutral nitrogen compounds. The latter column contains a small zone (approximately 10% of column length) of anionic exchange resin (OH^- form) at its bottom. The effluent from this column contains only hydrocarbons and nonpolar nonhydrocarbons.

B. Desorption Steps

1. Bases

Basic compounds are retained on cation exchange resins by at least two mechanisms. In the first one, hydrogen transfer to a nitrogen atom forms a positively charged ammonium ion which reacts with the sulfonate anion on the resin; this represents true ion exchanging and has been demonstrated with strongly basic aliphatic amines, anilines, and certain heterocyclics, such as pyridines and quinolines. These compounds can only be desorbed by chemical displacement with a stronger base. The other mechanism entails chemical adsorption of moderate to weak bases when transfer of the hydrogen to a basic site may be secondary to dipole-dipole interaction or hydrogen bond formation between the electrostatic surface of the aromatic sulfonic acids on the ion exchange resins and highly polarized multifunctional nonhydrocarbons. The strength of these bonds is solvent dependent. It is greatest in aliphatic hydrocarbons and decreases with increasing solvent polarity. In practice very weak bases can be desorbed with benzene and moderately strong bases with methanol or tetrahydrofuran. Therefore, the preferred desorption sequence for petroleum bases is exhaustive elution with (a) benzene, (b) methanol, and (c) methanol(i)propylamine or methanol-ammonia. Step c is followed by benzene elution to assure complete solubilization and recovery since many strongly basic high molecular weight components are only moderately soluble in methanol after they have been displaced from the ion exchange resin. Frequently water is desorbed from the resin matrix with anhydrous methanol-amines and must be removed from recovered strong bases by azeotrope distillation. Since this desorption step is very exothermic and accompanied by some swelling of the ion exchange resin, a slow elution rate is advisable for preparative separations.

2. Acids

Acidic components are retained on strong anion exchangers (with either HCO_3^- or OH^- counter ions) by mechanisms similar to those given for bases. Strongly acidic compounds such as carboxylic groups react irreversibly with quaternary ammonium ions on the resin surface to form ionic bonds, which can only be broken by displacement with a stronger acid; hydrogen transfer via abstraction is complete for carboxylic acids but incomplete for weak acids such as phenols, amides, and carbazoles. The preferred desorption sequence is exhaustive elution with (a) benzene, (b) methanol, and (c) methanol-carbonic or -formic acids; step c is again followed by benzene elution, to completely solubilize the free acids.

3. Neutral Nitrogen Compounds

a. Separation process. Many nonbasic and nonacidic (Brønsted) nonhydrocarbons can react with anhydrous ferric chloride, either in solution or as a coating on an inert support. Carbazoles [5], nitriles [3], and ketones [1] are well known examples of this reaction. These reactions form highly colored coordination compounds in which a pair of nonbonding electrons on a nitrogen or oxygen atom coordinate with the unfilled d-orbitals of the Fe^{3+}. The coordination compounds are insoluble in aliphatic hydrocarbon solvents, which affords their separation. A potential source of contamination is the tendency of ferric chloride to form weak complexes with high ring-number pericondensed aromatics (e.g., perylene, coronene) and certain methyl-substituted catacondensed aromatics (e.g., 9,10-dimethyl-benzanthracene) [8]. Many transition metal salts will also react with Lewis bases, including sulfides; however, generalized schemes for routine application to petroleum have not been evolved.

The coordinate covalent bond between ferric chloride and nitrogen (or oxygen) ligands is not dissociated by polar solvents but must be broken either thermally or chemically (e.g., NH_3, H_2S, or ligand exchange). The complexes are soluble in benzene (slightly) or chlorinated solvents and can be broken by contacting their solutions with a strong anionic macroreticular exchanger (quaternary ammonium hydroxide). This cleavage reaction is essentially one of ligand exchange, not unlike that commonly used for cleaving mercury-nitrogen complexes [6]. In practice, this is achieved by placing a small zone of the same anion exchange resin as described earlier for acid removal below the ferric chloride-clay adsorbent. When a maltene solution in n-pentane, previously freed of acids and bases, is passed through this adsorbent, the yellow surface slowly turns green or blue due to the formation of coordination compounds. When the color reaches the bottom of the column, the capacity of the adsorbent has been reached. This color change, therefore, is a convenient visual (qualitative) indicator for column activity. Exhaustive elution of the recycle column with n-pentane removes entrained slow-moving aromatic hydrocarbons. Desorption is then accomplished by exhaustive elution with chloroform, methylene chloride or 1,2-dichloroethane. As the colored solutions of complexes migrate through the zone of strong anion exchange resin, the iron-nitrogen bond is broken and the metal is retained by the resin. These reactions are not exothermic. Excess iron chloride, which did not coordinate, will remain on the clay with these solvents. Thus, upon reconditioning, the adsorbent can be reused with another sample of maltenes, which is especially convenient with repetitive separations.

b. Comparison of supports for ferric chloride. Since the reaction of ferric chloride with neutral nitrogen compounds proceeds rapidly in solution or on the surface of coated supports, the effect of different supports is important but has not been seriously studied. The ideal form of iron chloride for these reactions is completely anhydrous Fe_2Cl_6 [5]. Because of its ease of hydrolysis, this form cannot be maintained as a coating. Completely inert and dry supports are necessary. Since chromatographic-grade silica and alumina gels are rarely inert or dry, they make poor supports. Common clays are much better; they do not show the high chromatographic activity of alumina gels, are readily dehydrated, have high mechanical strength, and exhibit a high capacity for physically adsorbed metal salts. Cellulose is inert but has only a low coating capacity for the metal salt. Common gas chromatographic column supports, such as Chromosorb W, are good candidates for supporting iron chloride but are quite expensive for preparative applications. Oelert's group [8] has recently studied numerous parameters in this technique. Table 1 shows comparisons of different supports for ferric chloride at constant loading of metal. The capacity was determined by using a 3-g portion of each separation phase in a 4 by 200 mm column and pumping a 0.01% solution of carbazole in n-heptane through it until a breakpoint was detected by an ultraviolet monitor. A low surface area calcined kaolin clay exhibits the highest capacity. Oelert et al. also determined the capacity (for

TABLE 1. Comparison of Supports for $FeCl_3$ [8]

Support (size μ)	Bulk volume ($g\ cc^{-1}$)	Surface area ($m^2\ g^{-1}$)	Capacity with 6 wt. % Fe mg carbazole g^{-1} support
Kaoline schamotte (40-150)	1.316	<1	7.9
Clay 191-700°C (125-250)	1.096	22	20
Calcined kaoline (125-250)	0.962	3	20.8
Chromosorb-W	0.242	1-3.5	7.2
Attasorb LVM (Engelhardt) (177-297)	0.714	125	15.6

carbazole) of the various supports as a function of the iron content. Figure 1 shows this variation for two different clays. A load of 5 to 7% iron gives a support for optimum capacity for neutral nitrogen compounds. Experiments with natural mineral oils, as well as coal liquids, affirmed that calcined kaolin with 5 to 6% iron is the best combination because of its good mechanical stability and high capacity by volume. The commercial availability of this support in the United States, as well as in Europe, would be highly desirable.

c. Complexation of various model compounds with iron chloride. Although coordination chromatography has been discussed primarily as a technique for isolating neutral nitrogen compounds, limitation to this group is not implied. If basic and acidic compounds are <u>not</u> selectively preremoved by ion exchange resins, they too react with transition metal salts. Oelert [9] suggested the possibility of using only one column of metal-coated calcined kaolin for the quantitative retention of all polar <u>nonhydrocarbons</u> as a rapid analytical approach for their determination. As future studies progress in this area, one should expect the development of universal precolumns for the quantitative removal of nonhydrocarbons <u>before</u> the separation of hydrocarbon types in preparative LC or HPLC.

In order to demonstrate the variety of nonhydrocarbons that react with ferric chloride-kaolin, Oelert and co-workers have determined the reactivity and some common properties of the resulting complexes. The data are tabulated in Table 2. They were obtained by percolating small quantities of

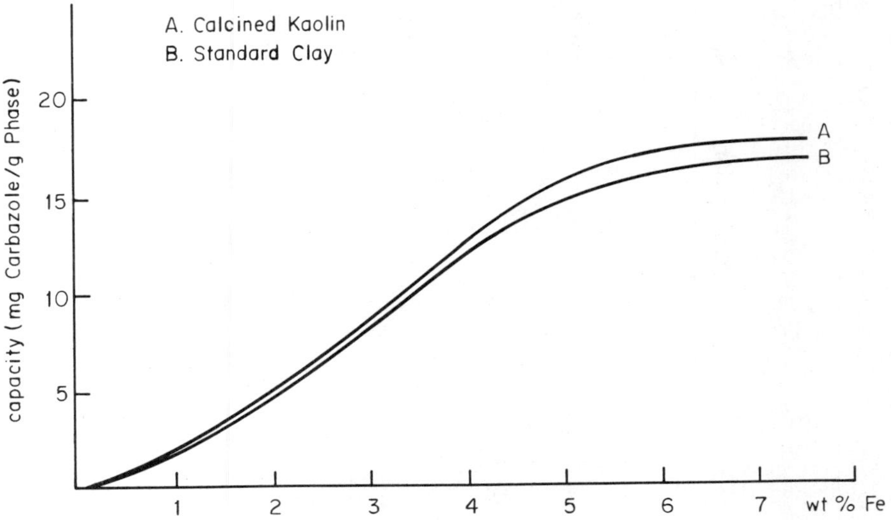

FIG. 1. Variation of capacity (carbazole) with iron content and support.

TABLE 2. Reactivity and Properties of Iron-Chloride Complexes [8]

Compound	Complexation	Color on support	Color of complex in 1,2-DCE[a]	Characteristic absorption (nm): Pure compound in n-heptane	Characteristic absorption (nm): Complex in 1,2 DCE[a]
$FeCl_3$		yellow		358, 320	
Carbazole	+	blue	green	290, 231	615, 300, 293, 238
1,2,3,4-Tetrahydrocarbazole	+	brown-orange	orange	273, 225	530, 480, 363, 290
Dodecahydrocarbazole	+	orange	orange	221	560, 520, 364, 315
N-n-Butylcarbazole	+	green	blue	293, 236	643, 575, 430, 358, 316
5,6,7,8-Tetrahydro-3,4-benzocarbazole	+	green	red → green	329, 258, 251, 246, 226	red: 540, 496 green: 650, 500, 364, 301, 290, 280
Indole	+	green	violet	266, 246, 220	<u>598</u>, 546, 500, <u>364</u>, 315, 287, 280, <u>269</u>
2,3 Dimethylindole	+	brown	red	265-285, 227	490, 364, 305, 280
Phenanthridine	+	yellow	colorless	259, 253, 248, 241	<u>365</u>, 319, <u>268</u>
Acridine	+	yellow	yellow	358, 253, 247, 241	<u>358</u>, 340, 325, <u>263</u>
Phenazine	+	yellow	yellow	361, 251, 246, 241, 236	<u>386</u>, 368, 318, <u>264</u>

2-Hydroxyquinoline	+	orange	yellow	330,318,267,236	358,320
Naphthylamine	+	green-blue	blue	346,340,336,315	580,362,310
α-Naphthol	+	green-blue	green	320,290,228,221	360,321,308,295, 260
p-Aminophenol	+	yellow	yellow	345,329,325	358,320
2-Pyrrole carboxylic acid	+	brown	green	267,257,251,245	358,320
Anthracene-9-carboxylic acid	+	yellow	yellow	259,253,248,242	357,324
α-Naphthalene-carboxylic acid	+	brown-orange	red	305,223	580,515,350,324, 305
Benzene	-				
Naphthalene	-				
Anthracene	-				
Naphthacene	-				
Chrysene	-				
Pyrene	-				
9,12-Dimethyl-1,2-benzanthracene	+	green	gray violet	361,296,285,222	590,450,360,310
Perylene	+	brown	violet	433,406,350,320	525,515,460,450
Coronene	+	green	green	303	—
1,3,6,8-Tetraphenyl-pyrene	+	green	reddish	293,279,259,252, 247	520,358,319

[a] 1,2-Dichloroethane.

each test compound in n-hexane through a 4 by 200 mm column of 1.57% Fe-calcined kaolin until a distinct color change occurred. The complex was then eluted with 1,2-dichloroethane and subjected to visible-ultraviolet absorption spectroscopy. Besides all nitrogen compounds, many phenols and carboxylic acids, as well as several large polycondensed aromatic hydrocarbons, were found to react. The polyaromatics not only form charge transfer complexes but several also undergo polymerization reactions. These side reactions, which are pronounced on supports with high iron content and low surface area, may be a source of artifacts with natural mixtures. Side reactions (polymerization) also occur with certain nitrogen compounds. They require more comprehensive study.

With modern HPLC instruments coupled with stop-flow scanning spectrophotometers, this list should readily be extended to document (a) retention volumes, (b) complexation, and (c) spectral properties of many additional structures. Since the complexes are frequently sensitive to air and moisture, flow-through instruments are preferred since they circumvent the isolation and handling problems.

It may be noted that this coordination chromatographic technique offers a convenient approach to purifying many commercial aromatic compounds which are cross-contaminated with either hydrocarbons or nitrogen compounds.

IV. INTEGRATION OF TECHNIQUES FOR STUDYING RESINS

The comprehensive analysis of petroleum should combine the various techniques just discussed in order to optimize the effectiveness of each reagent. Figure 2 shows a scheme that has been found very effective for isolating discrete classes of polar nonhydrocarbons (resins). This scheme provides 10 groups of resins which, when obtained preparatively and studied in depth, begin to demonstrate the great complexity of petroleum. If a particular class of compounds is desired, this scheme may provide a direct route to its isolation. Table 3 illustrates several structural types that have been isolated by this scheme; this list is not meant to be all inclusive. Overlap occurs in certain types due to relative positions and multiplicity of functional groups and their molecular weight ranges. Although Webster et al. [11] showed with several relatively simple mixtures that discrete separations are easily achieved, experience demonstrates that some overlap must be expected when progressing from naphthas to residuals. Pyrrolic structures are the most ubiquitous nitrogen functional groups in petroleum; various forms of indoles and carbazoles thus appear in several categories. A potential disadvantage of ion exchange resins lies in the fact that they may cause a GPC effect with large molecules such as asphaltenes. If the molecular

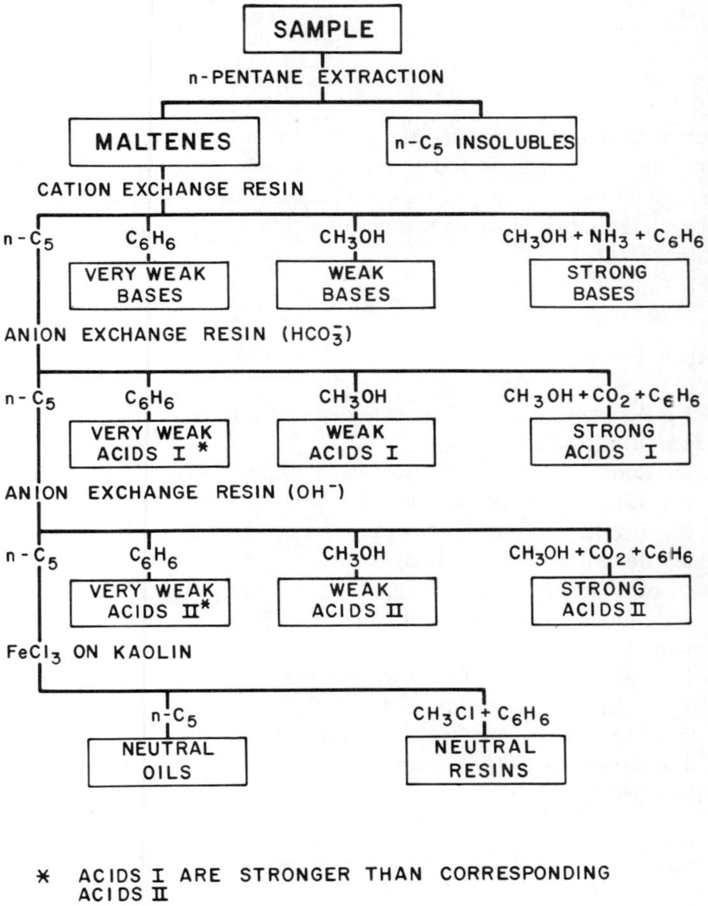

FIG. 2. Stepwise separation of resin.

TABLE 3. Structural Types Isolated by Figure 2.

Classification	Structures
Very weak bases	Phenolic amines, sulfoxides
Weak bases	Phenolic pyridines, alkyl indoles
Strong bases	Alkyl pyridines, quinolines, anilines, aliphatic amines
Very weak acids	Alkyl phenols, naphthols, carbazoles, indoles
Weak acids	Quinolones, amides, phenolic carboxylic acids
Strong acids	All simple alkyl carboxylic acids (benzoic, naphthoic, steroid)
Neutral resins	Carbazoles, benzocarbazoles, phenazines, aromatic ketones, nitriles

dimensions approach or exceed the pore dimensions, an exclusion effect will compete with the expected chemical reaction. However, this problem is encountered only in special circumstances, that is, with cyclohexane solutions of asphaltenes or of residuals which contain high molecular weight asphaltenes.

REFERENCES

1. O. F. Boberov, V. N. Tolnachev, and V. F. Lavrushir, J. Gen. Chem. USSR 35:2119 (1965).
2. B. E. Buell, Anal. Chem. 39(7):756, 762 (1967).
3. G. K. Hartung and D. M. Jewell, Anal. Chim. Acta 27:219 (1962).
4. D. M. Jewell and R. E. Snyder, J. Chromatog. 38:351 (1968).
5. D. M. Jewell and J. W. Bunger, J. Heterocyclic Chem. 8:333 (1971).
6. W. Jordon and P. J. Scheuer, J. Chromatog. 19:175 (1965).
7. W. A. Munday and A. Eaves, Proc. Fifth World Petrol. Congr., sec. V, paper 9, New York, 1959.
8. H. H. Oelert, personal communication, 1975.
9. A. W. Sangster, J. Chem. Educ. 37:454 (1960).
10. J. W. Vogh and J. E. Dooley, Anal. Chem. 47:816 (1975).
11. P. B. Webster, J. N. Wilson, and M. C. Franks, Anal. Chim. Acta 38:193 (1967).

Chapter 12

GEL PERMEATION CHROMATOGRAPHY (GPC)

Klaus H. Altgelt
Fuels and Asphalts Research Department
Chevron Research Company
Richmond, California

I. OVERVIEW

Gel permeation chromatography is now widely used in the analysis of petroleum and other fossil fuels. Since the introduction of microgels and HPLC equipment, analytical GPC runs can be completed in just a few minutes (see Chap. 14). Although the interpretation of analytical gel chromatograms is not as straightforward with fossil fuels as with polymers, the method allows at least relative determinations of molecular size and, with certain provisions, molecular weight (mol. wt.). Preparative GPC, in combination with molecular weight determinations, will not only provide mol. wt. distributions of a sample but also information on the compactness of the molecules and, thus, on their basic architecture.

GPC has the advantage of separating mixtures by molecular size only and not by chemical functionality as long as no interfering mechanism (sorption) competes with it. Thus, it lends itself to two-dimensional separations. By combining it with adsorption chromatography or any other separation method that discriminates mainly by chemical structure, one can separate a mixture into a group of (mol. wt.) fractions, each of which is subdivided into fractions of different chemical composition. This principle is illustrated in Figure 1. Altgelt first made use of such a scheme in his analysis of asphalt [4, 5].

The question arises, which separation should come first, GPC or the complementary one. The answer will depend on several factors. In principle, GPC should be first since it is the "cleaner" separation. It ordinarily discriminates only by molecular size, while most other methods (fractional precipitation and dissolution, as well as adsorption and partition chromatography) separate by both functionality and molecular size. While the dual response is an intrinsic feature of these methods, there can be, in addition, a GPC effect in chromatography resulting from the porosity of the adsorbents.

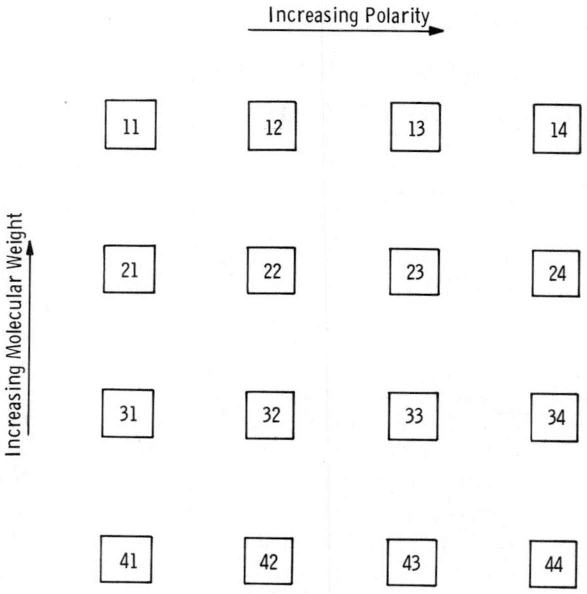

FIG. 1. Two-dimensional separation. A mixture is separated first by one method. Each resulting fraction is then further separated by a complementary method (e.g., GPC-adsorption chromatography).

The GPC effect operates simultaneously and opposite to the adsorption principle, thereby rendering the separation less effective.

On the other hand, regular preparative GPC columns may not be large enough for the separation of several grams of material that may be desirable for a two-dimensional matrix. In this case, it may be advantageous to begin with the complementary separation method, which usually can accommodate larger samples.

With petroleum and other fossil fuels, it is often advisable to extract the sample with n-pentane or other suitable light solvents since some very polar resins, asphaltenes, and especially benzene-insoluble components, tend to be irreversibly adsorbed on column packings. This happens not only on alumina, silica, and porous glass, but even on polystyrene gels. The adsorption is usually less serious with a whole (asphaltene-containing) petroleum residuum than with the asphaltene fraction itself because some of the maltene components strongly peptize the asphaltenes [3]. If asphaltene fractions of different molecular weight are desired, it may sometimes be necessary to chromatograph by GPC the whole residuum and then to separate the asphaltenes from each fraction by precipitation with n-pentane.

Sorption can seriously interfere with GPC in case of condensed polynuclear aromatics [58, 60]. This will be discussed in Sec. III in some detail. Here we shall only mention three aspects of this interference. First, gross effects have only been observed with unsubsituted aromatic model compounds and with relatively simple, purely aromatic (unsubstituted) species in petroleum distillates and pitches. Second, a group of workers at the Energy Research Centers in Bartlesville, Okla. and Laramie, Wy. made use of this effect for determining the ring structure of aromatic compounds in petroleum distillates by mass spectrometry and GPC [23, 43, 61, 72]. Third, with regular alkyl-substituted aromatic materials, this sorption can be overcome by using potent solvents [12, 34a] and was found to be minor with heavy residuals in the presence of chloroform or benzene containing 5 to 10% methanol [1, 2, 4, 5, 69]. Working at higher temperatures (50°C or more) also helps to minimize problems caused by association or sorption.

With proper precautions, analytical GPC will give the approximate mol. wt. distribution of a heavy petroleum sample. Characteristic elution patterns and changes in the mol. wt. distribution due to aging, weathering, and oxidation are easily and rapidly established by GPC. Examples of these applications are the elution patterns of petroleum pitches and coal tars that Edstrom and Petro [32] established, those of coal extracts determined by Limpert and Obermiller [50], the fingerprinting of crude oils described by Done and Reid [30], and by Albaugh and Talarico [13], the aging studies of asphalts by Minshull [51], Bynum et al. [20], Dougan [31], Haley [36, 37], and Hayes et al. [38, 39], investigations of the effect of air blowing on asphalts [12, 51], and that of desulfurization on petroleum residuals [14, 43], and the change of asphaltene molecular weights after steam stimulation of oil wells [71].

Jewell and Ruberto et al. [44] demonstrated the power of GPC for heavy petroleum samples by comparing the elution patterns of various fractions, differing by chemical composition as well as by mol. wt., with that of the parent residual (Figures 2 to 4). In all cases, the proper mol. wt. sequence is observed.

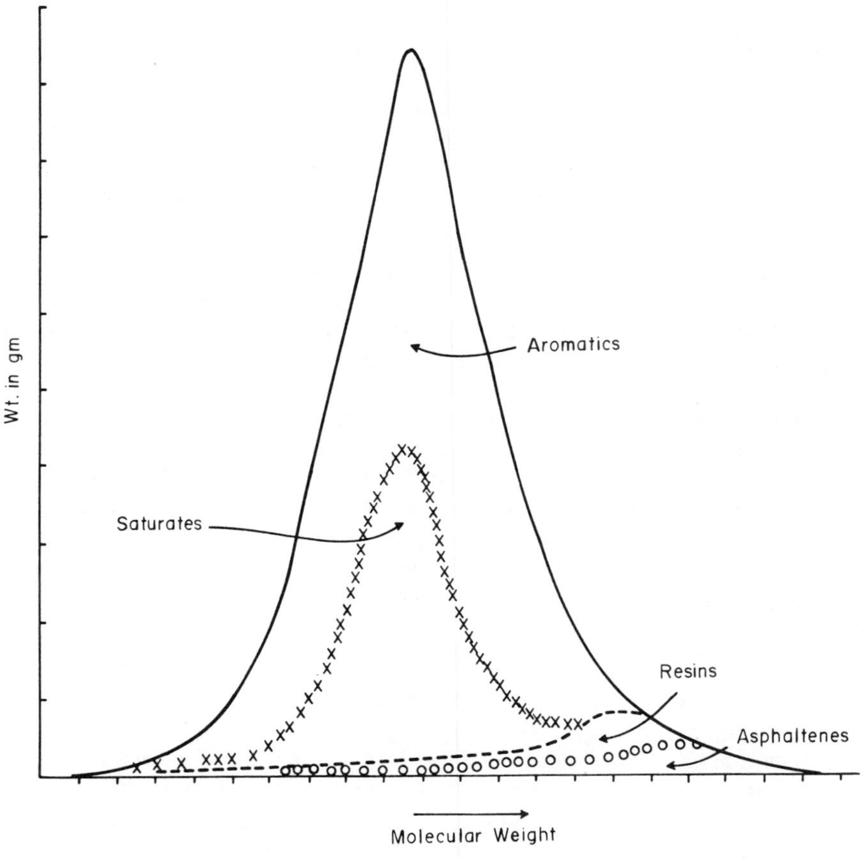

FIG. 2. Elution patterns of a Kuwait residual and its main constituents: asphaltenes, resins, aromatics, and saturates [43]. (Courtesy ACS, Ind. Eng. Chem., Fundam.)

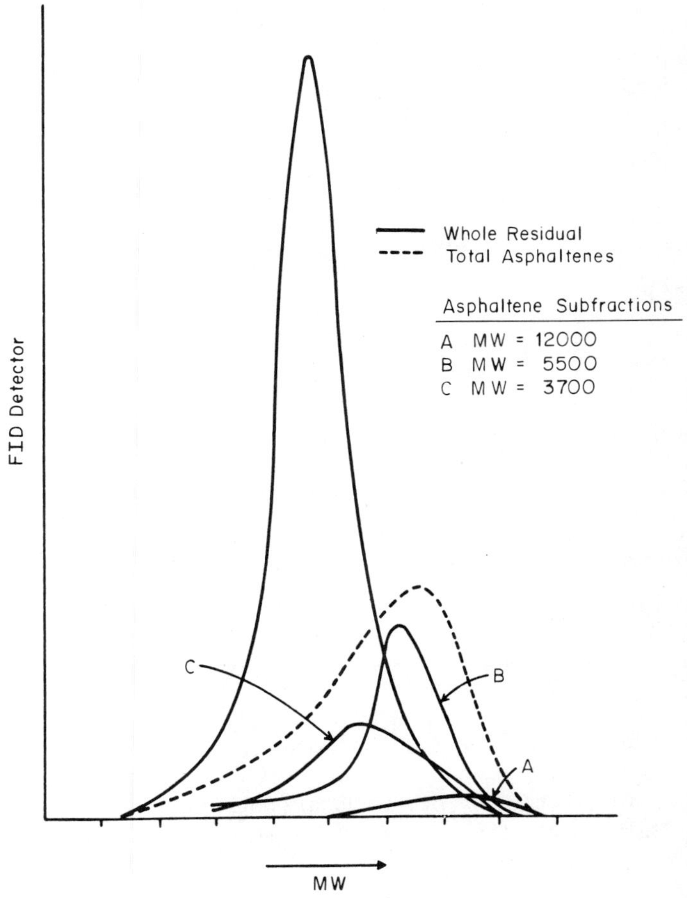

FIG. 3. Elution patterns of a residual, its asphaltenes, and three asphaltene fractions of different MWs [43]. (Courtesy ACS, Ind. Eng. Chem., Fundam.)

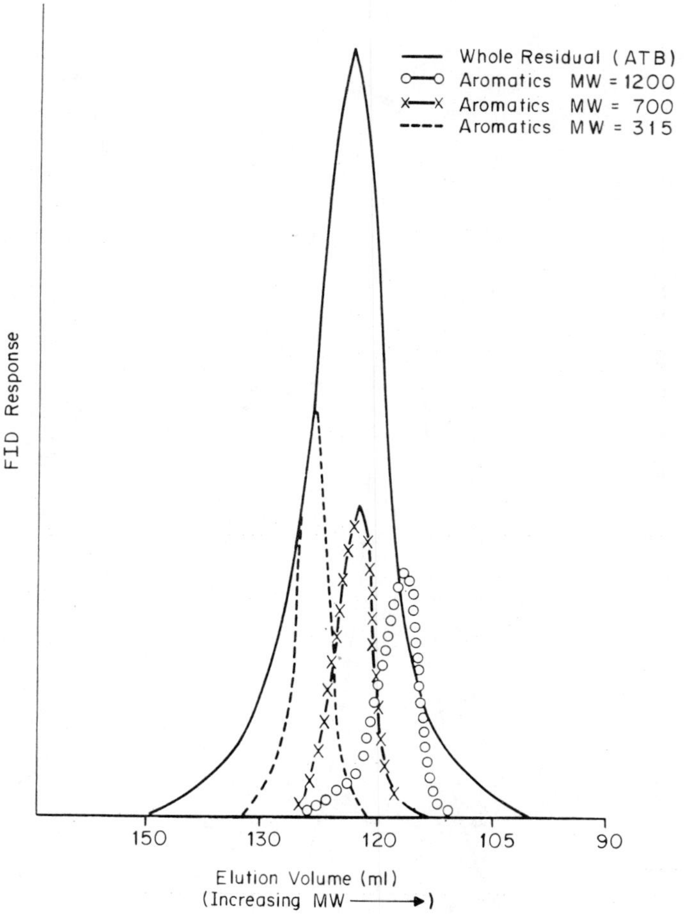

FIG. 4. Elution patterns of a residual (same as in Figure 3) and three aromatics fractions of different MW [43]. (Courtesy ACS, Ind. Eng. Chem., Fundam.)

II. METHODOLOGY

General GPC technique and apparatus have been adequately described in the literature [7, 9, 17, 21, 22, 46, 64]. Here we want to point out a few factors in methodology which have special application in the GPC of heavy petroleum samples.

A. Analytical GPC

Here the procedure is straightforward, but the interpretation may be misleading. First of all, a weight balance seems to be rarely obtained or, at least, is rarely reported. With strongly adsorbing samples, such as coal asphaltenes or, even worse, benzene insolubles, and also various petroleum asphaltenes and resins, it is easy to miss 50 to 80% of the material without even realizing it. Tailing may be so weak, even with gross adsorption, that one is tempted to dismiss it as being caused only by a minor portion of the sample and, therefore, to ignore it. Thus, getting a weight balance is imperative when working with a new fossil fuel sample.

A second problem can occur in the calibration of the GPC column. Compared to polymers, the components in petroleum and coal liquids are much more compact. Also, they can have widely differing mol. wt. size relations. This aspect will be discussed further in Sec. IV. See also Ref. 3.

On the other hand, much valuable information can be gleaned even if the true mol. wt. distribution is not attainable. Figure 5 shows an example of Done and Reid's fingerprinting technique [28] that uses gels of small pore size. The high mol. wt. portion of a sample is unresolved and appears as a spike at the beginning of the fingerprint. The bulk gives distinctive patterns determined by the average mol. wt., its distribution, and—with use of a refractive index detector—its aromaticity (positive peaks) or paraffinicity (negative peaks). Combination of several detectors as practiced by Albaugh and co-workers [12, 13] further increases the discriminatory power of the method. Figure 6 illustrates the different responses from flame ionization, which gives nearly the true mass distribution (especially with the Pye LCM2 detector) [13], from uv, which is sensitive to the aromatics, and from refractive index detection, which senses the aromatics-saturates distribution. Benzene as solvent gives very distinctive patterns for both ir and uv. The latter, at wavelengths of >320 nm, then sees only the condensed polynuclear aromatics and not the monoaromatic hydrocarbons. Specific detectors, for instance, for sulfur [14], can further increase the usefulness of analytical GPC for petroleum samples.

With analytical GPC the choice of solvent is critical, as discussed in Sec. III. Gioumatzidou and Oelert [34a] found chloroform + 10%, 1,2,4-trichlorobenzene to be a good compromise between low sorption effects and convenience (viscosity, detection).

The conversion of elution curves to molecular weight distribution is discussed in Sec. IV.

B. Preparative GPC

Many GPC runs of residua or asphalts are made on a preparative scale. Gram quantities are often needed when the fractions are used for further analysis by chemical or physical methods or when they are further separated by other techniques. True mol. wt. distributions of heavy petroleum fractions are obtained by direct mol. wt. measurements on all or some of the GPC fractions. Preparative GPC does not have to be inferior to analytical GPC in terms of resolution, but it will take considerably more time.

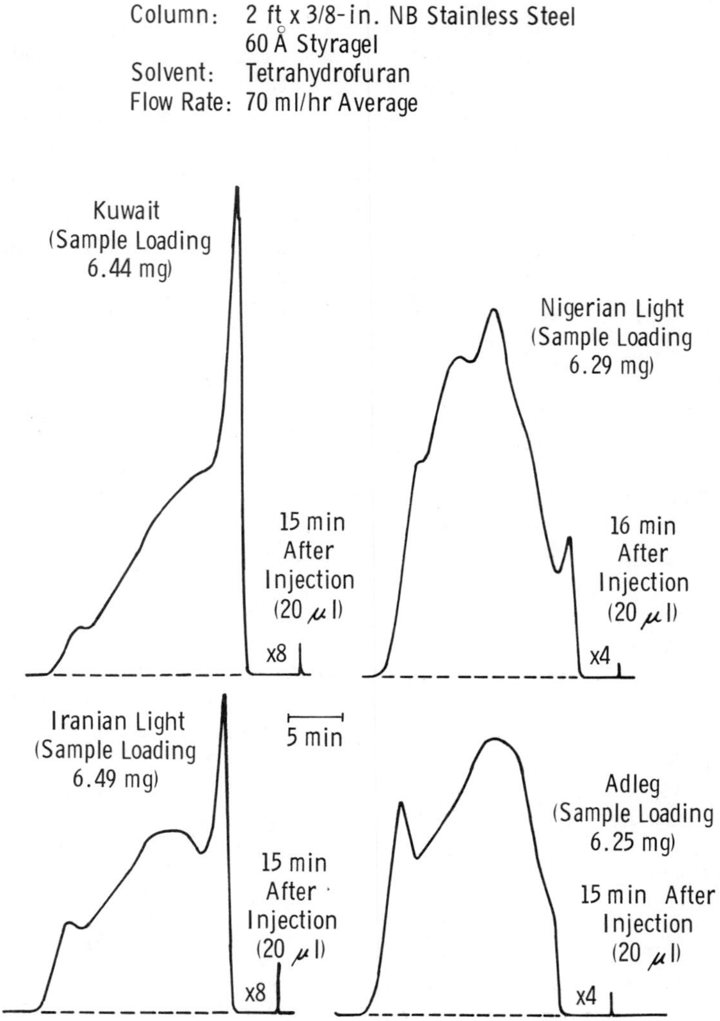

FIG. 5. GPC "fingerprints" of different crude oils [30]. (Courtesy Separ. Sci.)

1. Exclusion of Air

Because of the longer exposure of the sample in preparative GPC, even greater care must be taken here than in analytical runs, to avoid oxidation. In our laboratory we purge the solvent reservoirs with nitrogen. Before the solvent enters the pump, it passes through a 4-in.-long heated section of 1/4-in. stainless steel tubing attached to the bottom of the reservoir in

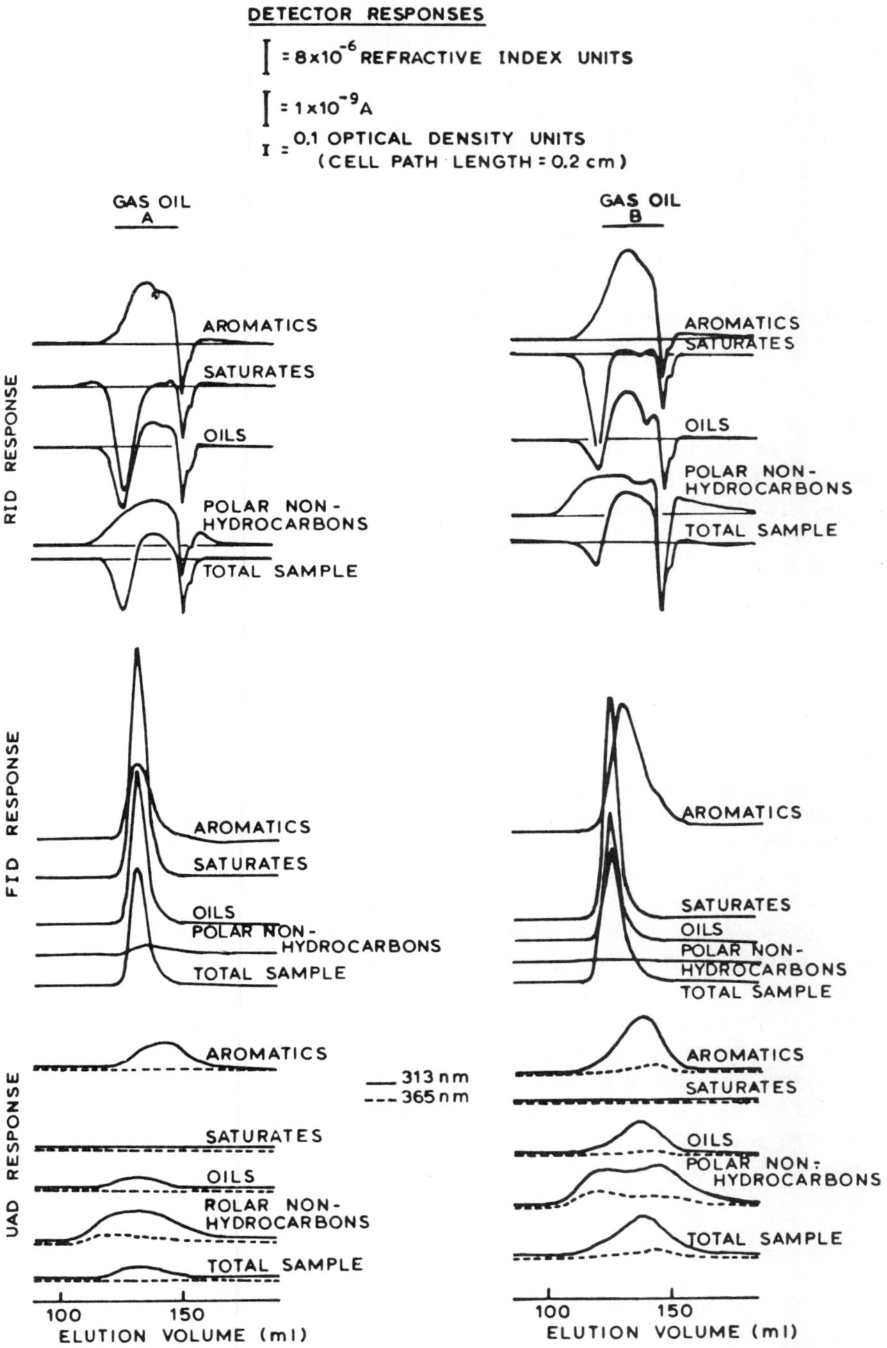

FIG. 6. GPC traces from refractive index (RID), flame ionization (FID), and uv absorption detectors (UAD) of two gas oils and their main fractions [13]. (Courtesy J. Chromatog.)

such a way that vapor and air bubbles rise through the bulk of the solvent into the nitrogen atmosphere. If the temperature of the tubing is adjusted to give very light boiling, e.g., one vapor bubble every 2 to 3 sec, no significant change in solvent composition (in case of mixed solvents) is observed in 1 or 2 days. The solvent composition is easily checked by determining the refractive index and restored to the original composition if necessary. A reflux condensor attached to the top of the reservoir may be required for very low boiling solvents. Ordinarily, the solvent composition is not critical in GPC as long as adsorption is avoided.

2. Gels

We prefer the rigid, heavily cross-linked polymers offered by Waters Associates, Bio•Rad, Merck, and others, to the softer, lightly cross-linked gels. They give more stable packings and allow considerably faster separations although at the expense of resolution. Porous glass and silica, on the other hand, strongly adsorb asphaltenes and other heavy polar compounds. Silylation of these packings does not seem to alleviate or even lessen this problem for fossil fuel samples. Some adsorption occurs also with the organic gels, but usually much less than with the inorganic ones.

3. Solvents

Very important for good results with petroleum components is the use of the proper solvent. In our experience, chloroform + 5% methanol or benzene + 5% methanol give the best results in terms of solvency, suppression of adsorption, and ease of evaporation (see also Ref. 14). There are more powerful solvents such as pyridine, trichlorobenzene, chloroform + 10% trichlorobenzene or α-methylnaphthalene + 5% methanol, which, indeed, better suppress any interfering sorption effects (see Sec. III). However, these solvents have the disadvantage of a high boiling point, which makes recovery of the fractions difficult. Since these solvents cannot be evaporated, they have to be separated from the solute on a short, small-pore GPC column. Methylene chloride containing 5% methanol would also be a good choice, except for its very low boiling point (33°C) which would make it necessary to refrigerate the solvent reservoir or use a cold finger to avoid excessive losses.

4. Columns

Very helpful for preparative GPC of heavy petroleum fractions is the use of glass columns because they allow us to see what is going on. Heavy petroleum compounds are usually black or brown and can, therefore, be observed in their progression along the column. Uneven flow or the presence of air, in case of leakage, is easily recognized, and immediate corrective action can be taken. Also, we can see early in the separation process when a sample is badly adsorbed. In such a case we may decide to backflush the sample out in order to save the column; or we may want to clean up the column right after the run by rinsing it with a more powerful solvent, e.g., a mixture of α-methylnaphthalene and pyridine. Digressing for a moment, we want to point out that whenever strong adsorption is suspected, a precolumn should be used, preferably also one made of glass. It is easier and cheaper to clean or discard a small precolumn than the larger main column.

With glass columns, pump rates are restricted in that only limited pressure can be applied, depending on column diameter and wall thickness. M. W. Lame's formula

$$P = 2S\,\frac{(OD - ID)^2}{(OD + ID)^2} \sim 2S\,\frac{W}{ID}$$

where OD = outer diameter, ID = inner diameter, W = wall thickness, and S = 68.05 atm (1000 psi) has been suggested [52] for estimating a safe upper pressure limit for Pyrex columns (safety factor of 10). Under the assumption of medium wall thickness, this means that the pressure should not exceed 150 psi (10 atm) for 1/2-in. (12.7-mm) column diameter and 100 psi (7 atm) for 1-in. (25.4-mm) diameter. Another precaution with glass columns is that they should not be exposed to bright sun or other light containing a substantial amount of uv frequencies. This helps to prevent photoreactions which might change the sample and aggravate adsorption on the gel.

Even with the best precautions, some adsorption of asphaltenes cannot be avoided; and, after a number of runs, the columns take on brown or gray shades. In our experience, this gradual discoloration usually does not impair their performance, especially if they are regularly and generously rinsed between runs. Eventually, however, heavily stained column packings will increase the adsorption tendency of a sample and must then be discarded. Again, a glass column makes it much easier to recognize when this point has been reached than when an opaque metal column is used.

5. Leaks in Columns

Glass columns also make it easy to see leaks. It may be (and is) disconcerting to realize just how often columns leak and air get in, but at least one can take corrective measures if one sees it happen. As it turns out, the effects of minor leaks are overcome easily and without loss of column performance by pumping deaerated solvent at relatively high pressure through the system. This takes the air out in a few minutes and restores the column to its original efficiency unless the gel had been so dry that it shrank. The procedure does not stop the leak, of course.

Leaks are usually caused by gel particles wedged between the column wall and the Teflon sleeve of the end piece. If tightening of the end piece does not help, one has a choice between two evils. One possibility is taking the end piece out for cleaning. This may not only be time consuming, but it may change the calibration of a column and even its performance. The other evil is often the lesser one: Instead of letting the column sit between runs, one keeps pumping solvent through it at a very slow rate. Also, the solvent should leave the system at a point which is located higher than the leak or, better yet, higher than the whole column. Then there is always positive pressure in the column and no air can enter. The leaks are usually small enough that only air but no solvent or sample will go through.

6. Packing of Glass Columns

Glass columns cannot be packed by the usual high-pressure method. However, very good results are obtained with most polystyrene gels in the following way. An auxiliary column fitted with a large funnel or an open flask on top is attached to the main column. The latter is filled to about one-half of its length with acetone. A slurry of the gel in enough acetone to have about one-half as much supernatant liquid as swollen gel is poured all at once into the two columns. The valve at the bottom is opened to allow the solvent to drain. While the gel is settling, the columns are turned back and forth around their longitudinal axis about once every 2 to 3 sec. The turning action agitates the gel when it settles on the packed surface, thus helping to pack it more tightly. Furthermore, it serves to even out any irregularities arising from a slight deviation from the vertical position of the columns or, more importantly, from the eddies caused by the gel falling through the solvent. When the gel has settled, the level of the solvent is lowered to the top of the gel. The solvent flow is then stopped, excessive gel and the auxiliary column are removed, and the main column is closed with its upper end fitting, which is connected to the solvent pump. Now about one column volume of acetone is pumped rapidly downwards. If some settling should occur, either the fitting is readjusted or more gel is added. When no more settling takes place, the solvent is changed stepwise from acetone to the intended solvent, usually chloroform + 5% methanol. The polystyrene gel swells slightly more in the good final solvent than in the poor initial solvent acetone. As a consequence, the packing becomes tight and quite regular to judge from the high efficiency usually achieved, viz., 800 to 3000 theoretical plates (tp) per foot (2600 to 10,000 tp m^{-1}) as measured with benzene. The backpressure remains reasonably small (30 to 40 psi) at a flow rate of 600 cc hr^{-1} in a column of 1 in. diameter and 36 in. length. With this packing technique, it is impossible for the bed to shrink when changing from chloroform to a poorer solvent. In no case will a solvent suitable for petroleum components be poorer for the polystyrene gel than acetone.

III. POLYNUCLEAR RING SYSTEMS

We stated above that GPC separates by molecular size and not by chemical structure. The discriminating parameter is, in good approximation, the hydrodynamic volume [25, 26, 53, 73], or, for small molecules, the molar volume [41, 47]. The hydrodynamic volume of polymers and other large molecules can be expressed as the product of intrinsic viscosity $[\eta]$ and weight average molecular weight, mol. wt. The plot of molar volume or of hydrodynamic volume against elution volume renders one and the same curve for a given column irrespective of solvent and sample type [17, 35]. This is very important in general and for petroleum compounds in particular because of their great variety of chemical structure. However, in the case of polynuclear ring systems, this principle is seemingly violated by an anomaly, the cause of which has not been clearly identified yet but is probably related to partitioning or adsorption of these compounds on the gel [16, 57, 59, 60, 72]. According to Oelert et al. [57, 59, 60, 72], the plot of molar volume versus elution volume yields three different relations for the three groups: paraffinic and monoring hydrocarbons, catacondensed hydrocarbons, and pericondensed aromatic hydrocarbons (see Figure 7). The catacondensed compounds follow a curve with decreasing mol. wt. for increasing V_e as is expected in GPC, but with a considerably steeper slope than the curve for the paraffins. The pericondensed hydrocarbons, however, <u>increase</u> in mol. wt. with increasing V_e indicating a predominant sorption or partitioning rather than an exclusion effect. Oelert et al. [60] established similar relations for nitrogen- and oxygen-containing compounds. It should be pointed out here that the samples used in these studies were model compounds, presumably synthetic ones, and generally without alkyl substituents. This means that they constituted extreme cases which are not necessarily representative of the majority of polynuclear aromatic compounds present in petroleum. Whatever alkyl-substituted samples were included in the comparison eluted at lower volumes than their parent compounds. This is to be expected not only because of their higher solubility but also because of a shielding effect both of which serve to minimize the adsorption.

Bergmann et al. [16] observed the same elution anomalies with numerous model compounds including n-alkanes, cata-, and pericondensed aromatics, hydropyrenes, and heterocyclics. However, they could practically eliminate the differences in elution behavior by carrying out the GPC with trichlorobenzene (TCB) as solvent. They performed most runs at 140°C, but they found that TCB overcame the anomalies even at room temperature.

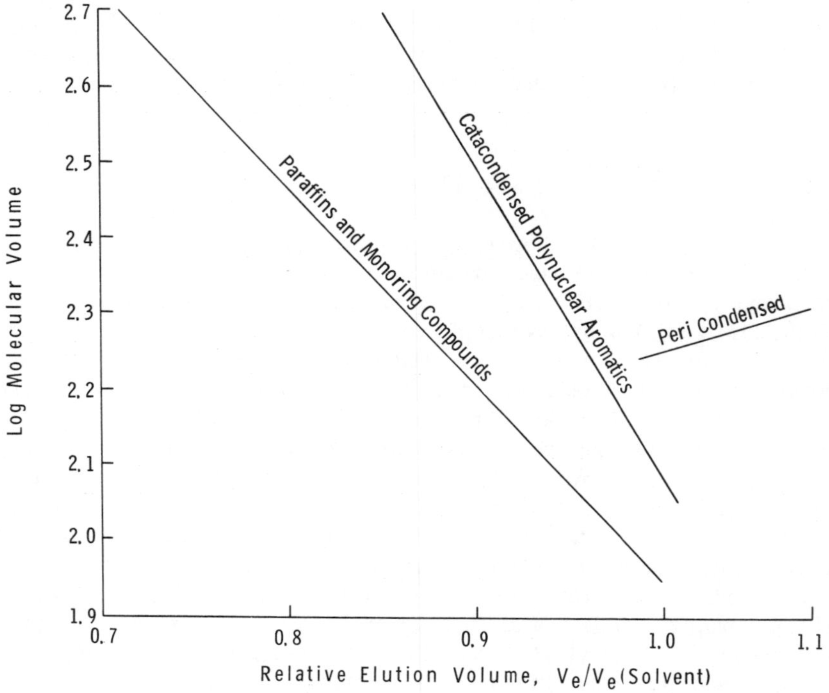

FIG. 7. Calibration curves, log V_m versus V_e of paraffinic, catacondensed and pericondensed hydrocarbons [59]. (Courtesy Separ. Sci.)

This was the case even though their model compounds did not contain any alkyl substitution and included molecules as large as coronene (300 mol. wt.) and ovalene (396 mol. wt.). Generally, the elution anomaly decreased with increasing sample solubility. Gioumatzidou and Oelert [34a] found that 10% TCB in chloroform was sufficient to suppress sorption effects on styrene gels. Lewis and Petro [49] reported successful GPC separations of petroleum and coal tar pitches containing fused aromatic ring systems with up to 68 atoms. Their solvent was toluene. Of petroleum constituents, the asphaltenes contain the greatest amounts of condensed aromatic rings. Yet, the higher mol. wt. portions elute before rather than after the other, less condensed, components. Figure 2 shows an example of such a separation. Also, the mol. wt. sequence of different asphaltene fractions is right as seen in Figure 3. On the other hand, the calibration curves of the components richer in condensed aromatic ring systems (asphaltenes, resins) are distinctly shifted toward higher elution volumes (or toward lower mol. wts.) compared to those with fewer or no polyaromatics (see Figure 8). Whether this shift is caused by adsorption or other factors is not clear in this particular case.

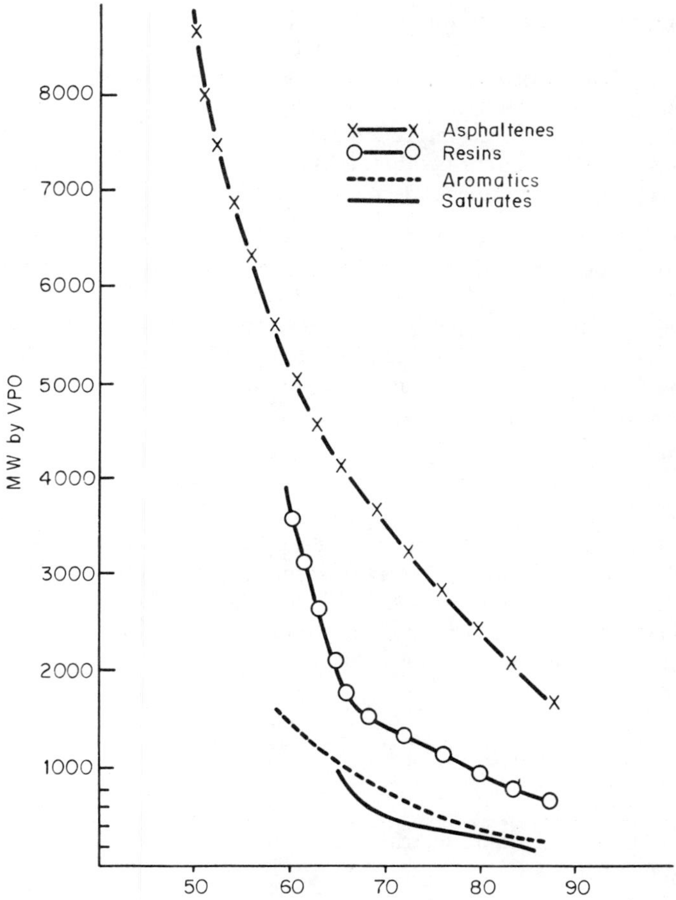

FIG. 8. Calibration curves, molecular weight (MW) versus V_e of asphaltenes, resins, aromatics, and saturates of a Kuwait residual [43]. (Courtesy Ind. Eng. Chem., Fundam.)

However, Altgelt and Harle [10] have provided evidence from viscosity data that the greater compactness of the asphaltene molecules would result in such a shift of calibration curves.

Recently, Popl et al. [65] published a comprehensive study of the adsorption effect in GPC separation of polycyclic aromatic hydrocarbons and some alkyl derivatives. For unsubstituted ring systems, they found a set of parallel straight lines in the plot of molecular size (they used the parameter $\overline{L}$ suggested by Giddings et al. [34] versus elution volume. Each line had the usual negative slope—lower V_e for greater $\overline{L}$—but the curves for molecules with the greater number of aromatic carbon atoms N_c were located above these with the smaller N_c. Furthermore, the distance between curves different by $N_c = 2$ was constant over the range from $N_c = 10 - 24$. Alkyl substitution shifted V_e to much lower values. These results confirm our previous contention [5, 11]. Increasing N_c increases adsorption and thus V_e; for a given N_c, increasing size due to different configuration decreases V_e in the regular fashion; alkyl substitution greatly diminishes adsorption and with it the special behavior of the polynuclear aromatics.

Heitz [40] and Dawkins and Hemming [27] took sorption into consideration in describing GPC. Heitz and Platt et al. [40] observed sorption effects with OH-substituted polyphenylenes on dextran and, in referring to the potential dual nature of the separation mechanism in GPC, called it a network-limited partition process. Dawkins and Hemming [27] extended this concept to include adsorption and introduced the solute-gel interaction term K_p into the equation for elution volume V_e

$$V_e = V_0 + K_D K_p V_i$$

where V_0 is the void volume, V_i is the internal pore volume, K_D is the distribution coefficient for steric exclusion, and K_p is the distribution coefficient for solute-gel interactions. If there is no solute-gel interaction, $K_p = 1$; otherwise $K > 1$. Dawkins and Hemming [27] also modified Grubisic's et al. [35] relation to incorporate their K_p:

$$\frac{V_e - V_0}{K_p} = V_i[-A \log [\eta]M + B]$$

Here A and B are constants characteristic of the solute-solvent system. Rearrangement leads to

$$\log [\eta]M = \frac{B}{A} - \frac{V_e - V_0}{AV_i K_p}$$

or

$$\log [\eta]M = \frac{C}{K_p} - \frac{C'V_e}{K_p}$$

These equations should describe the results of Popl et al. [65] at least qualitatively (a quantitation has not come to this author's attention). We can assume that for unsubstituted polynuclear aromatics K_p depends primarily on the number of aromatic rings, or, put in other words, for polynuclear aromatics of equal ring number but different size K_p remains nearly constant as, indeed, Popl and co-workers found. Alkyl substitution should decrease K_p and bring it closer to unity. For $K_p = 1$ no solute-gel interaction is left and Grubisic's et al. original formula for steric exclusion holds.

$$\log [\eta]M = C - C'V_e$$

Feurer and Gourdenne [33] made another observation of interest in this context. They found that aromatic ring systems are strongly solvated by certain polar solvents, e.g., dimethylformamid. As a result, the ring compounds increase in size and V_e decreases. Such solvation would counteract the effect of adsorption in GPC not only indirectly because of the size increase but also directly by competition for the interactive aromatic rings.

In summary, adsorption and partition effects of polynuclear ring systems in GPC have been observed and mathematically described. With petroleum samples they are usually not dominant; but, since these principles counteract the exclusion effect, we must be aware of their existence and try to suppress them. Their effect will be smaller with increasing alkyl substitution of the molecules—which is known to increase solubility and would also tend to shield the aromatic rings—and with increasing solvent power and temperature. Oelert [62] suggested taking advantage of the sorption effect for separations based on chemical structure by using carefully selected solvents and gels. This can lead to simple and effective fractionations [59, 62] and offers another option in addition to a combination of regular GPC with adsorption chromatography on alumina or silica.

GPC of asphalts and crude oil residua has generally yielded fractions of decreasing molecular weight and color intensity [1, 2, 4, 5, 8, 12-14, 18, 20, 28, 30, 31, 36-39, 44, 49, 51, 59, 67-69, 71]. Yet, investigation of high and low mol. wt. asphaltenes by NMR, elemental analysis, and density measurements indicated higher degrees of aromaticity and ring fusion in the low mol. wt. fraction, quite possibly as a consequence of some adsorption. In plots of these properties and of S, N, and O contents versus mol. wt. or elution volume, often V-shaped curves are obtained with distinct minima in the medium ranges [5, 29, 69].

IV. CALIBRATION AND DETERMINATION OF MOLECULAR WEIGHT DISTRIBUTIONS

Most GPC runs of nonpetroleum samples are performed in analytical columns with milligram or microgram amounts. Calibration curves are used to convert, often by computer, elution volumes into molecular weights. This technique depends on the knowledge of the relation between hydrodynamic volume, in case of polymers, or molar volume, in case of small molecules, and molecular weight. The relation of hydrodynamic volume [η]M with M, or of intrinsic viscosity, [η], with M depends on the configuration of the molecules [7]. Once this relation has been determined for a polymer across the whole mol. wt. range, the mol. wt. distribution of any sample of this polymer type can be determined by GPC by means of a calibration curve without any further measurement.

However, the situation is different with heavy petroleum samples. These are not homologous mixtures differing only in mol. wt. Even within a given crude oil or asphalt, we have a tremendous variety of different configurations ranging from largely paraffinic wormlike molecules to compact, nearly spherical or disk-shaped polynuclear ring systems [10]. We would not expect all these constituents to follow the same [η]M relation. Likewise, petroleum samples of different origin have different [η]M relations [5, 8]. In Figure 9, [η]M is plotted versus M for various asphaltene fractions deriving from four crude oils with different chemical composition. Three curves result reflecting the different compactness of the molecules though not necessarily their paraffinic, naphthenic, and aromatic character. Differences between the asphaltenes and the resin fractions of one asphalt are seen in Figure 10. These differences are, of course, paralleled in the mol.

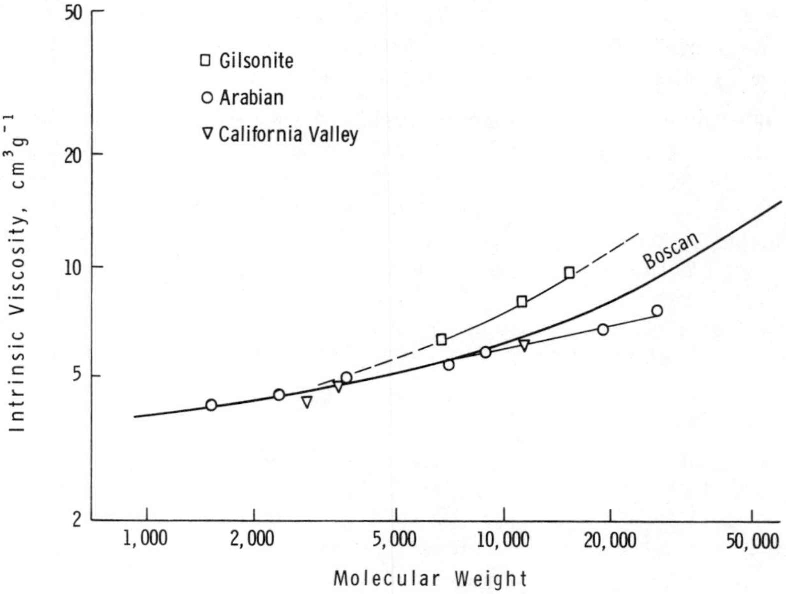

FIG. 9. [η]-MW relation of various asphaltenes [5]. (Courtesy Bitumen, Teere, Asphalte, Peche.)

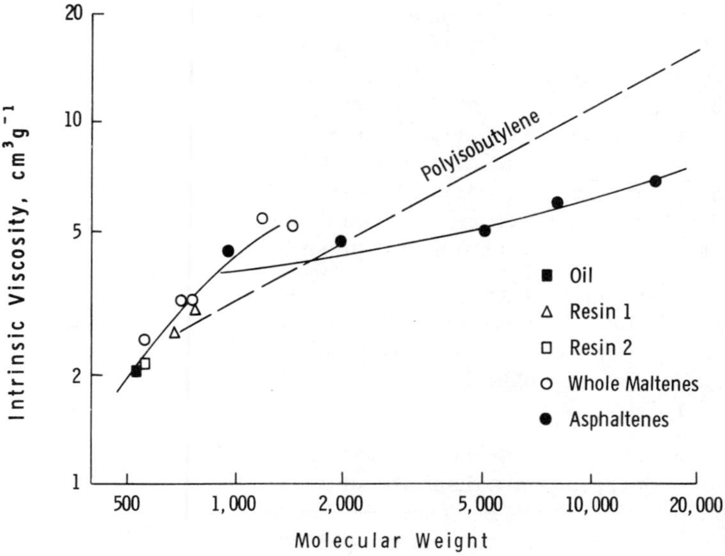

FIG. 10. η-MW relation of oil, resin, and asphaltene fractions of a steam-refined Boscan asphalt [5]. (Courtesy Bitumen, Teere, Asphalte, Peche.)

wt.-elution volume curves. The different calibration curves in Figure 8 for asphaltenes and other fractions from a petroleum residual confirm this view.

Another problem encountered with petroleum samples is the variation of the detector response with the type of material monitored. Refractive index, uv, and visible light adsorption all depend greatly on the aromaticity, the degree of ring fusion, and the size and nature of the polynuclear aromatic ring systems. Variations by factors of 100 and more have been observed in our laboratory [12] between the light absorption of different GPC fractions of asphaltenes as well as of whole asphalts. Snyder [69] reported similar results, although only in qualitative terms. Even ir absorption in the C—H stretch vibration range at 2900 cm^{-1} becomes unreliable for highly aromatic samples.

We conclude that, except for routine measurements on a well-characterized sample, the mol. wt. distributions of heavy petroleum fractions as determined by GPC cannot be based on analytical runs and on the use of a universal calibration curve without additional measurements. The only safe method is to determine the mol. wt. distribution independently and plot the measured mol. wts. versus elution volume. This has been done in two ways: by preparative GPC combined with vapor phase osmometry (VPO) [2, 5, 8, 19, 44, 48, 70] and by ultracentrifugation [66]. In the first method enough material (about 0.5 g or more) was fractionated on a preparative GPC column. The mol. wts. of several fractions, measured by VPO and plotted against the elution volume, resulted in specific calibration curves which could be used for the interpolation and, to some extent, the extrapolation to all remaining fractions of a sample and, if necessary, to elution volumes between fractions. Corrections for axial dispersion (peak broadening) are unnecessary in this case as the method is self-correcting.

In the second method [66] the mol. wt. distribution was measured by ultracentrifugation. Comparison of the mol. wts. plotted versus the cumulative weight fraction with the elution volumes from a GPC run plotted versus the cumulative weight yields the calibration curve. The relation is illustrated in Figure 11.

For quick and easy assessment of the quantitative mol. wt. composition of a sample, the results are displayed in terms of an integral mol. wt. distribution curve such as the one shown in Figure 12. Here the cumulative weight percentages of the fractions are plotted versus mol. wt. or log mol. wt. Often the differential mol. wt. distribution is preferred because it gives a more graphic and easily comprehended illustration of the mol. wt. composition. An example is Figure 13, which shows the differential distribution curve (DDC) of the same sample as the one depicted in Figure 12. The

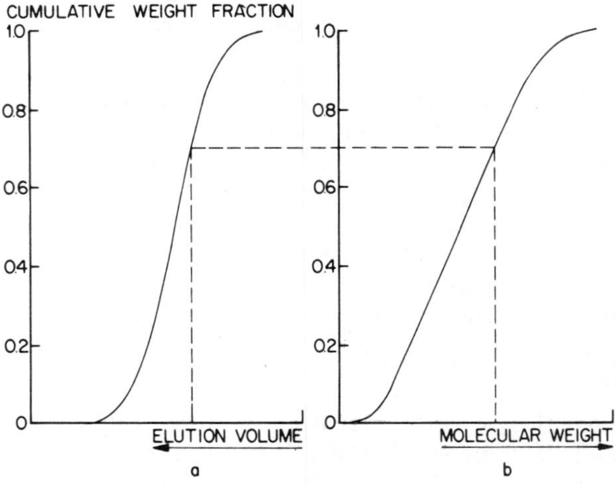

FIG. 11. Determination of MW from V_e via MW distribution from ultracentrifugation [65]. a. Cumulative weight fraction versus V_e. b. Cumulative weight fraction versus MW. (Courtesy ACS, Anal. Chem.)

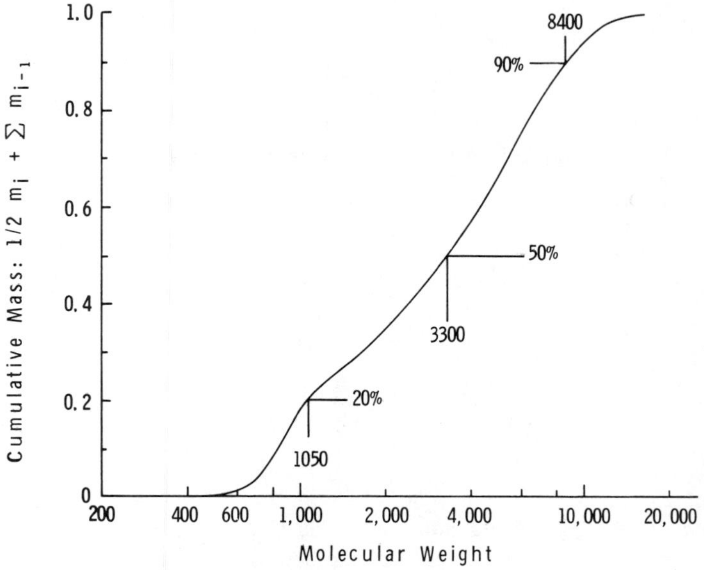

FIG. 12. Integral MW distribution of a model mixture of three components [5]. (Courtesy Bitumen, Teere, Asphalte, Peche.)

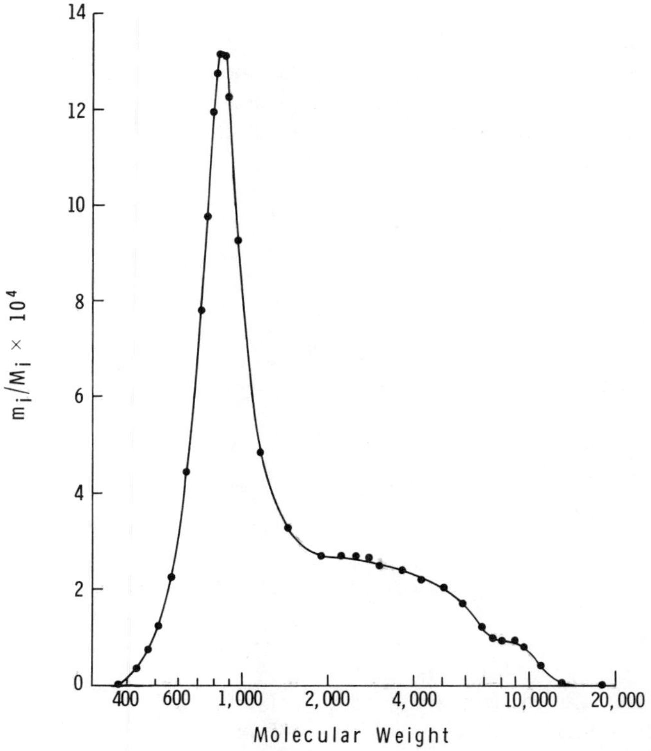

FIG. 13. Differential MW distribution of a model mixture of three components (same as in Figure 11) [15]. (Courtesy Bitumen, Teere, Asphalte, Peche.)

ordinate in Figure 13 is related to the slope of the curve in Figure 12, i.e., the differential of the ordinate over the (linear) mol. wt. It can be calculated from the equation [5, 6]

$$\frac{d\sum_{i=1}^{j} m}{dM_i} = \lim_{M_{j-1} \to M_j} \left(\frac{m_j}{M_j - M_{j-1}} \right) = \lim_{V_{e(j-1)} \to V_{ej}} \left(\frac{m_j}{V_{ej} - V_{e(j-1)}} \right) \times \lim_{M_{j-1} \to M_j} \left(\frac{V_{ej} - V_{e(j-1)}}{M_j - M_{j-1}} \right)$$

by use of a computer or even a desk calculator. Here m_j is the mass percentage of fraction j, M_j its mol. wt., and V_{ej} its elution volume. The first factor on the right-hand side of the equation defines the concentration of fraction j, the second one the reciprocal slope of the calibration curve at point j. Thus, the DDC is a plot of the product of (normalized) concentration and reciprocal slope of the calibration curve of fraction j against its mol. wt.

It is interesting to compare the DDC of a sample with its elution curve. In Figure 14 we see the elution curve of the same sample as that used for Figures 12 and 13. The peaks and valleys observed in the elution curve (Figure 13) are reflected in the DDC although in a grossly distorted fashion.

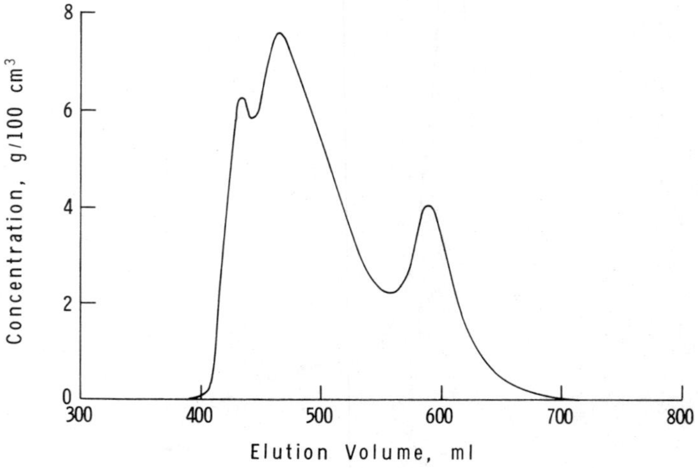

FIG. 14. Elution curve of a model mixture of three components (same as in Figures 11 and 12) [5]. (Courtesy Bitumen, Teere, Asphalte, Peche.)

This comparison demonstrates how difficult it is to estimate the mol. wt. distribution from the appearance of an elution curve without the proper evaluation. The primary reason for the distortion is the logarithmic relation between mol. wt. and V_e. An additional contribution to the distortion is the nonlinearity of the calibration curves. The greatest uncertainty in the shape of the DDC is near the peaks and the valleys. These regions, therefore, require the most data points. Here it is often good practice to interpolate between fractions while in other regions one may even leave one or several fractions out of the calculation.

V. GPC and Structural Analysis

Done and Reid [30] first utilized GPC as a rapid tool for the classification or recognition of petroleum samples. Albaugh and co-workers [13, 14] expanded this method by employing three or more detectors in series for the analysis of the eluate. Qualitative and semiqualitative information on the mol. wt. distribution and the chemical composition can be obtained rapidly and automatically. The amount of useful data is multiplied by applying this technique to fractions obtained by alumina chromatography or by one of the more elaborate separation schemes (see Figure 6).

More quantitative results are derived when fractions obtained by GPC and possibly other methods are subjected to structural analyses. Distillates and the light portion of residua can be analyzed by mass spectrometry; the nonvolatile compounds require NMR, preferably in combination with other measurements. Coleman et al. [24] give examples for both approaches. Oelert [15, 54-57, 63], Altgelt and Hirsch [6, 42], and Haley [36, 37] applied group-type analysis to GPC fractions of petroleum and coal and studied their chemical structure as a function of mol. wt.

Dickson et al. [29] characterized GPC fractions of a crude oil residuum with NMR and elemental analysis of C, H, and S. They determined molecular weights by vapor-phase osmometry and compared them with "unit weights" obtained from NMR. Altgelt and Hirsch [6] fractionated a whole asphalt by GPC. After further separation by n-pentane extraction and alumina chromatography (see Chap. 9), the fractions were characterized by "integrated structural analysis." By this term the authors describe the combination of NMR, elemental analysis, mol. wt., and density determinations with a computer program that reconciles all experimental data and evaluates them in terms of structural parameters [42] (see also Refs. 45 and 46).

Coleman et al. [23, 43] used GPC data to assist them in the interpretation of mass spectral mol. wts. in terms of molecular structures. The GPC data allowed them to distinguish the larger straight-chain compounds from the more compact ring compounds. For example, naphthenologs, i.e., saturated fused 6-membered ring compounds are higher in mol. wt. by 1.0 to 1.5 carbons per ring number than the basic alkyl substituted molecules eluting at the same volume. The combination of GPC elution volumes with mass spectral mol. wts. allowed this team to make structural assignments more easily and with greater certainty.

REFERENCES

1. K. H. Altgelt, J. Appl. Polymer Sci. 9:3389 (1965).
2. K. H. Altgelt, Makromol. Chem. 88:75 (1965).
3. K. H. Altgelt, 1966, to be published.
4. K. H. Altgelt, Separ. Sci. 5:855 (1970).
5. K. H. Altgelt, Bitumen, Teere, Asphalte, Peche 21:475 (1970).
6. K. H. Altgelt and E. Hirsch, Separ. Sci. 5:855 (1970).
7. K. H. Altgelt, Theory and Mechanics of GPC, Adv. in Chromatog. 7:3 (1970).
8. K. H. Altgelt, Preprints, Div. Petrol. Chem., ACS, 13(3):37 (August 1968).
9. K. H. Altgelt and L. Segal, eds., Gel Permeation Chromatography, Marcel Dekker, New York, 1971.
10. K. H. Altgelt and O. L. Harle, Ind. Eng. Chem. Prod. Dev. 14:240 (1975).
11. K. H. Altgelt and T. H. Gouw, Chromatography of Heavy Petroleum Fractions, Adv. in Chromatog. 13:71 (1975).
12. K. H. Altgelt, unpublished work.
13. E. W. Albaugh and R. C. Talarico, J. Chromatog. 74:233 (1972).
14. E. W. Albaugh and R. C. Query, Anal. Chem. 48:1579 (1976).
15. G. Alberts, L. Lenard, and H. H. Oelert, Fuel 53:47 (1974).
16. J. G. Bergmann, L. J. Duffy, and R. B. Stevenson, Anal. Chem. 43: 131 (1971).
17. K. J. Bombaugh, The Practice of Gel Permeation Chromatography, in "Modern Practice of Liquid Chromatography" (J. J. Kirkland, ed.), Wiley-Interscience, New York, 1971.
18. J. F. Branthaver and J. M. Sugihara, Preprints, Div. Petrol. Chem., ACS, 11(2):B-105 (1966).
19. E. H. Burk and J. Y. Sun, "Report of the NSF Workshop on the Fundamental Organic Chemistry of Coal," University of Tennessee, Knoxville, July 1975.
20. D. Bynum, Jr., R. N. Traxler, H. L. Parker, and J. S. Ham, J. Inst. Petr. 56:147 (1970).
21. M. J. R. Cantow, R. S. Porter, and J. F. Johnson, Rev. Macromolec. Chem. 1:393 (1966).
22. J. Cazes, J. Chem. Educ. 43:567 (1966).

23. H. J. Coleman, J. E. Dooley, D. E. Hirsch, and C. J. Thompson, Anal. Chem. 45:1724 (1973).
24. H. J. Coleman, D. E. Hirsch, and J. E. Dooley, Anal. Chem. 41:800 (1969).
25. H. Coll, Separ. Sci. 5:273 (1970).
26. J. V. Dawkins, J. W. Maddock, and D. Coupe, J. Polymer Sci. A-I(5): 1391 (1967).
27. J. V. Dawkins and M. Hemming, Makromol. Chem. 176:1795 (1975).
28. J. P. Dickie and T. F. Yen, Anal. Chem. 39:1847 (1967).
29. F. E. Dickson, R. A. Wirkhala, and B. E. Davis, Separ. Sci. 5:811 (1970).
30. J. N. Done and W. K. Reid, Separ. Sci. 5:825 (1970).
31. C. E. Dougan, Res. Proj. 175-210, Conn. Dept. Transport., Weathersfield, Conn., 1970.
32. T. Edstrom and B. A. Petro, J. Polym. Sci. C(21):171 (1968).
33. B. Feurer and A. Gourdenne, Preprints, Div. Polymer Chem., ACS, 17(2):495 (1976).
34. J. C. Giddings, E. Kucera, C. P. Russell, and M. N. Myers, J. Phys. Chem. 72:4397 (1968).
34a. P. Gioumatzidou and H. H. Oelert, Erdoel und Kohle 30:229 (1977).
35. Z. Grubisic, P. Rempp, and H. Benoit, J. Polym. Sci. B(5):753 (1967).
36. G. A. Haley, Anal. Chem. 43:371 (1971).
37. G. A. Haley, Anal. Chem. 44:580 (1972).
38. M. B. H. Hayes, M. Stacey, and J. Standley, Fuel 51:27 (1972).
39. M. B. H. Hayes, M. Stacey, and J. Standley, Fuel 57:32 (1972).
40. W. Heitz, K. L. Platt, H. Ullner, and H. Winau, Makromol. Chem. 102:63 (1967). See also W. Heitz, Ber. Bunsenges. Phys. Chem. 77: 210 (1973) and W. Heitz, Z. Anal. Chem. 277:323 (1975).
41. J. G. Hendrickson and J. C. Moore, J. Polymer Sci. A-1(4):167 (1966).
42. E. Hirsch and K. H. Altgelt, Anal. Chem. 42:1330 (1970).
43. D. E. Hirsch, J. E. Dooley, and J. H. Coleman, API RP 60, Published Report No. 27. Bureau of Mines Report on Investigations, Bartlesville, Okla., 1974.
44. D. M. Jewell, R. G. Ruberto, A. W. Albaugh, and R. C. Query, Ind. Eng. Chem., Fundam. 15:206 (1976).
45. Y. Katayama, T. Hosoi, and G. Takeya, Nippon Kagaku Kaishi 1:127 (1975).
46. Y. Katayama, T. Hosoi, and G. Takeya, Nippon Kagaku Kaishi 4:629 (1975).
47. A. Lambert, Brit. Polym. J. 3:13 (1971).
48. A. Lambert, Anal. Chem. Acta 53:63 (1971). Many other references given there.
49. I. C. Lewis and B. A. Petro, J. Polymer Sci. 14:1975 (1976).
50. R. J. Limpert and E. L. Obermiller, Sixth International GPC Seminar, 285, 1969.

51. J. A. Minshull, Proc. Fourth Conf. Aust. Road Res. Board 4(2):1468 (1968).
52. H. M. Nutt, Fusion 5:9 (1958). See also K. E. Bett and D. M. Newitt, in "Chemical Engineering Practice," vol. 5 (H. W. Cremer and T. Davis, eds.), Academic, New York, 1958, pp. 205-206.
53. A. C. Oanu, A. Broido, A. Barrall II, and A. C. Javier-Son, Preprints, Div. Polymer Chem., ACS, 12 (September 1971).
54. H. H. Oelert, Brennstoff-Chem. 11:3 (1967).
55. H. H. Oelert, Brennstoff-Chem. 12:1 (1967).
56. H. H. Oelert, Z. Analyt. Chem. 231:81 (1967).
57. H. H. Oelert, Z. Analyt. Chem. 231:105 (1967).
58. H. H. Oelert, Erdöl und Kohle 22:19 (1969).
59. H. H. Oelert, Z. Anal. Chem. 244:91 (1969).
60. H. H. Oelert, D. R. Latham, and W. E. Haines, Separ. Sci. 5:657 (1970).
61. H. H. Oelert and J. H. Weber, Erdöl und Kohle 23:484 (1970).
62. H. H. Oelert, J. Chromatog. 53:241 (1970).
63. H. H. Oelert, Adv. in Org. Geochem., 1971, p. 629.
64. E. P. Otocka, Acc. Chem. Res. 6:348 (1973).
65. M. Popl, J. Fähnrich, and M. Stejskal, J. Chromatog. Sci. 14:537 (1976).
66. H. Reerink and J. Lijzenga, Anal. Chem. 47:2160 (1975).
67. W. B. Richman, Assoc. Asphalt Technol. Paving Proc., 1968, p. 106.
68. R. J. Rosscup and H. P. Polhmann, Preprints, Div. Petrol. Chem., ACS 12:A-103 (1967).
69. L. R. Snyder, Anal. Chem. 41:1223 (1969).
70. H. W. Sternberg and C. L. Delle Donne, Fuel 53:172 (1974).
71. C. A. Stout and S. W. Nicksic, Separ. Sci. 5:843 (1970).
72. J. H. Weber and H. H. Oelert, Separ. Sci. 5:669 (1970).
73. W. W. Yau, J. Polymer Sci. A-2(7):483 (1969).

Chapter 13

SUPERCRITICAL FLUID CHROMATOGRAPHY

T. H. Gouw and R. E. Jentoft
Chevron Research Company
Richmond, California

I. INTRODUCTION

Consider a compound which is confined in a space of such volume that it is always present as a liquid and gas phase in equilibrium with each other. When heat is applied to this system, the intensive properties of these two phases will converge until, at the critical point, these two phases become identical. (Physical properties, which are independent of the amount of material involved, such as the density, viscosity, refractive index, thermal conductivity, etc., are called intensive properties.) This critical point marks the highest temperature on the gas-liquid coexistence curve. Further heating will result in the formation of a supercritical phase.

One of the many unusual properties associated with a compressed fluid at its critical point is its extremely large compressibility. Densities at and around the critical point are very sensitive to small variations in temperature and pressure. Gravity forces alone can set up appreciable density gradients. At the critical point of CO_2, which has a critical pressure of 73.8 bar, a pressure increase of only 1 mm Hg (one part in 50,000) will cause a density increase of more than 10%. Many of the unusual properties of supercritical fluid chromatography (SCF) are, as we will perceive, related to this phenomenon of extreme pressure dependence.

SFC differs from the other, more well-known chromatographic techniques in that the mobile phase is a substance occupying this particular region of the phase diagram. In this method, the carrier fluid is ordinarily maintained at 1 to 4 times the critical pressure and at 1 to 1-1/2 times the critical temperature in °K. This modus operandi is of special interest for the separation of substances with low GC migration rates and/or low thermal stabilities. Other major advantages of this technique are related to the extreme selectivities which can be attained, to the ease with which wide molecular weight range mixtures can be analyzed and, more recently, to the potential use of this technique in conjunction with a mass spectrometer for detection and further elucidation of the eluting solutes.

Because of the intense dependence of the density on the pressure of the system, the solvent power of the mobile phase and, hence, the rate of

TABLE 1. Properties of Selected Mobile Phases in Supercritical Fluid Chromatography

Compound	Atmospheric boiling point (°C)	Critical point data		
		T_c (°C)	P_c (bar)	d_c (g ml^{-1})
Nitrous oxide	-89	36.5	72.3	0.457
Carbon dioxide	-78.5[a]	31.3	73.8	0.448
Methanol	64.7	240.5	79.9	0.272
n-Propane	-44.5	96.8	42.5	0.220
n-Butane	-0.5	152.0	38.0	0.228
n-Pentane	36.3	196.6	33.7	0.232
Dichlorodifluoromethane	-29.8	111.7	39.9	0.558
Trichlorofluoromethane	23.7	196.6	42.2	0.554

[a]Sublimation point.

migration of the solutes can be varied by as much as one order of magnitude or more by relatively small changes in the operating pressure. In contrast to gas chromatography (GC), the supercritical mobile phase has the advantage of possessing a much higher solubilizing power than a gas, allowing higher molecular weight and even ionic compounds to be chromatographed at relatively low temperatures. In comparison to liquid chromatography (LC), the supercritical phase shows more desirable transport properties. This results in a significant reduction in the resistance to mass transfer in the mobile phase. Some popular mobile phases and their properties at the critical point are listed in Table 1. The suitability of various liquids as the mobile phase in SFC has recently been tested by Asche [17]. For a comprehensive treatment of this technique, we refer to the most recent review papers on the subject [3].

II. INSTRUMENTATION

Figure 1 shows a schematic diagram of an assembly for SFC. One may note the similarity of many components to those used in modern, high performance LC.

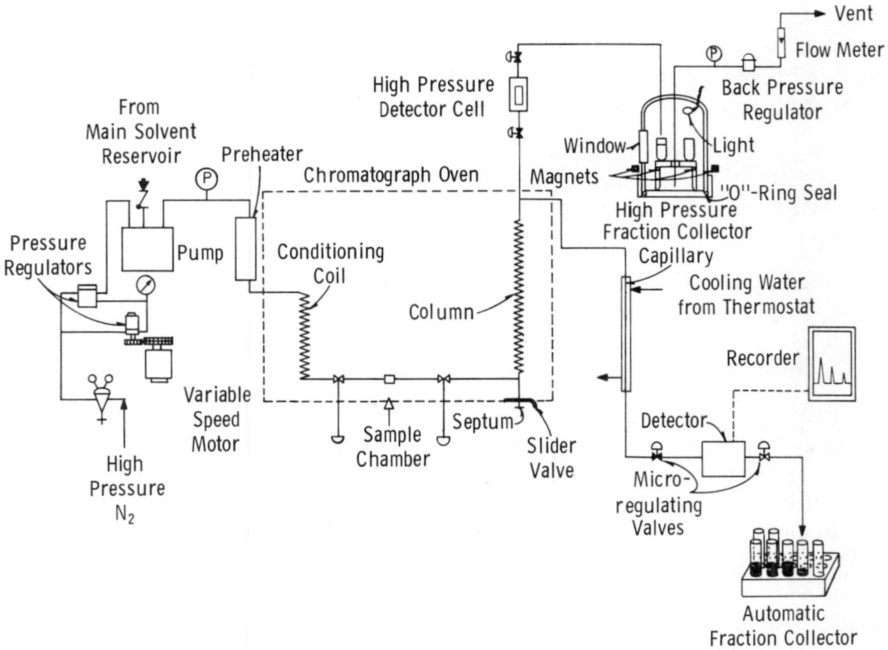

FIG. 1. Schematic diagram of an apparatus for supercritical fluid chromatography. (See also Ref. 8.)

The solvent reservoir depends on the mobile phase used. If the material is liquid at ambient conditions, a simple glass container with reflux condenser suffices; continuous degassing is recommended by, e.g., heating the solvent to gentle boiling. If a compound such as CO_2 or N_2O is used, the mobile phase container is commonly a standard steel cylinder with a dip tube reaching into the bottom of the vessel. This ensures that liquid instead of gas is being fed into the pump.

This mobile phase is compressed and brought to the desired pressure by a high-pressure pump, the output pressure of which can be controlled and programmed. In this particular layout, high-pressure nitrogen or air is used as the driving force. Pressure regulators, one of which is connected to a motor drive, are used to control and program the output pressure. To attain the desired high pressures, these pumps operate on the pressure intensification principle whereby the driving force is directed toward a piston of much larger cross section than the connecting piston on the pressure output side. The syringe, the diaphragm, and the reciprocating piston pump currently in use for most high-resolution LC all operate on a regulated output flow rate basis; the output pressure is observed as a dependent variable. In the gas-driven pressure transfer pump shown here, the output pressure is the variable being controlled; the flow rate is the dependent parameter. A differential pressure transducer can be used to measure the pressure drop over a piece of capillary tubing between the outlet of the pump and the injector to obtain an indication of the flow rate, and to provide the signal to a feedback loop to control the flow rate.

A preheater and/or a preconditioning coil is used to elevate the temperature of the mobile phase to the desired level. Because of the extreme sensitivity to small variations in temperature and pressure, long preconditioning coils are advisable to ensure adequate time for equilibration.

The high pressures in the system and the strong solubilizing power of the compressed liquid present special requirements in the injector design. Only a few of the several commercially available high pressure injectors for LC adequately meet these prerequisites.

After the sample has been fractionated on the chromatographic column, the stream is introduced into a detector and then directed to a sample collection device. Most of the work described with this technique employ a uv absorption detector. Even though this is not a universal- but a specific-type detector, it is suitable for a surprisingly large variety of compounds. SFC cannot compete against GC for the analysis of lower molecular weight compounds which can be volatilized without decomposition. Its superior capabilities lie in the analysis of larger molecules where the probability of a uv absorbing chromophoric group being present in the molecular structure is quite large. The mobile phases employed, on the other hand, are of sufficiently low molecular weight to be transparent in the uv wavelength region of interest.

Two options are shown. For mobile phases which are liquid at ambient conditions, the pressure is reduced to atmospheric levels over a section of capillary tubing and a microregulating valve. Detection and sample recovery are carried out at ambient conditions. When the mobile phase is a gas such as CO_2 or N_2O, detection and sample recovery are carried out under pressure. Commercially available detectors for high-resolution LC are ordinarily not designed to handle high pressures, and specially designed detection systems are required. Ultraviolet detector cells which have been used successfully to several hundred bars have been described [5, 10]. If pressure reduction is carried out at the outlet of the detector, sample collection can be performed by one of the many procedures reported for trapping fractions in preparative GC. An alternative possibility would be to employ a high-pressure fraction collector. In an arrangement reported earlier [5], the effluents are gathered in small test tubes from which the solvent can be carefully evaporated by adjusting the pressure in the collection vessel. If sample recovery from the effluents is not required, one can resort to a modified flame ionization detector to monitor the effluent stream for the presence of solutes [1, 16].

III. SELECTIVITY

The high density and the resultant large intermolecular interactions in the mobile phase in SFC and LC are the main causes for the more pronounced selectivity compared to what can be attained in GC. In the latter technique, selectivity considerations are based essentially on the composition of the stationary phase only. Due to the low density of the gas, the amount of sample components in the mobile phase is determined primarily by the vapor pressure of these solutes.

In both LC and in SFC, the carrier fluid competes actively for the sample molecules. Selectivity can, hence, be controlled by varying the composition of the moving phase. In SFC selectivity can, in addition, quite dramatically be modified by varying the temperature and/or the pressure of the system. Adjusting the latter parameters results in a high degree of operational flexibility.

Type selectivity in SFC is emphasized by the use of a nonpolar mobile fluid and/or a polar stationary phase. Sie et al. [13] report, e.g., that with n-pentane as the mobile phase on alumina as the stationary substrate, phenylnaphthalene elutes at approximately the same time as hexacosyl naphthalene which has 20 additional methylene units. On Polyethylene Glycol 6000 on Sil-O-Cel, the highly paraffinic di-n-tricontylphthalate is eluted well before the diphenylphthalate, which has 48 less carbon atoms [14]. Extreme type selectivities can, therefore, be induced between paraffins and aromatic hydrocarbons, especially condensed aromatic hydrocarbons.

We have mentioned earlier that operating pressures are ordinarily in the 1 to 4 P_c range, where P_c is the critical pressure. Obviously, it is not essential to adhere strictly to this region. By reducing this parameter to well below P_c, the decrease in density of the mobile phase transforms the technique into something more closely related to GC where molecular weight considerations prevail. Rijnders [11] notes, e.g., that with n-pentane and with Polyethylene Glycol 6000 on Sil-O-Cel, the compound 2-n-decylnaphthalene (BP 387°C) elutes before 1,3-dimethylnaphthalene (BP 263°C) at pressures above 39.2 bar. If the pressure is reduced to below 39.2 bar, a reversal in the elution order is noted. Molecular weight effects dominate the separation process in this case. The change in the relative retention times becomes more pronounced as the pressure is further decreased. This phenomenon is obviously accompanied by increasingly larger retention volumes.

Selectivity control can also be achieved by the appropriate choice of solvents. With alumina as the substrate, diethyl ether will cause condensed polyaromatic hydrocarbons to elute before monoaromatic hydrocarbons of the same molecular weight; with n-pentane as the mobile phase, the elution order is reversed [11].

A class of compounds which has been studied in some detail by SFC are the polynuclear aromatic hydrocarbons (PNAH). Some of these compounds have become the focus of considerable attention in the last few years (see Chap. 18). The carcinogenic properties of several specific PNAHs have been known for some time; and, with the increased emphasis on environmental protection, good analytical methods are necessary to quantitatively determine these substances in a wide variety of matrixes. In addition, the petroleum industry is also interested in PNAHs from a process point of view. Many of these compounds have a deleterious effect on certain catalysts; they reduce product yields and quality and shorten the life of the catalyst. Plugs in many process lines can be traced to the high melting point of these compounds. An interesting observation is that these products can often be present in fairly large concentrations in distillate fractions with end points considerably lower than the boiling point of these PNAHs, indicating a strong susceptibility to codistillation with a wide variety of materials.

PNAHs can be chromatographed at slightly above room temperature with supercritical CO_2 as the mobile phase. With Du Pont's Permaphase-ETH as the stationary phase, PNAHs are separated according to ring type. Permaphase-ETH is an ether polymer substrate, chemically bonded to superficially porous microbeads of 30-μ nominal diameter. The parent PNAH and its alkyl-substituted derivatives elute close to each other with the more highly substituted derivatives tending to elute earlier. This ring-type selectivity is illustrated by the chromatogram in Figure 2, which shows the segregation of four PNAHs of 252 molecular weight. This chromatogram is noteworthy because of the known carcinogenicity of benzo(a)pyrene and

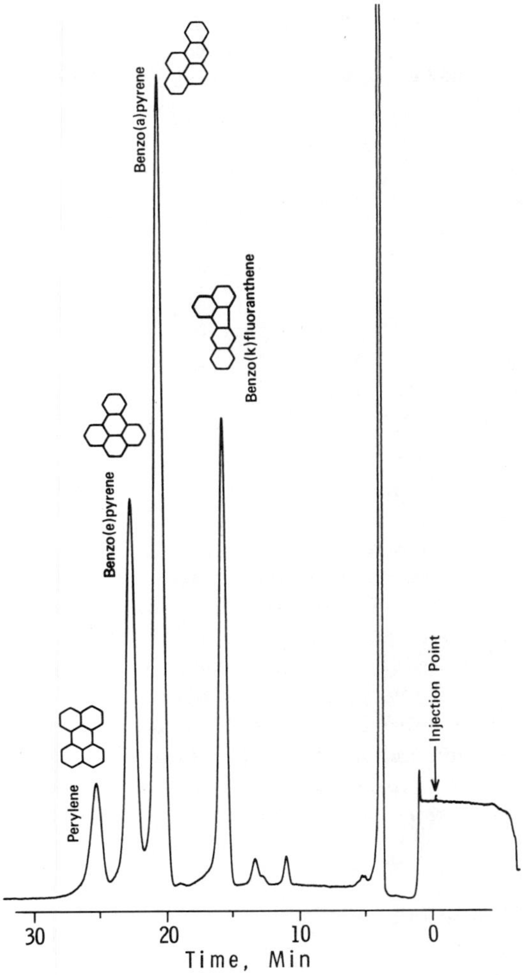

FIG. 2. Separation of 5-ring 252 MW polynuclear aromatic hydrocarbons [7].

because various current and pending legislation refers to the level of this particular substance in background materials. It is interesting to note that the separation of benzo(a)pyrene from benzo(e)pyrene by GC can only be carried out on a column with a substantially larger number of theoretical plates. On most common isotropic phases, the relative retention time for this pair is close to unity. The separation of these two compounds is important because of the widely different carcinogenic properties ascribed to each of these two substances.

When a VYDAC Reverse Phase column is used, however, elution takes place more or less according to molecular weight. Good separations are observed between a parent PNAH and its alkyl derivatives. Alkyl substitution increases the retention times. VYDAC Reverse Phase consists of 30 to 44 μ superficially porous bead packing to which (nonpolar) octadecyl groups are bonded.

IV. APPLICATIONS

In this section we will cover a number of the most interesting applications of SFC. The selection is weighted toward those subjects of possible interest to workers in the petroleum industry.

A. Wide Molecular Weight Range Samples

Because the density of the mobile phase is strongly influenced by the pressure of the system, pressure programming can most expediently be used to process mixtures present in a wide molecular weight range. Figure 3 was published in 1969, but it is still remarkable by current standards. The chromatogram displays as separate peaks the oligomers of a polystyrene mixture with molecular weights ranging from 266 to 1930. Chromatography has been carried out with pentane:methanol (20:1) as the mobile phase on a stationary sbustrate of Porasil C to which n-octyl groups have been bonded. The molecular structure of each oligomer can be described by $CH_3(CH_2)_3-(CH_2CHC_6H_5)_n-H$, where n is the number of styrene molecules in the polymer [6]. More recent work on these systems has been described by Klesper and Hartmann [8].

B. Polynuclear Aromatic Hydrocarbons in Automobile Exhaust

SFC has been found to be an excellent instrument for the isolation of very pure fractions of benzo(a)pyrene [B(a)P] and benz(a)anthracene [B(a)A] from automobile exhaust. This technique is used for the unambiguous determination of the level of these compounds in the original sample. In the actual process, a known fraction of the automobile exhaust is diverted into an analytical train where all the organic material is recovered by a combination of cooling, scrubbing with cold methylcyclohexane, and filtration through a Millipore filter [9]. After reduction of the sample size to a few hundred milliliters by distilling of the excess solvent, known amounts of tritiated B(a)A and B(a)P are added as internal standards [7]. At this point, the sample is still much too complex and the levels of the B(a)A and the B(a)P are still very low. Concentration of these PNAHs is carried out by extraction

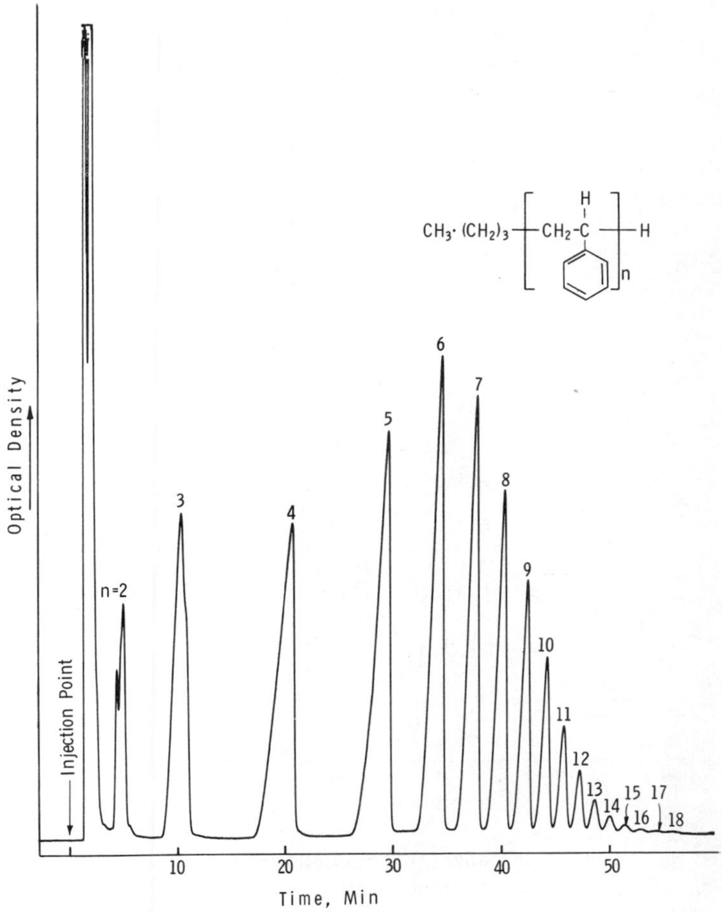

FIG. 3. Chromatogram of a "monodisperse" polystyrene of 600 nominal MW [6].

with dimethyl formamide (DMF), followed by silica gel column chromatography. The B(a)A- and B(a)P-rich fraction is subsequently subjected to SFC for the final purification step.

Figure 4 shows a typical chromatogram of this material. The detector is first set at 280.5 nm to obtain maximum sensitivity for B(a)A; the wavelength is then shifted to 372 nm for the optimum response to B(a)P. B(a)A and B(a)P are isolated and recovered in a high degree of purity by collecting the effluents in a high-pressure fraction collector and by carefully evaporating off the solvent. Quantitation is by uv spectroscopy and

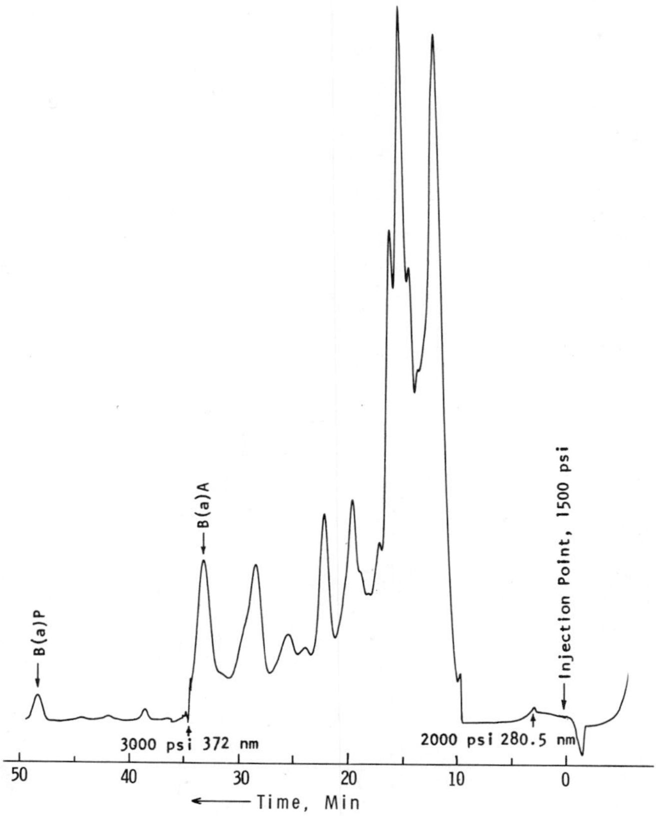

FIG. 4. Isolation of benzo(a)pyrene and benz(a)anthracene from a PNAH-rich fraction from automobile exhaust [7].

by scintillation counting. Ultraviolet spectroscopy determines the absolute amounts of B(a)A and B(a)P in the recovered fractions; the radioactive levels assayed by the latter technique define the aliquot of these PNAHs recovered in the sample work-up. The lower detection limit under regular conditions is a few parts per billion in the original liquid sample.

C. Polynuclear Aromatic Hydrocarbons in Refinery Solid Waste

In another related application, SFC is also employed as the final purification stage in the analysis of B(a)P and B(a)A in refinery solid waste. This matrix is an undefined mixture of sludge, hydrocarbon-water emulsions, and organic and inorganic dirt from various sources such as from settlers, holding tanks, recovered oil spills, etc. This material is difficult to utilize in regular refinery operations because of its inhomogeneous composition. The constitution of the PNAH components in this mixture is much more complex than in the products from automobile exhaust. Because of the high temperatures in the combustion process, only parent and simple alkyl substituted polynuclear aromatic hydrocarbons are observed to be present in the latter case.

The work-up procedure as described for the previous subject is inadequate to yield sufficiently simple mixtures of PNAHs to allow for an isolation of B(a)A and/or B(a)P in high purity by SFC. One major drawback is the relatively large proportion of alkylated PNAHs in the original sample. Another encumbrance is the relatively high total concentration of these compounds.

The successful approach to this problem again utilizes tritiated B(a)A and B(a)P as internal standards to monitor the fate of these PNAHs during the development of the many steps involved in the procedure. Without these radioactive tracers, development of an appropriate analytical method would have been extremely difficult. After addition of these internal standards, the sample is first subjected to column chromatography over a bed of specially deactivated clay. This step removes the resins and the particulate material. Concentration is carried out by DMF extraction, followed by chromatography over porous polystyrene beads (Poragel). This step is necessary to remove a major portion of the substituted alkyl derivatives from the parent PNAHs. The fractions high in radioactivity are then subjected to chromatography over alumina, and SFC is again relied upon to carry out the final isolation of pure B(a)A and B(a)P. A quantitative measure of the amount of these two hydrocarbons in the original mixture is obtained by assaying the recovered fractions by uv spectroscopy and by scintillation counting.

D. SFC of Coal Tar Pitch

Sie and Rijnders [15] obtained the chromatogram shown on the left in Figure 5 with isopropanol as the supercritical mobile fluid and alumina as the substrate. The investigated sample is a coal tar pitch commonly used in bitumen blending. The expected location of various condensed polynuclear hydrocarbons are indicated. These substances are known to be present in coal tar pitch. The chromatogram shows a series of distinctly separated peaks. These would roughly correspond to groupings of molecules of increasing molecular weight and ring number. The last two broad peaks probably represent 7- and 8-ring aromatic compounds.

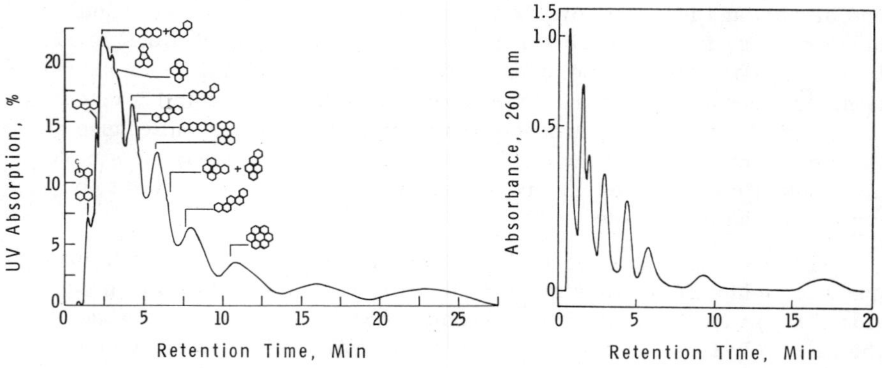

FIG. 5. Chromatograms of coal tar.

The chromatogram on the right in Figure 5, on the other hand, was obtained by Fuzita and Shimokobe et al. [2] with n-hexane on Alumina W-200 (neutral). Even though a nonpolar solvent is used, it is interesting to note that the chromatograms from the two different runs are essentially similar. The results obtained by the Japanese workers do show a somewhat improved resolution.

E. SFC of a Lube Oil Fraction

Rijnders described the chromatographic separation of a narrow, high boiling distillate fraction (345 to 350° C) of a mineral oil fraction on alumina with n-pentane as the supercritical mobile phase [12]. Although no single component peaks were observed, the procured chromatogram was sufficiently expanded to display a fair number of broad composite peaks. Ten different cuts were isolated in this process. Ultraviolet spectroscopy on the individual fractions showed that a distinct separation between different molecular structural types can be attained by this technique. The composition of the recovered products were as follows:

Fraction 1 Alkylbenzenes

Fraction 2 Alkylbenzenes and substituted alkylbenzthiophenes

Fraction 3 Alkyl substituted benzthiophenes and alkylnaphthalenes

Fractions 4 and 5 Alkylnaphthalenes

Fraction 6 Alkyl substituted dibenzthiophenes

Fractions 7 and 8 Alkyl substituted dibenzthiophene and phenanthrene

Fraction 9 Alkyl substituted phenanthrenes

F. SFC-Mass Spectrometry (MS)

One of the most recent developments in this field is to interface a SFC to a mass spectrometer [4]. This approach extends the applicability of GC-MS, an established and highly successful technique, to high molecular weight and/or thermally labile compounds. The potential of this combination method is almost equivalent to the extension of the scope of chromatography from GC to GC plus high-resolution mass spectrometry. We may note that LC-MS has not been very successful so far. Principally this is because the relatively high molecular weight of the mobile phase has hampered construction of a practical and effective interface to separate the solutes from the excess of solvent. In SFC-MS, one can choose a mobile phase which is gaseous at ambient conditions.

Figure 6 shows a so-called spectrum map, obtained by SFC-MS, of a synthetic mixture of n-paraffins and a number of miscellaneous other compounds. Chromatography has been carried out on a 3-1/2 m by 1/8-in. OD stainless steel column packed with Du Pont's Permaphase ETH packing. CO_2 is used as the mobile phase. The column temperature is maintained at slightly above ambient levels. Pressures are indicated on the spectrum map. The interface consists of a splitter with a swaged down 0.01-in. ID capillary tube inserted into one leg of a tee attached to the outlet of the column. The other end of the capillary tube leads to the mass spectrometer inlet.

The spectrometer is a Nuclide 90° magnetic sector instrument, operated under control of an INCOS computer system. In this particular case, the spectrometer is operated in the multiple ion detection mode, which allows for more sensitivity and resolution. In this specific method, the magnetic field is maintained at a constant value; and masses are scanned by programming the accelerating voltages. Scan cycle times are usually one second. In this particular run, the spectrometer is set to scan only m/e values of 128, 138, 142, 148, 170, 178, and 184. The spectrum map displays, in essence, the various chromatograms at these m/e values superimposed on each other and slightly offset to prevent parent peaks and fragments from overlapping. The spectrum map can also be considered as an orthogonal projection of a three-dimensional diagram.

An interesting note is that this is the first chromatogram by SFC showing the presence of n-paraffins; uv detectors have almost exclusively been used previously to monitor effluents from a SFC. These results show that under the chromatographic conditions employed, the C_9-C_{13} n-paraffins are barely retained. Saturated cyclic compounds also show very short elution times. Polycyclic aromatic hydrocarbons, on the other hand, are retained very strongly on this column.

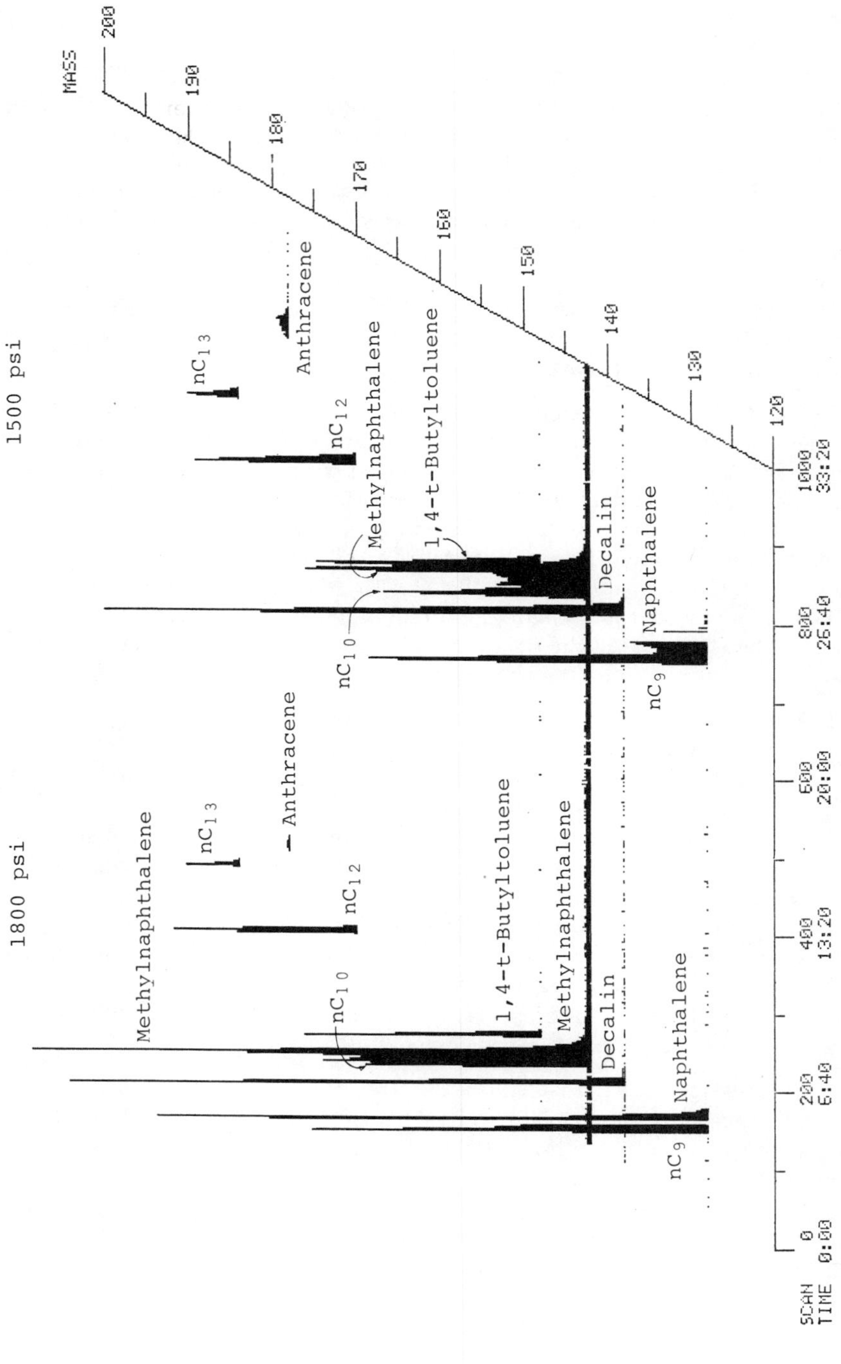

FIG. 6. Spectrum map of synthetic mixture of n-paraffins and other compounds by SFC-MS [4].

ACKNOWLEDGMENTS

The authors gratefully acknowledge the courtesy of the following copyright holders for permission to include these figures in their chapter: The American Chemical Society for Figures 2 and 4 from Analytical Chemistry; John Wiley and Sons, Inc., for Figure 3 from Polymer Letters; and Plenum Press for Figure 6 from the Proceedings of the 6th AIRAPT International High Pressure Conference in Boulder, Colo., 1977.

REFERENCES

1. D. Bartmann, Ber. Bunsenges, Phys. Chem. 76:336 (1972).
2. K. Fuzita, I. Shimokobe, and F. Nakazima, Nippon Kagaku Kaishi 8: 1348 (1975).
3. T. H. Gouw and R. E. Jentoft, J. Chromatog. 68:303 (1972); Advances in Chromatography (J. C. Giddings, ed.), 13, 1 (1976).
4. T. H. Gouw, R. E. Jentoft, and E. J. Gallegos, Proc. Sixth AIRAPT High Pressure Conf., Boulder, Colorado, I, 583 (1977).
5. R. E. Jentoft and T. H. Gouw, Anal. Chem. 44:681 (1972).
6. R. E. Jentoft and T. H. Gouw, J. Polym. Sci. B(7):821 (1969).
7. R. E. Jentoft and T. H. Gouw, Preprints, Div. Petrol. Chem., ACS, 20(4):797 (1975); Anal. Chem. 48:2195 (1976).
8. E. Klesper, W. Hartmann, Polymer Letters 15:9 (1977); 15:707 (1977).
9. H. K. Newhall, R. E. Jentoft, and P. R. Ballinger, SAE Paper No. 730834, presented at SAE Meeting, Milwaukee, Wisconsin, September 10-13, 1973.
10. J. A. Nieman and L. B. Rogers, Separ. Sci. 10:517 (1975).
11. G. W. A. Rijnders, Fifth International Symposium on Separation Methods: Column Chromatography, Chimia Suppl., 1969, p. 192.
12. G. W. A. Rijnders, Chem. Ing. Tech. 42:890 (1970).
13. S. T. Sie, W. van Beersum, and G. W. A. Rijnders, Separ. Sci. 1:459 (1966).
14. S. T. Sie and G. W. A. Rijnders, Separ. Sci. 2:729 (1967).
15. S. T. Sie and G. W. A. Rijnders, Separ. Sci. 2:755 (1967).
16. O. Vitzthum, P. Hubert, and M. Barthels, U.S. Pat. 3,827,859, August 6, 1974.
17. W. Asche, Chromatographia 11:411 (1978).

Chapter 14

THIN LAYER, PAPER AND HIGH-PERFORMANCE LIQUID CHROMATOGRAPHY

R. Amos
Analytical Department
Esso Petroleum Company Ltd.
Esso Research Centre
Abingdon, England

I. INTRODUCTION

Thin layer (TLC), paper (PC), and high-performance liquid chromatography (HPLC) are three forms of liquid chromatography which are becoming increasingly important in the analysis of petroleum and its associated products.

The three techniques should be discussed together because they are complementary to one another. TLC and PC are invaluable for the qualitative analysis of lipophilic and hydrophilic materials, respectively, and are useful for setting up and monitoring HPLC systems. HPLC is being developed into a reliable technique for routine high-speed quantitative analyses as well as a preparative tool for the isolation of unknown fractions for further analysis. As well as being complementary to TLC and PC, HPLC is also complementary to gas chromatography (GC). It should be recalled that only 15 to 20% of known organic compounds are sufficiently volatile to enable them to be analyzed by GC, and consequently the petroleum laboratory of the future may well contain more liquid chromatographs than gas chromatographs.

In this chapter, we will first describe the apparatus, materials, and technique of TLC, PC, and HPLC in turn, and then discuss some of their applications. Two of their most important applications so far, i.e., the determination of additives in lubricants and the examination of polynuclear aromatic hydrocarbons, are discussed in Chap. 17 and 15, respectively. This chapter will conclude with an appreciation of the relative merits of the three techniques and a brief discussion about their future.

II. HISTORICAL

Curiously enough, the chronological order in which the three techniques have achieved their popularity within the petroleum industry is almost in the reverse order in which they were invented. Prior to about 1960, paper chromatography was the most dominant of the liquid chromatographic techniques to be used, albeit not on a wide scale. The notable exception, of course, was the FIA analysis (see Sec. V.B). Paper chromatography was invented in 1944 by Consden and Gordon et al. [1]. This technique in which the sample is partitioned between an aqueous mobile phase and water held by the cellulose support, is most suitable for the separation of hydrophilic substances. It enabled rapid and substantial advances to be made in the biological, medical, and pharmaceutical fields, but had little impact on petroleum chemistry where most samples are hydrophobic in character.

Prior to this, Izmailov and Shraiber [2] separated plant extracts into concentric zones by applying drops of various solvents to the samples which had been spotted onto adsorbent-coated microscope slides. But TLC did not come into popular use until the efforts of Stahl [3, 4] resulted in standardized adsorbents and equipment becoming commercially available. In contrast to PC, TLC is eminently suitable for the separation of lipophilic components and Killer and Amos [5] demonstrated the usefulness of this technique in the petroleum industry and described a wide range of applications.

Finally, the last 10 years have seen the rebirth of liquid chromatography in columns originated by Tswett [6]. "Modern" liquid chromatography in columns, variously known as high-speed, high-resolution, high-efficiency, high-pressure, or high-performance liquid chromatography (HPLC) has emerged from the fusion of various chromatographic techniques which had originally developed along independent lines.

The development of gel permeation chromatography and its associated equipment, the development and use of linear elution adsorption chromatography by L. R. Snyder [7] and his colleagues in the determination of the composition of the heavier petroleum fractions, the potential and limitations of thin layer chromatography, and, finally, applications of the theories and practices derived from gas chromatography have all contributed to a renaissance of liquid chromatography in columns.

III. THIN LAYER AND PAPER CHROMATOGRAPHY

It is appropriate to consider TLC and PC together because it is probable that paper sheets as such are no longer utilized within the petroleum industry, except perhaps, in "blotter-spot" tests. Where the use of cellulose as stationary phase is desirable, as, for example, in the separation of certain polynuclear aromatic hydrocarbons, cellulose-coated glass plates or plastic sheets are now preferred. As a consequence, the apparatus and techniques of PC and TLC have become indistinguishable, even though the mechanisms of the separation differ. It must be respected, however, that PC is primarily used for the separation of hydrophilic substances and is extensively practiced with great success in other fields of chemistry and biochemistry.

The main attraction of TLC is that it utilizes inexpensive and portable equipment which is simple to use and a typical experiment only takes 20 to 30 min. A further advantage over column chromatography is that after separation, the whole of the sample is available on the plate for inspection (assuming the chromatogram has been adequately revealed and that volatile components are absent). This is particularly important when sample components may be strongly adsorbed and remain at or near the start of the chromatogram. Such components would be missed in column chromatography, since they would not have eluted from the column and passed through the detector. A further advantage of TLC is the use of specific chemical revealing reagents which react with certain compound classes to yield characteristically colored products. This can give a valuable insight into the composition, as well as complexity, of an unknown mixture.

The technique involves applying, by micropipette, 3 to 5 μl portions of 0.1 to 1% solutions of the components of interest as spots 15 mm above the lower edge of a plate coated with a layer of adsorbent 0.1 to 0.3 mm thick. The plate is then placed in a closed glass jar (developing chamber) just big enough to hold the plate in a vertical position. The plate dips 10 mm deep into the mobile phase which then ascends through the adsorbent by capillary action. In doing so, the solvent takes sample components at different speeds through the layer according to their distribution coefficients and thus separates them (Figure 1).

Development is stopped by removing the plate from the developing tank when the mobile phase has risen to a predetermined position (the solvent front). The plate is allowed to dry and the positions of the spots are located (if not already visible) either by viewing under uv light or by spraying the plate with a chromogenic reagent. The number of chromogenic reagents useful to the petroleum chemist is fairly limited. Lists and further details will be found in Sec. V of this chapter and Sec. III in Chap. 17.

Currently, precoated TLC plates are popular because they can be used straight from the packet without any further preparation. Plate sizes are normally 20 cm by 20 cm, 20 cm by 10 cm, and 20 cm by 5 cm, which

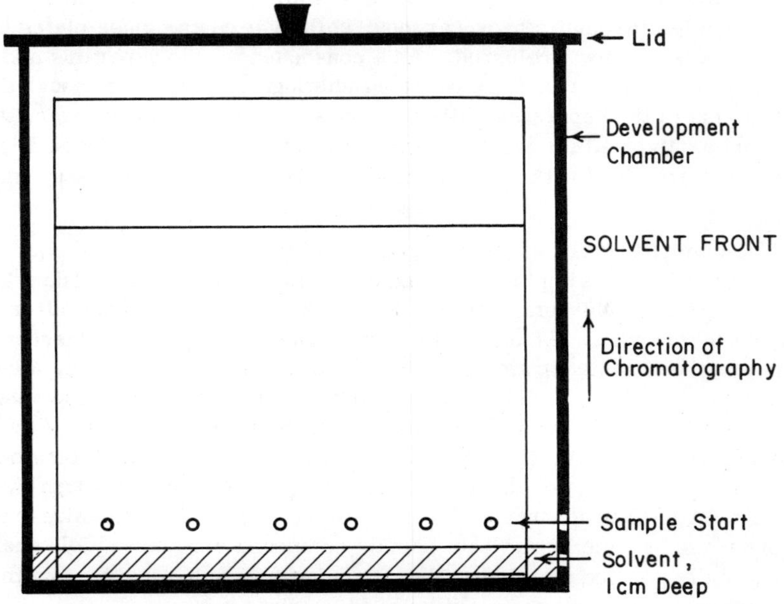

FIG. 1. 20- × 20-cm TLC plate in development tank.

accommodates about 16, 8, and 4 samples, respectively. Stationary phases may be supported by metal foil, plastic sheet or glass-plate. The latter is preferable on account of its rigidity and its ability to accommodate thicker layers (0.1 mm to 1.5 mm). Some types contain an organic binder to make them abrasion resistant (i.e., they can be written on by pencil) and some have a fluorescent indicator incorporated to aid visualization of colorless spots under uv light. The presence of such additives may, on occasion, interfere with certain visualization methods. Some organic binders react with aggressive revealing reagents, e.g., concentrated sulfuric acid, while zinc phosphors will interfere with the revelation of zinc dialkyldithio-phosphates by zinc indicators such as dithizone.

Plates precoated with silica, alumina, kieselguhr, magnesium silicate polyamide, cellulose, and modified celluloses are readily available from a number of sources and have been summarized recently by Macek and Becvarova [8].

Precoated plates have largely eliminated the need to prepare one's own plates, which is a space- and time-consuming process and also requires considerable skill and experience. They also give more reproducible results than homemade plates. Methods of coating one's own plates can be gleaned from standard references text books on TLC such as Stahl [9] or Kirchner [10].

The choice of solvent as mobile phase depends on the separation mechanism to be exploited and on the chemical nature of the sample. The mobile phase should have a low viscosity so that it travels through the stationary phase quickly, and low volatility so that after development the plate can be quickly dried. The solvent should also be inert. (See Table 1 on page 341.)

To find the right mobile phase for adsorption chromatography, one would try out a series of solvents of gradually increasing polarity, using mixtures if necessary, until the components of interest lay within the range of R_f values of 0.3 to 0.8. A suitable series of solvents, in order of increasing eluent strength, is n-pentane, carbon tetrachloride, toluene, diethyl ether, chloroform, ethyl acetate, pyridine, methanol, acetic acid, and ammonia. Pentane and carbon tetrachloride will separate saturated hydrocarbons from aromatic hydrocarbons, while toluene, ether, and chloroform will carry hydrocarbon base oils as one spot along with the solvent front, leaving the more polar additives behind them. Those additives that are so polar that they do not move even in chloroform-ethyl acetate mixtures are best separated by a double development. The first development is carried out in toluene so that the base oil is carried right to the top of the plate, leaving the polar additive behind at the starting point. A second development is carried out until the solvent front has reached halfway up the plate using a mobile phase incorporating a very polar solvent, such as pyridine, acetic acid, or aqueous ammonia. Care must be taken when using polar solvents that the phenomenon of "solvent demixing" does not occur. For example, it might be thought that a suitable mixture for the separation of sulfonic acids would be methyl acetate containing small amounts of ammonia. However, during development, the mobile phase demixes into its two components to form an upper solvent zone rich in methyl acetate followed by a secondary, aqueous, solvent zone. This problem can be overcome by incorporating a mutually miscible bridging solvent, such ethanol, into the mobile phase. Plates developed in polar eluents such as those containing ammonia, acetic acid, or pyridine need to be dried by heating in an oven at 110°C for at least 30 min to completely remove the strongly adsorbed solvent before spraying with the revealing reagent.

The position of a separated spot on a developed chromatogram is described by R_f value, which is defined as

$$R_f = \frac{\text{Distance moved from point of application by solute}}{\text{Distance moved from point of application by mobile phase}}$$

i.e., the R_f value of all adsorbed solutes is less than 1.00.

For convenience, the R_f values are sometimes quoted as hR_f which are defined as $100 \cdot R_f$. An alternative method for measuring the position of a spot is to relate it to the position of a reference compound x, chromatographed simultaneously. Its position is then given by the R_x value defined as

$$R_x = \frac{\text{Distance moved from point of application by solute}}{\text{Distance moved from point of application by compound x}}$$

In contrast to R_f values, R_x values of greater than 1 are possible. There are many factors that can affect the measurement of R_f value and consequently reproducibility from laboratory to laboratory is rather poor; these factors include the chemical nature, surface area, particle size and activity of the stationary phase, layer thickness, purity of the mobile phase, degree of exposure of the adsorbent to solvent vapor in the developing tank, effect of temperature, position of starting point in relation to depth of mobile phase in developing tank, development time, mobile phase velocity, sample size, and complexity. Nevertheless, tables of R_f values are published regularly in, for example, Journal of Chromatography, and these are useful because they indicate whether or not it is possible to separate a particular group of substances, and they offer a guide to the use of the chromatographic system for attempting a particular separation for the first time.

Tables of R_f values still do not give the whole story. The shapes and sizes of spots and drawings of actual chromatograms obtained probably give the most useful information. Examples of both shall be given later.

The best uses of TLC and PC lie in qualitative and semiquantitative analysis and in combination with other ancillary techniques, particularly HPLC. Examples of the first two applications are in monitoring the synthesis of additives and, conversely, following the fate of additives during service. Experimental conditions are first established for obtaining thin layer chromatograms of the starting materials. Samples are taken at regular intervals and the disappearance of the starting materials and the appearance of fresh spots due to products are noted. If the samples and a series of standard solutions are chromatographed together on the same plate, then visual comparison of the spot size and intensities can give a semiquantitative indication of the amounts present to within about ±20%.

The literature on TLC and PC is replete with methods for improving the accuracy and precision of quantitative analysis by these techniques. The most prominent of these include densitometry and fluorimetry in which the amount of material on the layer is assessed directly, and spectroscopic analysis follows elution of the spot from the layer. The amount of extra work and instrumentation necessary to achieve only a moderate increase in accuracy detracts from the basic simplicity of TLC and PC so that HPLC is rapidly gaining recognition as being the primary tool for quantitative analysis in liquid chromatography. For the interested reader, further accounts of quantitative analysis in TLC and PC have been given by Shellard [11]. Similarly, the recovery of fractions for further analysis is much more simply achieved by HPLC. It can be seen then, that TLC is a valuable screening technique for HPLC analysis.

IV. HIGH-PERFORMANCE LIQUID CHROMATOGRAPHY

A. Background

HPLC is the modern development of classical liquid chromatography in columns invented by Tswett [6] and revived some 20 years later by Kuhn and Lederer [12]. The next 30 years produced an enormous number of applications, especially in the natural products field, but no real advance in technique. Meanwhile, Martin and Synge [13] invented liquid-liquid partition chromatography, conceived the theoretical plate model of chromatography "and predicted that high performance liquid chromatography would be achieved by using very small particles and a high-pressure difference across the length of the column." Such theories were developed and put into practice in the technique of gas chromatography with such an enormous effect in the petroleum industry that liquid chromatography was virtually forgotten during the 1950s. By 1960, however, the usefulness of TLC as a simple and rapid technique had emerged, but comparison with gas chromatography revealed its limitations. The tremendous success of gas chromatography had provided the impetus to apply its theories and practices to classical liquid chromatography and the result has been a massive revival of interest and activity in this, the original chromatographic technique.

Among the features of gas chromatography which are required by HPLC are high resolving power, short analysis times, ease and simplicity of operation, continuous monitoring of column effluent, precise identification based on accurate retention measurement, accurate quantitative measurement, repetitive analysis with the same column and automation of the complete analysis with data handling facilities. All these requirements can now be readily achieved in HPLC.

The essential features of a liquid chromatograph are outlined in Figure 2. Solvent, degassed by prior refluxing or sparging with helium if necessary, is gravity fed to a high-pressure pump, equipped with a pressure gauge. 1 to 10 μl sample solution is injected directly into the column, typically 10 to 50 cm long, 2 to 5 mm internal diameter, containing stationary phase of narrow particle size range, somewhere between 5 and 40 μm in diameter. Mobile phase is forced through at a constant flow rate, between 0.5 and 3.0 ml min^{-1}. Typical operating pressures are often between 200 and 1000 psig, although pumps capable of operating up to 6000 psig are now available. Separated components, as they emerge from the column are measured by a continuously-recording detector, usually a uv monitor or differential refractometer. The column effluent passes to waste or to a fraction collector if further identification work is required.

An excellent introduction to HPLC has been given by Snyder and Kirkland [14] to which both the beginner and more experienced are referred. Their book also gives recommendations for further reading. We shall confine ourselves to a brief outline of the technique, materials, and equipment before discussing some applications in the petroleum industry. Four separation mechanisms are possible: adsorption, partition, exclusion, and ion exchange. The first two mechanisms only are discussed in this chapter. Gel permeation chromatography is the subject matter of Chap. 12. As far as we are aware, high-performance ion exchange chromatography has not yet been used in the petroleum industry, although as Chap. 17 points out, a clear need exists.

B. Columns

The heart of the liquid chromatograph is the column and measurements of its performance are similar to those used in gas chromatography. The evaluation of a chromatogram can be considered in terms of just completely resolving the two most difficult-to-separate components, A and B, of a complex mixture, in the shortest possible time.

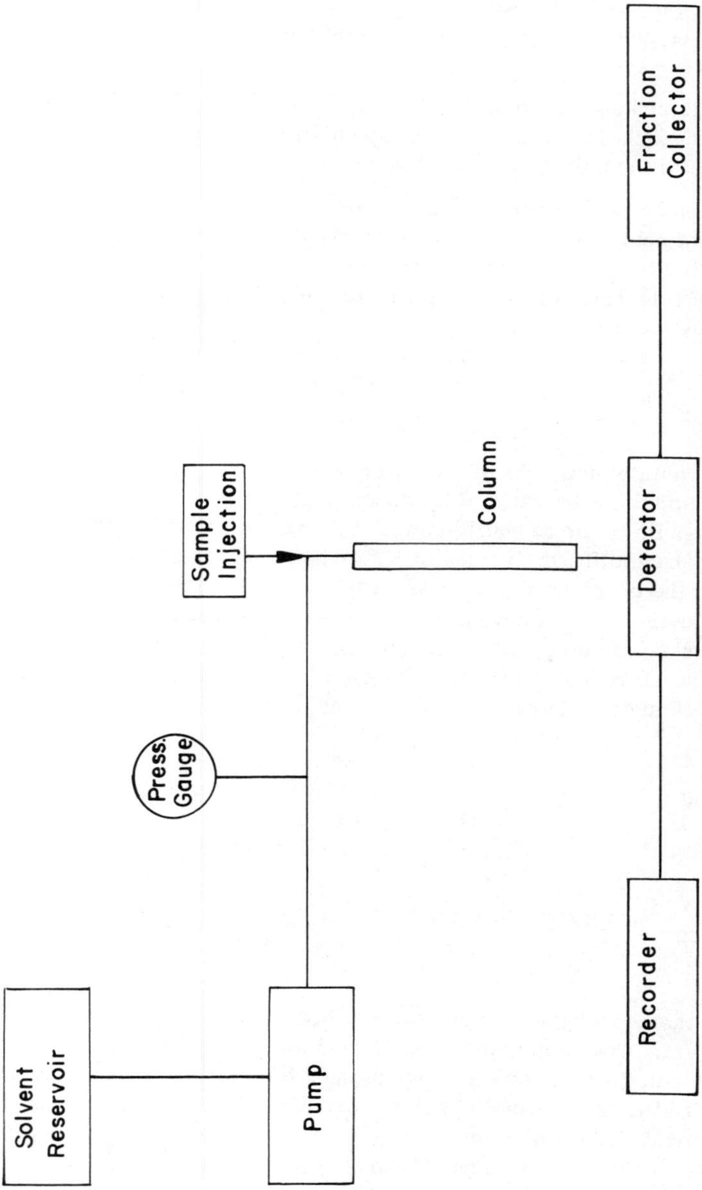

FIG. 2. Schematic of high-performance liquid chromatograph.

Suppose, in Figure 3, the sample is injected at point x and after time t produces the chromatogram shown, in which the ordinate indicates the detector response, and the abscissa is proportional to both the analysis time and the volume of mobile phase that has passed through the column after time t.

An unretained solute would emerge after volume V_o of mobile phase has passed through the column. This is equal to the total volume V_M of mobile phase in the column (the dead volume).

Component A is retained until a volume of mobile phase V_r^A (the retention volume of component A) has passed through the column and is conveniently measured by retention time or retention distance (i.e., mm of chart paper). The retention volume is related to the distribution coefficient K_A of solute A by the expression

$$V_r^A = V_M + K_A V_S \tag{1}$$

where V_S = volume or weight of stationary phase. The distribution coefficient K of a solute is the ratio of its concentration in the stationary phase to that in the mobile phase at equilibrium. Components A and B are separated because they have different values of K. It is a characteristic property of the solute in the phase pair chosen and varies in a systematic way with variation in structure. Measurement of V_r can therefore be used to tentatively identify an unknown compound, i.e., by chromatographing, under the same conditions, a reference compound suspected to be identical to the unknown component. Equation (1) can be rewritten as

$$V_r^A = V_M (1 + k') \tag{2}$$

where

$$k' = K \frac{V_S}{V_M} = \frac{\text{Amount of solute in stationary phase}}{\text{Amount of solute in mobile phase}} = \frac{V_r - V_M}{V_M}$$

k' is called the capacity factor and can be measured directly from the chromatogram. Optimum separations are achieved when k' lies between 2 and 6. If k' = 0, then the solute is unretained by the stationary phase and $V_r = V_M$. If k' is large then the solutes are strongly retained and emerge as broad diffuse bands which may not be seen by an insensitive detector. The expression $1/(1 + k')$ is identical to the R_f value in PC and TLC.

In HPLC, in contrast to gas chromatography, we have an almost infinite number of ways of adjusting k' values. Firstly, we can decide which of the

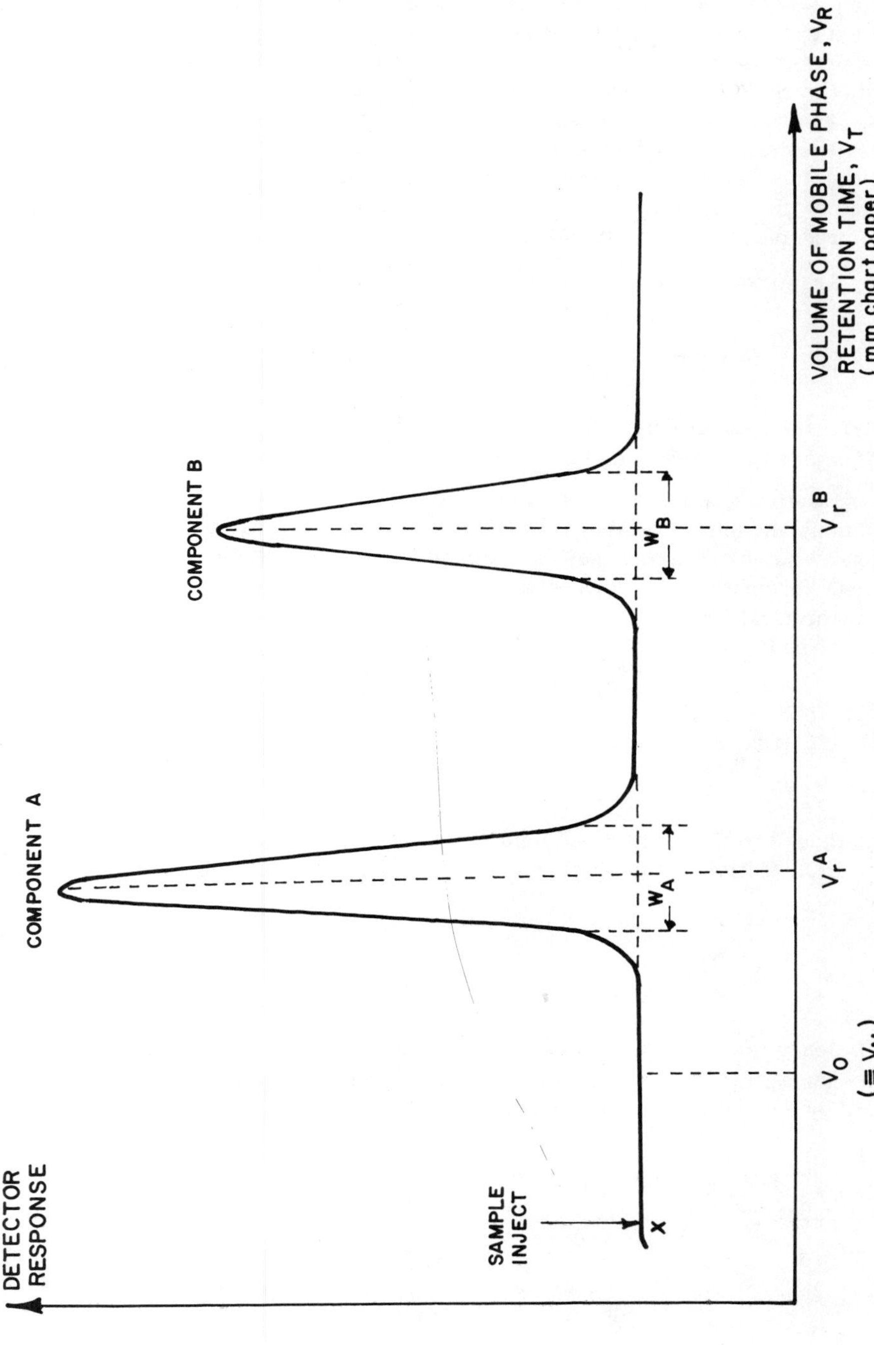

FIG. 3. Definitions for liquid chromatography.

four basic separation mechanisms, adsorption, partition, gel permeation, or ion exchange, is most likely to achieve the required separation; stationary phases are readily available for all four types of chromatography; second, the choice of solvent mixtures as mobile phases is infinitely large, so that, in practice, other considerations such as properties affecting column efficiency and detector sensitivity, chemical reactivity, and volatility can dictate the final choice of mobile phase composition to achieve the desired value of k'. A list of solvents, suitable for both HPLC and TLC of petroleum products, together with some significant properties, is given in Table 1.

From Eq. (2) we can derive an expression for the analysis time, t_r

$$t_r = \frac{L}{u}(1 + k') \tag{3}$$

where L = column length and u = mobile phase velocity measured by L/t_o and where t_o corresponds to V_o in Figure 3.

Equation (3) tells us that quicker analyses are obtained by using short columns and high flow rates. To be able to do this, we also need high efficiency and resolution. These two terms are measured in the same way as in gas chromatography. Efficiency is expressed in terms of N, the number of theoretical plates. N generated by the column for component A in Figure 3 is given by

$$N = 16\left(\frac{V_r^A}{W_A}\right)^2 \tag{4}$$

and therefore the length of column per plate, i.e., plate height, is given by H = L/N. Efficiency is often quoted in terms of "effective plates" measured by

$$N_{eff} = N\frac{(k')^2}{(1 + k')} \tag{5}$$

Efficiency tells us how sharp the peaks are, but gives no information as to whether a pair of components have actually been separated. This is measured by resolution, rs, defined as

$$rs = \frac{2(V_r^B - V_r^A)}{W_A + W_B} \tag{6}$$

TABLE 1. Eluotropic Series for Polar Adsorbents [7]

Solvent	$\epsilon_0(Al_2O_3)$	Viscosity, cP (20°C)	RI	UV cutoff (nm)	Boiling point (°C)
Fluoroalkanes	-0.25	—	1.25	—	—
n-Pentane	0.00	0.23	1.358	210	36
i-Octane	0.01	0.54	1.404	210	118
n-Heptane	0.01	0.41	1.388	210	98.4
Cyclohexane	0.04	1.00	1.427	210	81
Cyclopentane	0.05	0.47	1.406	210	49.3
Carbon tetrachloride	0.18	0.97	1.466	265	76.7
Amyl chloride	0.26	0.43	1.413	225	108.2
i-Propyl ether	0.28	0.37	1.368	220	69
i-Propyl chloride	0.29	0.33	1.378	225	43.8
Toluene	0.29	0.59	1.496	285	110.6
n-Propyl chloride	0.30	0.35	1.389	225	46.6
Benzene	0.32	0.65	1.501	280	80.1
Ethyl ether	0.38	0.23	1.353	220	34.6
Chloroform	0.40	0.57	1.443	245	61.2
Methylene chloride	0.42	0.44	1.424	245	41
Tetrahydrofuran	0.45	0.55	1.408	220	65
Ethylene dichloride	0.49	0.79	1.445	230	84
Methyl ethyl ketone	0.51	0.43	1.381	330	79.6
Acetone	0.56	0.32	1.359	330	56.2
Dioxane	0.56	1.54	1.422	220	104
Ethyl acetate	0.58	0.45	1.370	260	77.1
Methyl acetate	0.60	0.37	1.362	260	57
Nitromethane	0.64	0.67	1.394	380	100.8
Acetonitrile	0.65	0.37	1.344	260	57
Pyridine	0.71	0.94	1.510	305	115.5
i-Propanol	0.82	2.3	1.38	210	82.4
Ethanol	0.88	1.20	1.361	210	78.5
Methanol	0.95	0.60	1.329	210	65.0
Acetic acid	large	1.26	1.372	251	118.5

It will be noted that resolution includes both a selectivity contribution (in V_r) and a performance factor (in peak width). The value of rs indicates the maximum number of resolved components that could be inserted between components A and B. When rs = 1, two components are just completely separated, i.e., with only about a 2% overlap of one peak over another. An excellent account on the control of separation has been given by Snyder and Kirkland [14].

The factors affecting column performance can be expressed in some form of the well known van Deemter equation [15]. According to Done et al. [16] the form of plate height equation which now seems best to represent the available experimental data is

$$H = Au^{0.33} + \frac{B}{u} + Cu \qquad (7)$$

The A term represents flow path irregularities, the B term axial diffusion, and the C term interphase mass transfer. The aim, of course, is to minimize H while maintaining u as high as possible for fast analyses. In liquid chromatography, the B term is low and, for practical purposes, can be ignored. The C term in Eq. (7) tells us to use small, spherical particles with minimum size distribution. But smaller particles are more difficult to pack and a poorly packed column leads to an increase in the A term.

Equation (7) also tells us to avoid viscous or polymeric stationary phases and that supports or adsorbents should have pore sizes greater than about 30 A so as to reduce the mass transfer term. For the same reason, the mobile phase should have a low viscosity.

Progress in liquid chromatography over the years can be followed in terms of the capability of packing columns of higher and higher efficiency. The classical columns of Tswett, which are still widely used today, have efficiencies of only a few hundred plates, while 300 to 1000 plates can be generated in TLC. Stewart et al. [17] concluded that 4 mm ID columns filled with 40 μm silica particles and having efficiencies in the range 1000 to 2000 plates per meter, i.e., plate heights of 0.5 to 1 mm represented a practical optimum. Such columns were packed by adding sufficient dry adsorbent at a time to fill a 2-cm length of column. After each addition, the column was simultaneously bounced on the floor, rotated, and tapped at the level of the packing with a piece of copper tubing about a dozen times. This technique is very effective for particles down to about 30 μm diameter, but below this level column performance rapidly deteriorates.

Improvement in both plate heights and analysis times were achieved with the introduction of the so-called pellicular column packings. These consist of an inert solid core, about 40 μm in diameter, coated with a single or double layer of porous adsorbent. A list of some commerically available adsorbents is given in Table 2. These dense, spherical particles are

TABLE 2. Porous Adsorbents

Adsorbent	Particle diameter d_p, μm	Surface area ($m^2\ g^{-1}$)	Manufacturer/supplier
Silica			
Sil-x	36-45	300	Perkin-Elmer
Biosil	20-44	—	BioRad Labs
Merckosorb (lichrosorb)	5,10,20 30,40	350-350	Merck A.G., Darmstadt
Spherisorb	5,10,20	200	U.K.A.E.A., Harwell
Spherosil	5,10,30		Rhone-Poulenc Industries (Waters Associates)
Nucleosil	5,7.5,10	—	Machery-Nagel
Partisil	5,10,20	400	Reeve-Angel
Woelm	18,30		
Alumina			
Merckosorb Alox	5,10,30	70	Merck A.G., Darmstadt
Woelm	18,30	200	
Spherisorb	5,10,20	95-175	U.K.A.E.A.
Pellicular Silica			
Corasil II	37-50	14	Waters Associates
Pellosil HC	37-44	14	Reeve-Angel
Perisorb A	30-40	10	Merck A.G., Darmstadt
Vydac	30-44	12	Separations Group
Pellicular Alumina			
Pellumina	37-44	14	Reeve-Angel

free flowing and can be easily poured into 4 mm ID columns to give plate heights of 0.2 to 0.5 mm. They have rather flat H versus u curves, indicating that they can by operated at high flow rates (e.g., 3 to 5 ml min^{-1}) with little sacrifice in column efficiency.

The improvement in plate heights is due to rapid mass transfer of the solute between the mobile phase and the thin layer of adsorbent and to the elimination of stagnant pockets of mobile phase within the depths of a large porous particle. Pellicular packings suffer from the disadvantage of low capacity ratios. They can therefore only accept very small sample sizes, putting a high demand on detector sensitivity. Pellicular packings are also extremely expensive.

Current practice in HPLC column technology is to pack 10, 25, or 50 cm lengths of 4 to 5 mm ID stainless steel tubing with porous adsorbents, particle diameter 5 to 10 μm. This is an alternative method of achieving rapid mass transfer. It requires that columns be packed with a balanced-density slurry under high pressures. A number of spectacular procedures have been described, involving pressures of up to 10,000 psig or even greater. The author prefers a much simpler approach, using a slurry mixture recommended by Asshauer and Halász [18]. 4 g silica are thoroughly shaken with 25 ml of tetrabromoethane + dioxane + carbon tetrachloride (4 + 3 + 3) v/v. The slurry is poured into a precolumn (50 cm by 1 cm) attached to the column to be filled (25 cm by 4.6 mm precision-bore stainless steel tubing with a highly polished internal finish). After topping-up with high density solvent, the precolumn is attached directly to an Orlita diaphragm dosing pump (Type DMP-1515) operating at maximum output and fed with heptane solvent. Pressures of between 2000 and 3000 psig are developed during the expulsion of the high-density solvent from the column. When column filling is complete (10 to 15 min), pumping is continued at maximum stroke until 250 ml each of ethyl acetate, heptane + ethyl acetate (80 + 20) v/v and heptane has been passed through, at which stage the column is fully activated and ready for use. Typical plate heights of 0.01 to 0.06 mm for 10 μm silica, and 1/2 of this for 5-μm diameter particles, are routinely obtained. Table 2 lists some brands of silica and alumina which can be used to pack the columns in this way, at only a fraction of the cost of purchasing prepacked columns.

Such columns exhibit relatively high backpressures, but only short lengths are needed because they are so efficient. For example, a 25-cm by 4.6-mm column packed with 5-μm Partisil generates 15,000 theoretical plates when operated at a flow rate of 1.0 ml min^{-1} and a pressure of 400 psig, which is well within the capabilities of even the simplest liquid chromatograph. Such a column is highly reproducible and is capable of handling hundreds of repetitive analyses, which means the cost and effort expended in preparing the column is of minor consequence.

Conventional liquid-liquid chromatography, in which the stationary phase is an inert liquid impregnated on an inert support, is probably not used at all in the petroleum industry because, in practice, it is very difficult to maintain a stable chromatographic system. Problems can arise from loss of stationary phase due to solubility in the mobile phase, mechanical instability at high pump rates, and instability in temperature and solvent gradients.

These disadvantages have now been overcome by the use of chemically bonded stationary phases in which organic functional groups have been chemically combined with the surface of siliceous supports. Table 3 lists some commercially available stationary phases. Two types of partition chromatography can be recognized. The first is normal phase partition chromatography in which the bonded phase carries a polar group such as nitrile or ethoxy, and the mobile phase is a relatively nonpolar solvent, such as a hydrocarbon or halogenated hydrocarbon. Such chromatographic systems are useful for separating polar additives. The second type of partition chromatography that can be identified is reversed phase partition chromatography, in which the stationary phase is a silica surface to which has been bonded a nonpolar hydrocarbon group such as octadecyl. These are used with aqueous-organic mobile phases (e.g., various mixtures of lower alcohols, tetrahydrofuran, or acetonitrile with water).

Such chromatographic systems are being used with remarkable success in the separation of polynuclear aromatic hydrocarbons, particularly for the complete separation of benzo(a)pyrene, from benzo(e)pyrene, perylene, benzo(k)fluoranthene and 1,12-benzo(g,h,i)perylene.

Obviously, a wide variety of chromatographic selectivities can be obtained by preparing packings with different functional groups. This is potentially excitingly useful. Nevertheless, it is to be hoped that the liquid chromatographer will not (as the gas chromatographer has been), be subjected to a bewildering array of stationary phases, most of which are either unsuitable or lack novelty. We have already encountered commercial packings which were hydrolytically or thermally unstable and some with polymeric coatings, which have offered high resistance to mass transfer.

It is instructive to evaluate each new column by constructing an H versus u curve under standardized conditions. This gives an indication as to how well the column has been packed and is likely to perform. This is quite important because one of the particularly depressing facts of present-day commercial stationary phases is their wide batch-to-batch variation in product quality. The author routinely evaluates columns with a standard test solution of naphthalene and phenanthrene. Figure 4 shows the progress during the last 5 years in packing efficient silica columns. Repeat evaluations at two monthly intervals during the life of a column reveals any deterioration in its characteristics. The evaluation of a column should also

TABLE 3. Some Chemically Bonded Stationary Phases

Commercial name	Functional group	Manufacturer/supplier
Nonpolar		
Bondapak, Corasil, C18	octadecyl	Waters Associates
Bondapak, Porasil, C18	octadecyl	Waters Associates
Bondapak, Corasil, Phenyl	phenyl	Waters Associates
Permaphase ODS	octadecyl	Du Pont
Vydac 201 RP	octadecyl	Separations Group/Applied Sciences
Lichrosorb RP8	octadecyl	E. Merck A.G.
Spherisorb ODS	octadecyl	U.K.A.E.A.
Phenyl, Sil x-1	phenyl	Perkin-Elmer
Octadecyl Sil x-11	octadecyl	Perkin-Elmer
Octadecyl Sil x-1	octadecyl	Perkin-Elmer
Nucleosil C_{18}	octadecyl	Machery Nagel
CO-PELL ODS	octadecyl	Reeve Angel
Partisil 10 ODS	octadecyl	Reeve Angel
Polar		
Durapak, Carbowax	polyethylene glycol	Waters Associates
Durapak ODPN	oxydipropionitrile	Waters Associates
Permaphase ETH	aliphatic ether	Du Pont
Vydac 501 PP	nitrile	Separations Group/Applied Sciences
Fluoroether, Sil x-1	fluoroether	Perkin-Elmer
Nucleosil CN	nitrile	Machery-Nagel

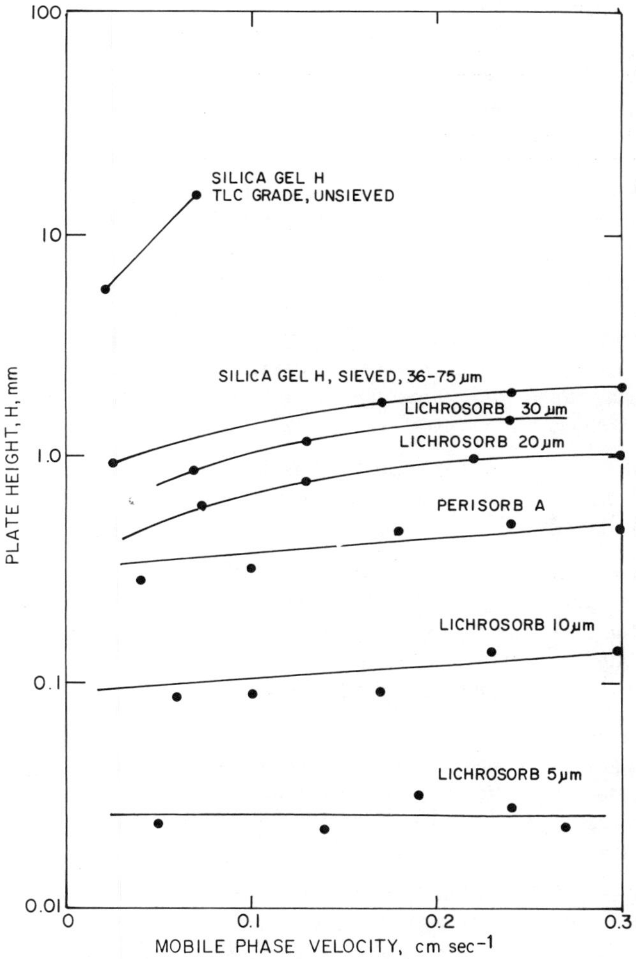

FIG. 4. Improvement of plate height with decreasing particle size.

include measurements of resolution, analysis times, and permeability (i.e., pressure drop to achieve a given mobile phase velocity or analysis time). To compare packings of different particle sizes, some authors prefer to plot reduced plate height h (equal to H/d_p) against reduced velocity v (ud_p/D_m), where d_p = particle diameter and D_m = diffusivity of the solute in the mobile phase. Further comparisons of column performance can be made by measuring the number of effective plates per second [Eq. (5)], a quantity which should be maximized. Karger [19] has shown typical values of N_{eff}/t in GC to be 10 and 25 for classical packed and open tubular columns, respectively, while typical values for LC are 0.02 for classical columns, 2 for 20-μm diameter silica and 8 for porous layer beads. We calculated the value of N_{eff}/t for the 5-μm Partisil column previously described to be 17, which approaches that of a GC capillary column.

C. Equipment

While the theories and techniques developed within GC provided the impetus for the revival of LC in columns, the analogy has often been pushed to ludicrous extremes by some manufacturers of so-called liquid chromatographs. Many of these instruments are unnecessarily complicated and appear to be gas chromatographs which have been converted, quite unsuitably, to liquid chromatographs.

It must be reemphasized that the heart of the liquid chromatograph is the column; no liquid chromatograph, however sophisticated, can perform better than the column it contains.

The essential features of a simple, but effective, liquid chromatograph include a solvent reservoir, pump, injection system, column, detector, and recorder. This is all that may be required for many routine quantitative analyses, but for analyses of complex mixtures with very high efficiency columns gradient elution is also necessary.

McNair and Chandler [20] have recently reviewed the commercially available liquid chromatographs and Snyder and Kirkland [14] have discussed equipment for liquid chromatography in greater detail. We shall summarize the apparatus required for the HPLC analysis of petroleum products based on our own experience over the last 8 to 10 years.

Figure 2 (p. 337) is a diagram of a typical chromatographic assembly we use. It should be stressed at the outset that all components are interchangeable and that mention of a specific piece of equipment does not necessarily imply it is the only or best available to do that job. What we do claim, however, is that the system is simple, easy to use, free from maintenance problems, and, most important of all, that it works.

The solvent reservoir is a glass bottle, 500 ml capacity, which accommodates sufficient mobile phase for a day's work. Nonaqueous solvents for adsorption chromatography do not require degassing (i.e., removal of dissolved air). Solvents containing water, however, in reversed phase chromatography, for the separation of polynuclear aromatics do need to be degassed, otherwise bubbles of air will accumulate in the cell of the detector rendering it inoperable. For casual work, this can be simply achieved by heating water on a boiling-water bath for 15 min and stirring to expel dissolved air. The water is then allowed to cool before making up the mobile phase to the desired composition, which can then be stored for several weeks in a glass-stoppered bottle.

Should an air bubble inadvertently form in the cell—and this is most likely to occur either at the start of a program of work after the chromatograph has been standing idle for some time, or immediately after changing columns—the bubble can usually be dislodged by raising the pressure in the cell for a few seconds by sticking one's finger against the outlet pipe of the detector.

The mobile phase is fed to a pump—in our case an Orlita Diaphragm dosing pump Type DMP-1515 (Orlita K.G., 63 Giessen/Lahn, Max Eyth Strasse 10, West Germany). This is a reciprocating pump capable of delivering a constant volume of up to 10 ml min^{-1} against a back pressure of 4000 psig.

Operating pressures are typically in the 200 to 1000 psig range, even with 5 to 10 μm porous adsorbents, which is well within the pump's capability. Although the pump delivers mobile phase in a series of pulses, these are effectively dampened down by inclusion of a Bourdon-tube pressure-gauge and a freely suspended 2-m length of 0.001-in. ID stainless steel capillary tubing connecting the pump outlet with the sample injection port. As reciprocating pumps have small internal volumes, they are suitable for gradient mixing.

It is good practice to include a pressure switch device when using mechanical pumps (e.g., from K.D.G. Instruments, Ltd., Crawley, Sussex, England) which is set to trip the power supply to the pump in the event of a pressure rise in excess of safe operation. Nevertheless, it is pleasing to report that our pressure devices, set at 3000 psig have never been activated in 8 years of operation. The few occasions on which blockages have occurred have been manifested by leaking column connections, but more usually by mobile phase dribbling from a split septum.

The way a sample is introduced to the column is of crucial importance, especially when operating a short, high-efficiency column packed with microparticulates. Ideally, the sample should be rapidly injected onto the first theoretical plate of a column. There are two basic methods of doing this, as in GC: syringe injection through a septum, or use of a microsampling valve. Syringe injection is the cheapest, most convenient but least accurate in quantitative analysis, especially in the hands of an unskilled operator. The injection port we use (Figure 5) is based on the original design of Scott et al. [21]. Similar injection ports are now available commercially (e.g., Jobling Laboratory Division, Stone, Staffordshire, England). Silicone rubber septa coated with a PTFE film (Scientific Glass Engineering Pty, Ltd., Melbourne, Australia) can withstand about 50 injections under inlet pressures of up to 1000 psig before they need changing.

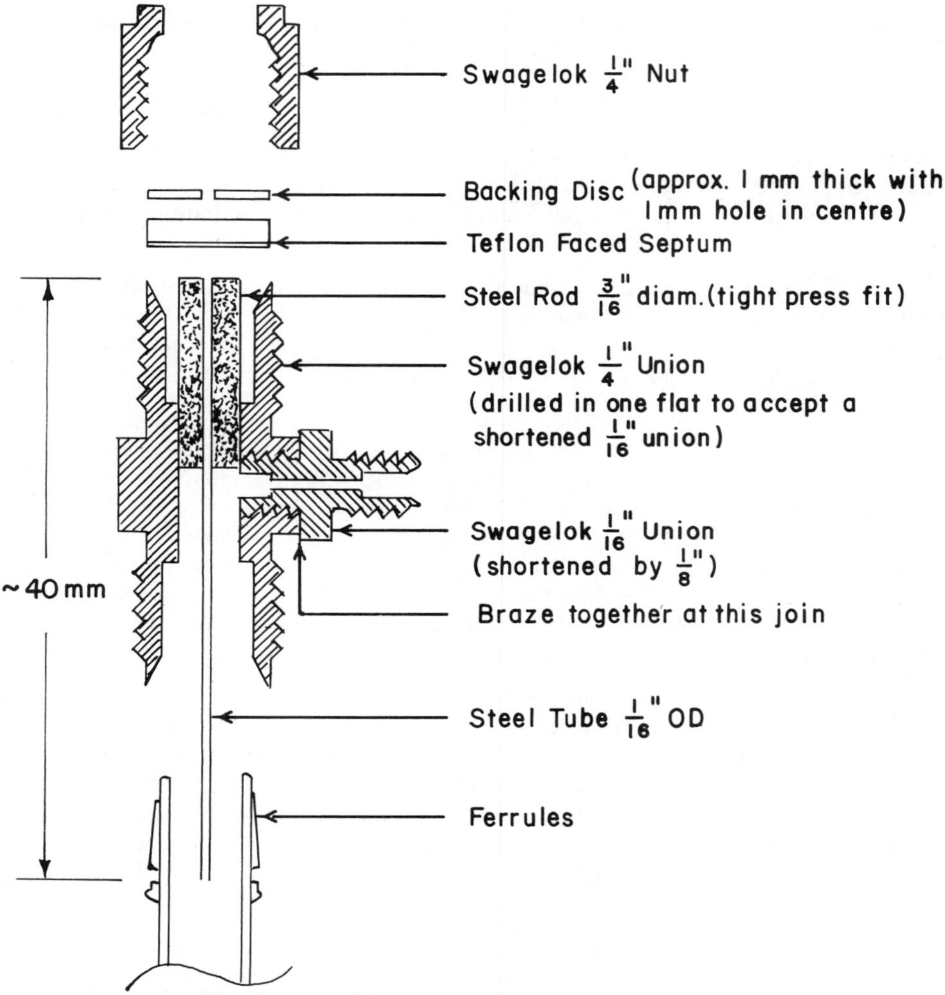

FIG. 5. Design of injection port for HPLC.

The sample is deposited by microsyringe just at the surface of a layer of glass beads (about 200 mesh) which occupy the top 5 mm of column length. The sample is therefore swept by the mobile phase into the stationary phase as a sharp band.

As we have already discussed, a typical column will consist of 10 to 50 cm length by 4 mm ID precision-bore stainless-steel tubing with highly polished inner walls. The packing is retained by a porous nickel or stainless steel plug and the column exit is connected to the detector cell by a short length of 0.001-in. ID tubing. Care must be taken to ensure that between the column exit and the detector cell there is no dead volume where remixing of separated components could occur. The detector cell itself should have maximum volume of 10 μl.

The petroleum analyst has the choice of only two types of detector—the uv monitor and the differential refractometer. (The moving wire detector can only be used with nonvolatile materials.) The most popular, by far, is the uv monitor, because it is cheap, reliable, easy to use, insensitive to small variations in temperature and flow rate, and can be used with gradient elution. There is now a wide variety of both fixed and variable wavelength detectors to choose from. A disadvantage of uv detectors is the enormously variable response of different compounds, which means that for quantitative work, pure reference compounds are essential. For example, whereas polynuclear aromatic compounds can be detected at the mg^{-1} level, saturated hydrocarbons, esters, and polymeric additives cannot be detected at all. To examine these compound types, a differential refractometer must be used, which presents several disadvantages: high detection limit (typically, 1% of a sample component can just be detected), high sensitivity to temperature and flow rate, and, in general, unsuitability for gradient elution. Its major advantage is that it is a universal detector. For further information on detectors, the reader is referred to Snyder and Kirkland [14].

The output from both types of detector is displayed on a 1- or 10-millivolt stripchart recorder, and the further evaluation of chromatograms is analogous to that of GC.

V. APPLICATIONS

The two most important areas in which PC, TLC, and HPLC have been most widely applied so far are in the determination of additives in lubricants and in polynuclear aromatics analysis. These two topics are discussed separately in Chaps. 17 and 15, respectively. In the remainder of this chapter we shall discuss their applications in hydrocarbon-type analyses and the determination of additives in fuels. These are fields in which neither PC nor even TLC has made a spectacular contribution, nor as yet, has HPLC. Nevertheless, the signs are that HPLC is about to make a significant impact.

A. General Hydrocarbon-type Analysis

The contribution of TLC to hydrocarbon-type analysis can be summarized by Figure 6, which shows a chromatogram of seven hydrocarbon oils of differing viscosities and aromatic content. The plate, coated with silica containing a uv fluorescent indicator (Merck Silica Gel GF 254), was activated by heating in an oven at 110°C for 30 min. 1 to 3 μl portions of 30% solutions of the oils in cyclohexane were spotted on the layer, and the plate developed in cyclohexane. After development, the plate was examined under uv light when spots corresponding to fluorescing and/or absorbing aromatic hydrocarbons could be detected. The plate was then sprayed with a solution containing 5 g potassium dichromate in 100 ml of 40% sulfuric acid and heated in an oven at 110°C for 5 min. Under these conditions both saturates and aromatics were revealed as light blue spots on an orange background.

It is apparent that all these oils are different, and contain different amounts and types of aromatic compounds but beyond this very little additional information can be gleaned.

The prospects for HPLC for hydrocarbon-type analysis looks much brighter but has yet to be exploited. Figures 7 to 11 show chromatograms of two different grades of gasoline, an aviation fuel, a middle distillate, and a lubricant base stock. The separations were obtained within 15 min each on a 25-cm by 4.6-mm stainless steel column packed with Reeve Angel's Partisil, 5 μm nominal particle diameter, using a Pye-Unicam uv 20 Liquid Chromatograph equipped with a fixed wavelength (254 nm) uv

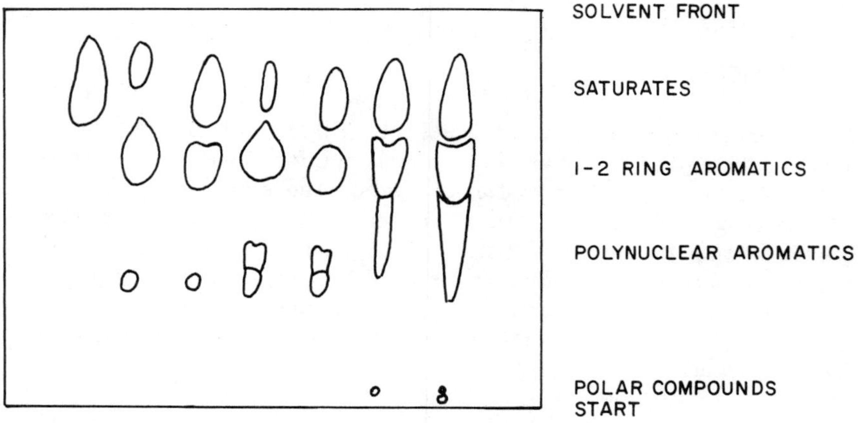

FIG. 6. TLC of lubricant basestocks. Adsorbent: Merck Silica G. Mobile phase: cyclohexane. Revealed by 5% potassium dichromate in 40% sulfuric acid.

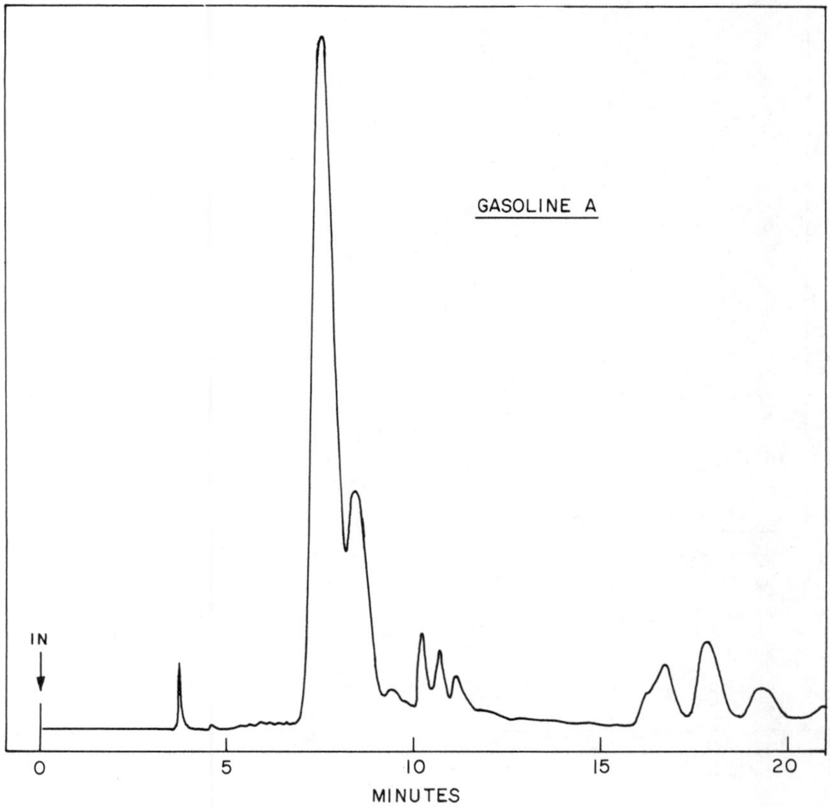

FIG. 7. HPLC of gasoline. Column: 25-cm × 4.6-mm i.d. stainless steel packed with Partisil, nominal particle diameter 5 μm.

monitor. The efficiency of this column, measured on the naphthalene peak at a flow rate of 1.0 ml min^{-1} was about 15,000 plates (H = 0.02 mm). It can be seen that, the column not only achieves a separation according to the number of aromatic rings, but also a considerable degree of separation within the members of each group. This separation can be improved by greater selectivity, e.g., using alumina instead of silica or by reversed phase chromatography, or it can be improved by increasing efficiency. No doubt, as in gas chromatography, we shall see columns of higher and higher efficiency being used for the separation of petroleum fractions.

The chromatograms reveal two limitations of uv detection. Different compounds may have widely different response factors. For example, the aviation fuel in Figure 9 contains about 18% alkyl benzenes and 0.5% naphthalenes and, as a result, the appearance of the chromatogram is misleading.

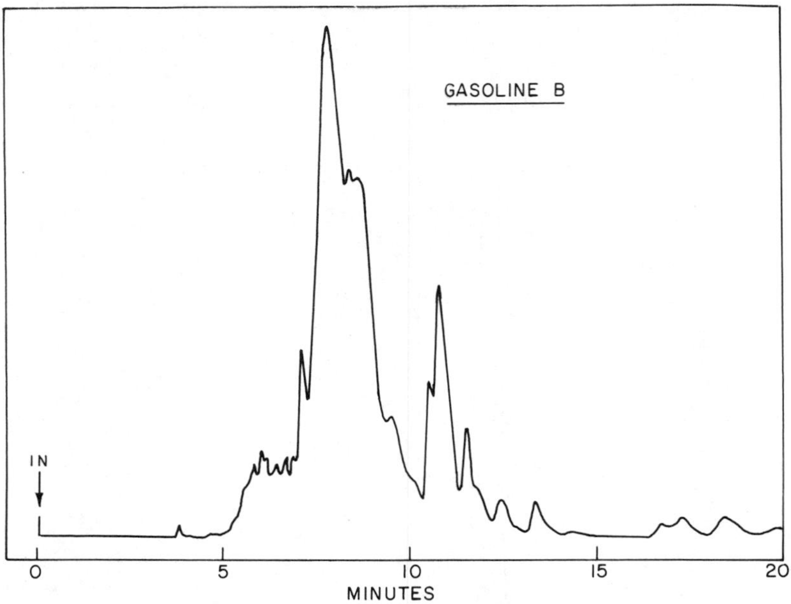

FIG. 8. HPLC of gasoline. Same conditions as Fig. 7.

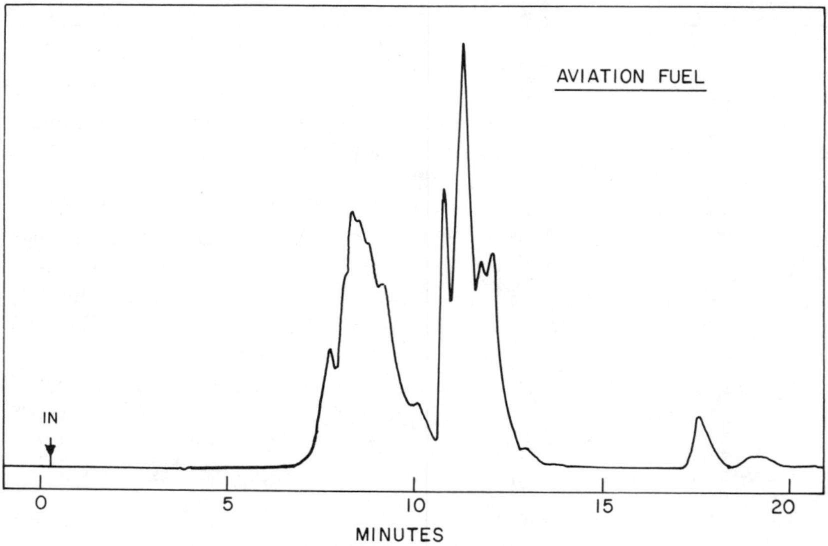

FIG. 9. HPLC of aviation fuel. Same conditions as Fig. 7.

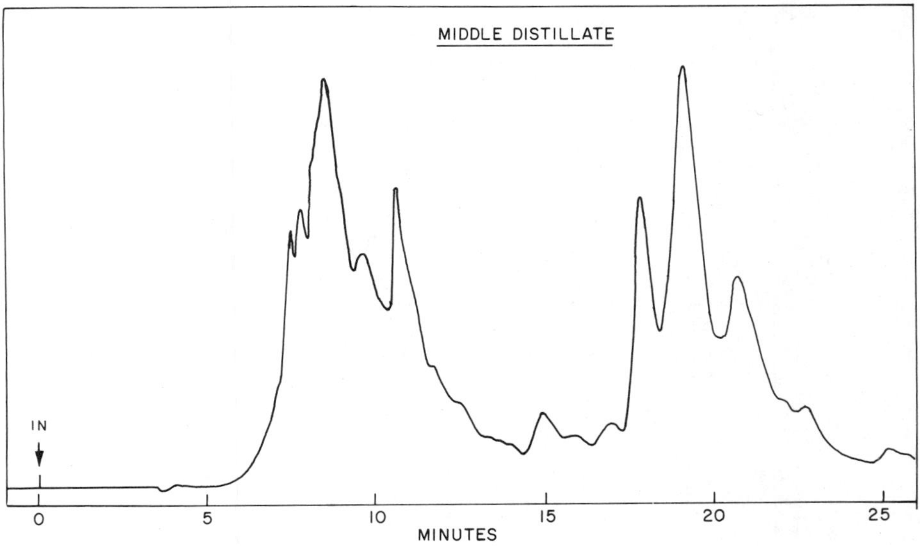

FIG. 10. HPLC of middle distillate. Same conditions as Fig. 7.

The second limitation is that 82% of the sample, i.e., the saturates, are not seen at all. Nevertheless, chromatograms of this type should be very useful for fingerprinting, especially as relative peak heights can be changed quite dramatically by rerunning at different wavelengths.

Differential refractometry offers the best hope for a quantitative method for hydrocarbon type analysis and the lead has been given in a recent paper by Suatoni and Swab [23], in which they measure total saturates, aromatic and polar compounds (and hexane insolubles) in total crude oils, gasolines, kerosenes, middle distillates, fuel oils, residuals, lubricating oils, and synthetic crudes derived from coal liquids, shale oils, and tar sands.

After removal of hexane-insolubles Suatoni and Swab separated their samples on a 30-cm by 4.0 mm μ-Porasil column (Waters Associates) using dried n-hexane as mobile phase at a flow rate of 2.0 ml min^{-1}. A preestablished time after the saturates peak has emerged, they reversed the direction of flow so that the aromatics also emerge as a single peak. The polar compounds are retained on the column and are determined by difference. The column is regenerated at weekly intervals by washing with dry dichloromethane. Calibration factors were obtained by isolating the saturates and aromatics from different boiling-range fractions from a number of crudes when it was found that smooth calibration curves could be constructed for each type of crude oil. The time taken for a total HPLC

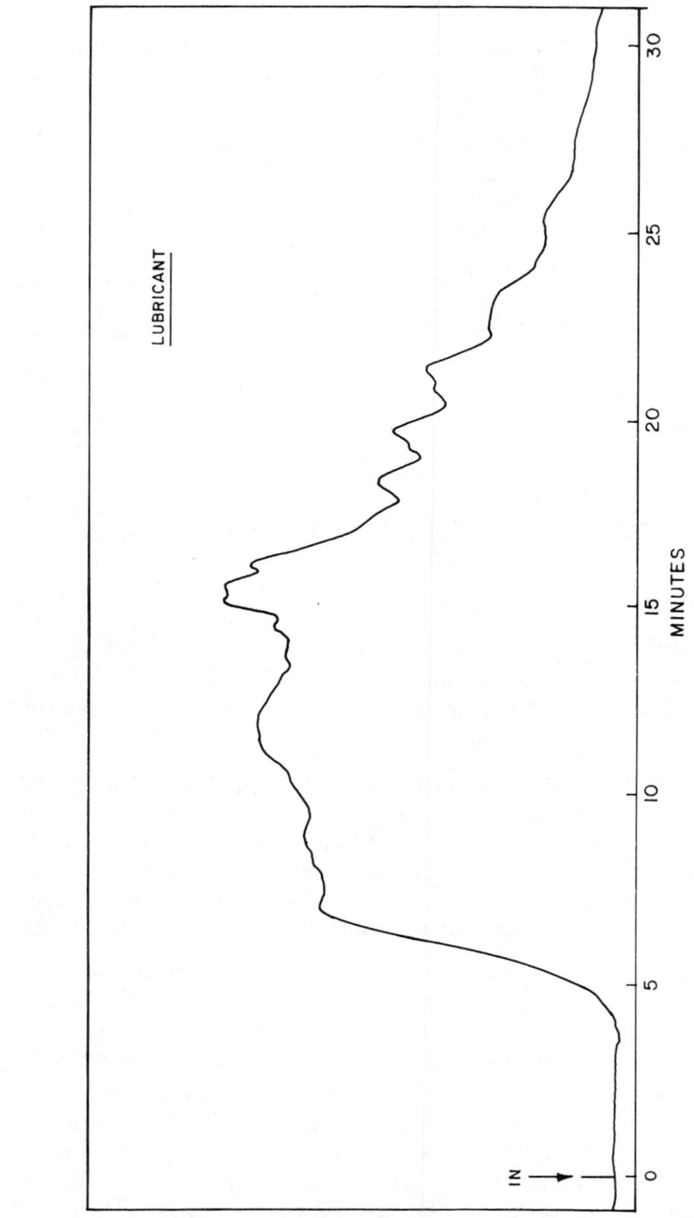

FIG. 11. HPLC of lubricant basestock. Same conditions as Fig. 7.

analysis is 8 min and samples can be injected at regular intervals so that the method could be adapted to onstream analysis. Typical standard deviations are ±0.4% for saturates and ±1.4% for aromatics at the 30 to 60% levels, which are similar to those obtainable by ASTM D 2007.

B. Hydrocarbon Type Analysis of Gasoline and Aviation Fuels

The most important liquid chromatographic method used in the petroleum industry for the last 20 years has been the fluorescent indicator adsorption (FIA) test to determine the saturate, olefin, and aromatic content of distillate fuels boiling below 315°C. The procedure, which has been widely adopted as a standard test (e.g., ASTM D 1319, IP 156, DIN 51791) involves the displacement of the hydrocarbon sample by isopropanol down a precision-bore glass column as originally described by Criddle and LeTourneau [24].

In doing so, the sample separates out into zones, the leading zone containing saturates, followed by a middle, olefinic zone, and a final, aromatic zone which also contains some diolefins, aromatic olefins, and sulfur- and oxygen-containing polar compounds.

The zone boundaries are located by inclusion of a uv fluorescent dye mixture, the components of which travel with each of the three zones and which are visible in uv light. The percentage, by volume, of the three hydrocarbon groups, is calculated from measurements of the lengths of the three zones.

The method has a number of disadvantages. It is rather a lengthy test (2 to 3 hr) and requires considerable operator skill. Reproducibility is frequently disappointing, notably for the olefin content. This is because the method is particularly sensitive to the properties of the silica gel being used as well as the variable product quality of the fluorescent dyes, especially during the last few years. Because of these disadvantages, there is much interest in any possible replacement for the FIA test.

Suatoni, Garber, and Davis [22] have recently described a rapid, accurate, and precise method by HPLC which could be adopted to onstream analysis. A 3-μl portion of sample is separated on a 30 cm by 4 mm ID stainless steel column packed with activated μ-Porasil (Waters Associates). Detection is by differential refractometry. The mobile phase used was Fluorinert FC-78 (3M Company) which has an extremely low solvent strength ($\epsilon° = -0.25$) compared with pentane ($\epsilon° = 0.00$) as well as a very low refractive index (approximately 1.25). Under these chromatographic conditions the unsaturates are separated from the saturates. After they have eluted, the direction of flow is reversed with a 4-way valve so that the aromatics are backflushed and emerge as a single peak. With a high flow rate of 3.5 ml min^{-1} a complete analysis takes about 10 min.

The amount of aromatics is calculated directly from the chromatogram since their refractive indices and thus their response factors are relatively constant (±2%) over the range of interest (C_6-C_{12} benzenes). The saturates and olefins, the response factors of which are less constant, are determined by difference by a weighted area fraction

$$\% \text{ Saturates} = (100\% - \% \text{ aromatics}) \frac{1.10 \text{ area (saturates)}}{1.10 \text{ area (saturates)} + \text{area (olefins)}}$$

and

$$\% \text{ Olefins} = (100\% - \% \text{ aromatics}) \frac{\text{area (olefins)}}{1.10 \text{ area (saturates)} + \text{area (olefins)}}$$

The standard deviation of an olefin measurement, at the 40% level, in a light gasoline distillate was calculated to be 1.07% by HPLC compared with an ASTM D 1319 reproducibility of 7.6%. The HPLC method is particularly well suited for aviation fuel analysis where a standard deviation for the aromatics content was found to be 0.32% at the 15% level compared with 3.2% by the standard FIA procedure.

C. Identification of Pollution Sources

TLC offers a rapid and simple way of obtaining some preliminary information on the nature of many organic materials including petroleum products. One specific area of interest is the identification of marine pollutants, and here it can be used to augment the information obtainable by GC (Chap. 7) and GPC (Chap. 12). As far as the author is aware, HPLC has not yet been used for this purpose, but it undoubtedly will or should be in the near future.

The TLC procedure to be described has now been adopted as an IP method (IP 314) and is based on the original work of Matthews [25], which has enabled the Greater London Council to greatly increase the number of cases in which it has successfully traced sources of oil pollution in London's sewer system, resulting in a considerable clean-up of the River Thames.

The varying chromatographic and fluorescence characteristics of petroleum and coal-tar derived oils, fats, and greases are combined to form a simple systematic approach to identify the source of marine pollution. Details of the scheme are given in Table 4. The sample is first chromatographed on an Alumina (T) F 254 plate with acetone as mobile phase, and the developed plate is examined under uv light at both 254 and 350 nm. This enables the sample to be placed in one of five classes. The sample is then chromatographed in a second system, details of which are

TABLE 4. Systematic Scheme for Differentiation of Oils by Fluorescence at 350 nm

System 1	Result	System 2	Result	Conclusion
Alumina T and acetone	Class 1—$R_f = 0$, blue, streak to front	Silica gel (6.5 cm) and ethyl acetate	(a) $R_f = 0$, streak to front (b) $R_f > 0$, slightly diffuse, streak to front	Petrolatums Motor and gear oils
	Class IIa—$R_f > 0 < 0.3$, strong blue streak to front	Use of system specific for a given oil	Specific for the oil	Heavy engineering oils
	Class IIb, c, d—$R_f > 0.3$, blue, may be diffuse or very pale at high R_f values	Alumina T and 1:1 ethanol/acetone	(b) $R_f < 0.4$ (c) R_f 0.3 - 0.7, long pale blue splot (d) Diffuse at solvent front, very pale blue; dark at 254	Light engineering oils Diesel fuel oils Solvent fuels
	Class IIIa, b—Orange or brown head at solvent front and brown streak	Band technique: Alumina T and petroleum ether, 40-60° C	(a) Multicolored bands up to solvent front (b) Some of bands missing when compared to crude coal tar	Coal tars Coal tar oils
	Class IVa, b, c, d—Yellow, brown, blue streak and head	Band technique: kieselguhr and acetone	(a) Very pale yellow brown head and band (b) Yellow-purple band ranging from pale to fairly strong, sometimes pale blue head (c) Intense yellow band (d) Very pale yellow band, blue yellow head	Lake asphalts Fuel oils Heavier crude oils Tanker sludges Bitumens Lighter crude oils
	Class Va—Fluorescing characteristics not listed above Class Vb—Very weak or no fluorescence	Examination by TLC or other techniques, e.g., saponification	Specific for the oil	Other non-petroleum fluorescing oils Non-fluorescing, non-petroleum oils or liquid paraffin

given in Table 4, which enables a further subclassification to be made. Hopefully, this will reduce the possible source of pollution to a narrow enough range to enable reference samples to be obtained for a final positive identification.

Classes I and II are characterized by the blue fluorescence at 350 nm due to polynuclear aromatics in refined petroleum products. Intensity of fluorescence increases as the midboiling point increases. At the same time, the product becomes less soluble in the mobile phase, so the R_f value is reduced. For example, a heavy lubricant will give an intense blue fluorescent spot near the starting line, whereas a lighter fuel distillate will give a less intense spot nearer the solvent front. Heavier lubricating oils are likely to contain additives which can be further characterized by TLC (Chap. 17), while distillates are best characterized by GC (Chap. 5).

Class III contains coal tar products which are easily recognized because they give a whole series of discrete, brightly-colored spots when chromatographed with the Alumina T/petroleum ether system. Improved sensitivity and reproducibility are obtained if the samples (10% solutions in toluene) are applied to the plate as a narrow band, 1.5 cm long.

Class IV contains residual petroleum products which are characterized by an intense yellow fluorescence due to the presence of asphaltenes. The color is modified by the presence of the blue fluorescence of polynuclear aromatics. Therefore an accurate distinction between different members of this class is difficult and much more information can be derived by gel permeation chromatography, e.g., as described by Done and Reid [26].

Class V contains those materials possessing fluorescent characteristics (usually no fluorescence at all) which are not covered by classes I-IV. With the exception of liquid paraffin, these materials are of nonpetroleum origin and are therefore beyond the scope of this book.

D. Additives in Fuels

Various chemicals are added to fuels, especially to gasolines, to improve either their performance or storage stability. Examples are icing inhibitors, antiknock additives, antioxidants, corrosion inhibitors, metal deactivators, spark plug deposit modifiers, and others. Liquid chromatography has been used to study two of these additive types: phosphate esters used as deposit-modifiers and antioxidants of the hindered phenol or amine type.

Such additives are usually present in the 1 to 100 mg l^{-1} concentration range and therefore need to be concentrated before examination by TLC. This can best be done by shaking 100 ml fuel with 5 g activated alumina for 5 min, during which time polar additives are adsorbed onto the alumina. Excess fuel is decanted off, the alumina washed twice with small portions of petroleum spirit and finally the additives are desorbed by shaking the alumina with acetone for 5 min. The solvent is removed at 40° C under vacuum in a rotary evaporator and the concentrated extract used for TLC.

Killer and Amos [5] examined the TLC behavior of phosphates (Table 5) and antioxidants (Table 6) and enabled studies of possible additive losses through distribution systems (i.e., from the refinery to the final customer) to be carried out. Detection of phosphate esters on a thin layer plate presents some difficulties. It is very difficult to convert triaryl phosphates to the characteristic Molybdenum Blue. One way of revealing them is to spray the plate first with calcium chloride (1% in ethanol) and second with Rhodamine B (0.05% in ethanol). Bromine vapors are then blown over the plate with the help of a hairdrier until the background has been bleached. After standing for 5 min the phosphate esters show up as white spots on a pink background.

In contrast to phosphate esters, antioxidants give characteristic colors with a number of revealing reagents and this specificity, taken in conjunction with R_f values enables a fairly confident identification to be made, particularly when reference compounds are available and are chromatographed alongside the sample under investigation. The color reactions with two reagents are given in Table 6. Reagent I is 2,6-dichloroquinone-4-chlorimine (1% in ethanol) and reagent II is prepared by mixing two volumes of antimony pentachloride with eight volumes of carbon tetrachloride. Further examples of characteristic color reactions with specific lubricant antioxidants are given in Chap. 17.

Amos [27] took the TLC analysis of antioxidants a stage further by developing a quantitative method for the routine determination of

TABLE 5. TLC of Phosphate Esters [5]

Phosphate ester	R_f
Trimethyl	0.36
Triethyl	0.43
Tri-n-butyl	0.67
Trichloroethyl	0.50
Triphenyl	0.80
Tritolyl	0.83

Adsorbent: Silica. Mobile phase: dichloromethane + methyl ethyl ketone (90 + 6)v/v. Revealed by Iodine vapor.

TABLE 6. TLC of Gasoline Antioxidants [5]

	A	B	C	D
2,6-di-tert-butyl, 4-methyl phenol	0.62	0.80	yellow	violet
4,4', bis (2,6-di-tert-butyl phenol)	0.82	0.88	yellow-green	brown
4,4' bis (2,6-methyl, 6-tert-butyl phenol)	0.22	0.55	yellow-brown	brown-red
2,4-dimethyl, 6-tert-butyl phenol	0.57	0.80	brown	black
2,6-di-tert-butyl-1-dimethylamino-p-cresol	0.00	0.08	blue-violet	yellow-brown
N,N' di-sec-butyl-p-phenylene diamine	0.00	0.24	red	light grey

A = Carbon tetrachloride mobile phase
B = n-hexane + diisopropyl ether + methanol (70 + 30 + 5) v/v mobile phase
C = revealed by 1% dichloroquinone-4-chloramine in ethanol
D = revealed by 20% antimony pentachloride in carbon tetrachloride

2,4-dimethyl,6-t-butyl phenol at the 10 mg l^{-1} level in aviation turbine fuels. The concentrated extract was chromatographed simultaneously with five standard samples covering the range 2 to 20 mg l^{-1}. The plate was developed with heptane + diisopropyl ether (80 + 20) v/v, sprayed with phosphomolybdic acid (5% solution in isopropanol) and heated in an oven at 110° C for 10 min when the antioxidant spots were revealed as intense blue-black spots on a pale yellow background. The area-intensities of the spots were compared by densitometry. A plot of this against the weights of additive per spot gave a smooth curve, from which the concentrations of additive in unknown samples were interpolated.

The extra cost in terms of both equipment and time involved in quantitative TLC are out of all proportion to the extra information involved. As a result, HPLC can be expected to completely replace TLC for quantitative analysis.

This determination [28] can now be carried out more simply, rapidly and accurately by HPLC using a 25-cm long column packed with Partisil, 10 μm particle size, with heptane-ethyl acetate (98 + 2) v/v as mobile phase. With a uv monitor at 280 nm the limit of detection is about 2 mg l^{-1}, which means no preconcentration of the additive is necessary. Points of comparison in these two approaches to quantitative analysis are summarized in Table 7 [29].

TABLE 7. Quantitative Analysis: HPLC versus TLC [29]

Item	Unit	HPLC	TLC
Cost of equipment	$	3625	3675
Cost of stationary phase: 1 sample	$	16.50	0.55
100 samples	$	16.50	5.50
Cost of mobile phase: 1 sample	$	0.32	0.32
100 samples	$	16.25	0.64
Cost of preparing 100 chromatograms	$	45.00	125.00
Detection limit	$\mu g\ l^{-1}$	2	2000
Initial sample size	ml	0.1	200
Analysis time, one sample	min	10	200
Analysis time, 100 samples	min	500	7000
Labor costs (100 samples)	$	100	1400
Error (2σ)		0.8	3.6

VI. STATUS OF LIQUID CHROMATOGRAPHY

The techniques described in this chapter are complementary to GC, which is a quantitative technique of high resolving power for the analysis of volatile mixtures, and GPC, which is also a quantitative, high efficiency (but low resolving power) technique suitable for the determination of molecular weight distribution of polymers of molecular weights in excess of 1000. The major role of liquid chromatography, as defined in this chapter, i.e., adsorption and partition, is clearly to fill the gap between GC and GPC.

Clearly, some degree of overlap is inevitable. Although it is tempting to argue that one technique has an advantage over the other, analysts should consider themselves extremely fortunate that they have several techniques available which are capable of contributing information that is either complementary or corroborative.

For example, GPC, while lacking the resolving power of HPLC, can give some useful information on the molecular weight of the major components in an unknown mixture, which can also be separated by HPLC by compound type; similarly GC gives information on the boiling range of a mixture. Now that the resolving power and speed of analysis by HPLC have approached those by GC, the analyst often has a choice which may be made on the basis of convenience. However, GC has something like a 15-year lead and there are now many instances where GC may have been overstretched. This is particularly true in additive analyses where a high temperature GC analysis can usually now be replaced by a quick HPLC compound-type separation at room temperature.

The choice between PC, TLC, and HPLC is now fairly clear cut. As far as petroleum chemistry is concerned PC and TLC have merged into one technique, adsorption chromatography on precoated silica plates being the most popular, but cellulose-coated plates having an application in certain circumstances, e.g., in polynuclear aromatic separations.

Amos and Perry [29] have appraised the relative merits of TLC and HPLC and have shown there is a genuine need for both. In TLC, the equipment is simple, cheap, and easy to use. In qualitative analysis, it can rapidly give information on the complexity of an unknown mixture, the whole of the chromatogram is open for inspection, and the use of specific chromogenic reagents enables something to be learned about the chemical composition of individual components. However, the technique has its limitations and these should be recognized. It has a low efficiency—only about 0.05 plates per second are generated—and attempts to improve this, e.g., by programming the mobile phase, only detract from the basic simplicity of the technique. In quantitative analysis, too, the additional effort, skill, and equipment required to obtain only a moderate accuracy is out of all proportion to the extra information obtained. Recovery of material from spots on a TLC also requires excessive sample handling, with apparently endless opportunities for contamination to occur.

In HPLC, by contrast, the equipment is more sophisticated and expensive, but it is amenable to automation, which makes it less susceptible to operator errors. Chromatographic conditions can be controlled to obtain high efficiency and resolving power. Quantitative analysis can be highly precise (less than 1% relative error) and sample recovery is straightforward, having none of the problems of either TLC or GC.

In conclusion, TLC should be used for preliminary examination of samples of unknown composition so that HPLC systems can be devised for a better separation and for recovery of pure components which can be further identified by infrared, ultraviolet, or mass spectroscopy. TLC can also be used for semiquantitative analyses, by visual comparison of spot sizes and intensities of large numbers of similar samples in which a precision of only ±20% of the amount present is sufficient. HPLC should be used for all routine high-speed quantitative analyses.

REFERENCES

1. R. Consden, A. H. Gordon, and A. J. P. Martin, Biochem. J. 38:224 (1944).
2. N. A. Izmailov and H. S. Shraiber, Farmatsiya 3:1 (1938).
3. E. Stahl, Pharmazie 11:633 (1956).
4. E. Stahl, Chemiker Zeit. 82:323 (1958).
5. F. C. A. Killer and R. Amos, J. Inst. Petrol. 52:315 (1966).
6. M. S. Tswett, Proc. Warsaw Soc. Nat. Sci. Biol., Sec. 14, Min. 6, 1903.
7. L. R. Snyder, "Principles of Adsorption Chromatography," Edward Arnold, London and Marcel Dekker, New York, 1968.
8. K. Macek and H. Bečvářová, Chromatog. Rev. 15:1 (1971).
9. E. Stahl, ed., "Dunnschicht-Chromatographie," Springer-Verlag, Berlin, Heidelberg, and New York, 1967 (English ed. 1969)
10. J. G. Kirchner, Thin Layer Chromatography, in "Technique of Organic Chemistry," vol. XII (E. S. Perry and A. Weissberger, eds.), Wiley-Interscience, New York, 1967.
11. E. J. Shellard, ed., "Quantitative Paper and Thin Layer Chromatography," Academic Press, London, 1968.
12. R. Kuhn and E. Lederer, Ber. Deut. Chem. Ges. 64:1349 (1931); Naturwiss 19:306 (1931).
13. A. J. P. Martin and R. L. M. Synge, Biochem. J. 35:1358 (1941).
14. L. R. Snyder and J. J. Kirkland, "Introduction to Modern Liquid Chromatography," Wiley, New York, 1974.

15. J. J. van Deemter, F. J. Zuiderweg, and A. Klinkenberg, Chem. Eng. Sci. 5:271 (1956).
16. J. N. Done, G. J. Kennedy, and J. H. Knox, in "Gas Chromatography 1972" (S. G. Perry, ed.), Applied Science Publishers, London, 1973, p. 145.
17. H. M. N. Stewart, R. Amos, and S. G. Perry, J. Chromatog. 38:209 (1968).
18. J. Asshauer and I. Halász, J. Chromatog. Sci. 12:139 (1974).
19. B. L. Karger, The Relationship of Theory to Practice in High-Speed Liquid Chromatography, in "Modern Practice of Liquid Chromatography" (J. J. Kirkland, ed.), Wiley-Interscience, New York, 1971, p. 47.
20. H. M. McNair and C. D. Chandler, J. Chromatog. Sci. 12:425 (1974).
21. R. P. W. Scott, D. W. J. Blackburn, and T. Wilkins, J. Gas Chromatog. 5:182 (1967).
22. J. C. Suatoni, H. R. Garber, and B. E. Davis, J. Chromatog. Sci. 13:367 (1975).
23. J. C. Suatoni and R. E. Swab, J. Chromatog. Sci. 13:361 (1975).
24. D. W. Criddle and R. L. LeTourneau, Anal. Chem. 23:1620 (1951).
25. P. J. Matthews, J. Appl. Chem. 20:87 (1970).
26. J. N. Done and W. K. Reid, Separ. Sci. 5:825 (1970).
27. R. Amos, J. Inst. Petrol. 54:9 (1968).
28. R. Amos, in "Gas Chromatography 1972" (S. G. Perry, ed.), Applied Science Publishers, London, 1973, p. 235.
29. R. Amos and S. G. Perry, J. Chromatog. 83:245 (1973).

Part V SPECIFIC APPLICATIONS

Chapter 15

ANALYSIS OF POLYNUCLEAR AROMATIC HYDROCARBONS

Ralph A. Brown and Thomas D. Searl
Analytical and Information Division
Exxon Research and Engineering Company
Linden, New Jersey

I. INTRODUCTION

The 1950s saw the initial major effort toward the development of methods to identify and measure polynuclear aromatic (PNA) hydrocarbons. Principal attention was directed toward air particulates, tobacco smoke, and tar. Benzo(a)pyrene (BaP) received major attention that continues to now. By the mid-1960s many papers had appeared on methods of analysis. Contributions came from laboratories of the United States, Canada, Australia, Europe, and Japan. In 1964, Sawicki [19] reviewed the analysis of PNAs based on a list of 244 references.

Most of the older literature describes the use of thin layer chromatography (TLC) to analyze these compounds. Although there is still an occasional reference to the use of this technique for the isolation and identification of PNAs, the use of TLC as the major tool to perform these analyses has been superseded by the more rapid techniques of gas and/or liquid chromatography.

The applicability of gas chromatography to PNA mixtures was apparently first demonstrated by Liberti et al. [15]. Since then, scores of publications have appeared which deal with techniques or methods based on the use of gas chromatography. An attractive feature of this separation is its ability to provide many enriched fractions in a single separation step. The best known procedure, called the GC/UV method, was developed and improved so as to handle tar from auto exhaust, gasoline, used crankcase oil [2], coke oven effluents [21], pitch volatiles in airborne particulates [23], and petroleum pitch volatiles [4].

There is a varying degree of difficulty involved in analyzing this range of samples and, consequently, there are some variations in complexity for given analyses. These variations will be discussed in Sec. III.

The GC/UV method is routinely employed to measure 18 individual PNAs, ranging from the 4-ring hydrocarbon, fluoranthene, to coronene, a 7-ring compound. The method is, however, not limited to this particular group of compounds. A list of the hydrocarbons that have been identified and measured is shown in Table 1. This list includes 42 hydrocarbons varying from 3- to 7-condensed ring compounds. Virtually every PNA of possible interest has been measured except for the dibenzopyrenes and the partially hydrogenated aromatics.

Table 2 summarizes the general applicability of the GC/UV method. This approach has been applied to the entire range of petroleum, tars, and particulates.

II. STORAGE OF SAMPLES

It is often necessary to store liquid samples, filters, and standard solutions. Liquids and filters are kept in containers covered with aluminum-lined caps. An initial blanket of nitrogen is provided. All samples are kept in the dark at room temperatures. Of particular concern in deterioration of samples is the possibility for photo-oxidation of some PNAs [13] or dropping out of solution due to limited solubility. Limited data indicate the storage procedures to be generally satisfactory; deterioration may occur, however, and must be prevented. Blends of PNAs in cyclohexane, for example, may show loss of benzo(g,h,i)perylene, coronene, or other compounds of low solubility. A preferred solvent is benzene or toluene, but even solutions of these solvents must be inspected prior to use after being in storage. Conversely, one must be alert to the possibility of losing solvent by evaporation so as to increase the PNA concentration.

TABLE 1. PNA Hydrocarbons Which Have Been Observed and Measured by GC/UV Method

3-Ring aromatics	4-Ring aromatics	5-Ring aromatics
Fluorene	Fluoranthene[a]	Benzo(g,h,i)fluoranthene[a]
1-Methylfluorene	Methylfluoranthenes	Benzo(b)fluoranthene[a]
2-Methylfluorene	Pyrene[a]	Benzo(j)fluoranthene[a]
2,3-Dimethylfluorene	Methylpyrenes	Benzo(k)fluoranthene[a]
Trimethyl and/or methylethylfluorenes	Dimethylpyrenes	Dibenz(a,h)anthracene
Anthracene	Cyclopenta(c,d)pyrene	Benzo(a)pyrene[a]
Methylanthracenes	5 H-Benzo(a)fluorene	Methylbenzo(a)pyrenes[a]
Phenanthrene	11 H-Benzo(f)fluorene	Benzo(e)pyrene[a]
1-Methylphenanthrene	Triphenylene[a]	Methylbenzo(e)pyrenes[a]
2-Methylphenanthrene	Chrysene[a]	Perylene[a]
Dimethylphenanthrenes	Methylchrysenes	
	Dimethylchrysenes	
	Benz(a)anthracene[a]	
	Methylbenz(a)anthracenes[a]	
	Dimethylbenz(a)anthracenes[a]	
	7,12-Dimethylbenz(a)anthracene	
	Benzo(c)phenanthrene	
	3-Methylcholanthrene	

6-Ring aromatics	7-Ring aromatics
Benzo(g,h,i)perylene[a]	Coronene[a]
Anthanthrene	

[a]Compounds which are routinely measured.

TABLE 2. Scope of GC/UV Method by Type of Sample and Measurement of 4- to 7-Ring PNA Hydrocarbons

Type of sample	Sample size	Range in concentration or quantity of individual PNA	Precision for individual PNA (1σ)	Accuracy for individual PNA
Gasoline	0.05-11	0.03-17 ppm	5-15 rel. %	
Heating oil / Diesel fuel	5-10 g	0.1-40 ppm	—	
Gas oil	1 g	—	—	
Lube oil, new	1 g	0.05-4 ppm	—	
Lube oil, used	1 g	0.5-62 ppm	—	
Heavy ends	10-100 g	10-240 ppm	—	
Crude oil	10-100 g	0.5-40 ppm	—	
Tar from auto exhaust gas	0.1-1.0 g	10-200 ppm	3-54 rel. %[d]	10 rel. %[a]
Tars from thermal cracking, etc.	0.1-2 g	100-15,000 ppm	—	
Particulates from ambient air, plant emissions	15 mg	0.1-20 μg	—	
Particulates from coke ovens	0.1-10 mg[c]	1-40 μg	—[b]	

[a]For BaA and BaP.
[b]See discussion.
[c]Cyclohexane solubles.
[d]See Table 3.

As described elsewhere [10a] a cyclohexane solution of a coke oven effluent was analyzed in duplicate, first in March 1969, and a second time in March 1970. Microgram quantities were, respectively, observed to be: pyrene 20, 21; BaA 22, 26; chrysene 15, 15; BaP 20, 21; and BeP 10, 9. Thus, no observed changes occurred during 1 year of sample storage.

III. DESCRIPTION OF METHOD

Because of the required sensitivity of the method, the possibility of error due to contamination is great. It is important that all glassware be scrupulously cleaned. Examination of all glassware, including stoppers and stopcocks, under ultraviolet light to detect any residual fluorescent contamination is required. As a precautionary measure it is recommended that all glassware be rinsed with purified isooctane immediately before use. No grease is to be used on stopcocks or joints. Tygon tubing must absolutely be avoided to eliminate di-isooctyl phthalate contamination, which interferes with uv measurements. Because some of the polynuclear hydrocarbons sought in this test are susceptible to photo-oxidation, the entire procedure should be carried out under subdued light. For laboratory lighting it is desirable to use yellow lights rather than the usual fluorescent lamps.

Radioactive materials must be handled and disposed of by accepted methods. Glassware used with radioactive species should be rinsed after use with chloroform-acetone solvent and washed in chromate-sulfuric acid cleaning solution.

The reagents and solvents employed in the method should be of high purity. The most important solvent, cyclohexane, is used to wash out trapping tubes and for the uv absorption measurements. Several sources are available, including Burdick and Jackson Laboratories, Inc., "Distilled in Glass" purity. To check purity, evaporate approximately 180 ml down to 5 ml. Run a uv scan on this residue in a 1-cm cell from 400 to 280 nm; the absorbance should not exceed 0.01 units. To purify, percolate through activated silica gel, Grade 12, in a glass column, 90 cm long and 5 to 8 cm diameter.

A. Internal Standards

Carbon-14 labeled BaA and BaP are ideal internal standards, with the preferred practice being to utilize both compounds in each analysis. They can be added at the beginning of any analytical procedure with the assurance that they will behave as the PNAs being measured. By using both compounds, insight is gained as to whether or not preferential loss might occur due to differences in volatility, thermal stability, or chemical stability.

In terms of the stability criteria, BaP is less stable than BaA. A difference between the two in recovery would reflect such an effect. On the other hand, equal recovery assures that the various PNAs have behaved similarly through the successive steps of the analytical procedure.

For some laboratories it may not be possible or necessary to use carbon-14 labeled internal standards. Compounds such as: 1,3,5-triphenylbenzene [22], o-terphenyl and bibenzyl [4, 21, 23] have been successfully used in samples that did not require a concentration step prior to injection in the gas chromatograph. Such compounds may not be satisfactory when prior concentration is required because they do not concentrate to the same degree as the PNAs being measured. Grimmer and co-workers [5, 6, 8] add internal standards such as dimethylphenanthrene and benzo(b)chrysene just prior to injection of the sample to a gas chromatograph. This practice does not provide a monitor for the entire analytical procedure.

B. Outline of Method

Figure 1 shows the individual steps of the method. Because of the variety of samples handled by the method there are a number of variations in individual steps. Indeed, some sample types are implicitly easier to analyze and do not require all of the steps as shown. Tars and pitches, for example, are relatively rich in PNAs and do not require separation on partially deactivated alumina. In order to describe the variations, the major steps of the procedure will be discussed in terms of the various sample types. While this descriptive material is quite detailed, a step-by-step description is available for coke oven effluent [10a] and tar from auto exhaust gas [2, 10b].

1. Preparation and Preliminary Sample Cleanup

a. Gasoline. A starting sample of 50 to 1000 ml is spiked with ^{14}C BaA and ^{14}C BaP and evaporated on a steam bath under nitrogen. As evaporation proceeds the sample is twice replenished with cyclohexane and finally reduced to 25 ml. The sample is then ready for the alumina column.

b. Heating Oil, Diesel Fuel. Standards are added to a weighed amount of sample (5 to 10 g). Because of the larger than normal size sample (1 g), the subsequent separation on partially deactivated alumina is in a scaled-up column.

c. Gas Oil, Lube Oil. Carbon-14 standards are added to a weighed amount of sample (0.5 to 1 g). The sample is then ready for charging to the partially deactivated alumina column.

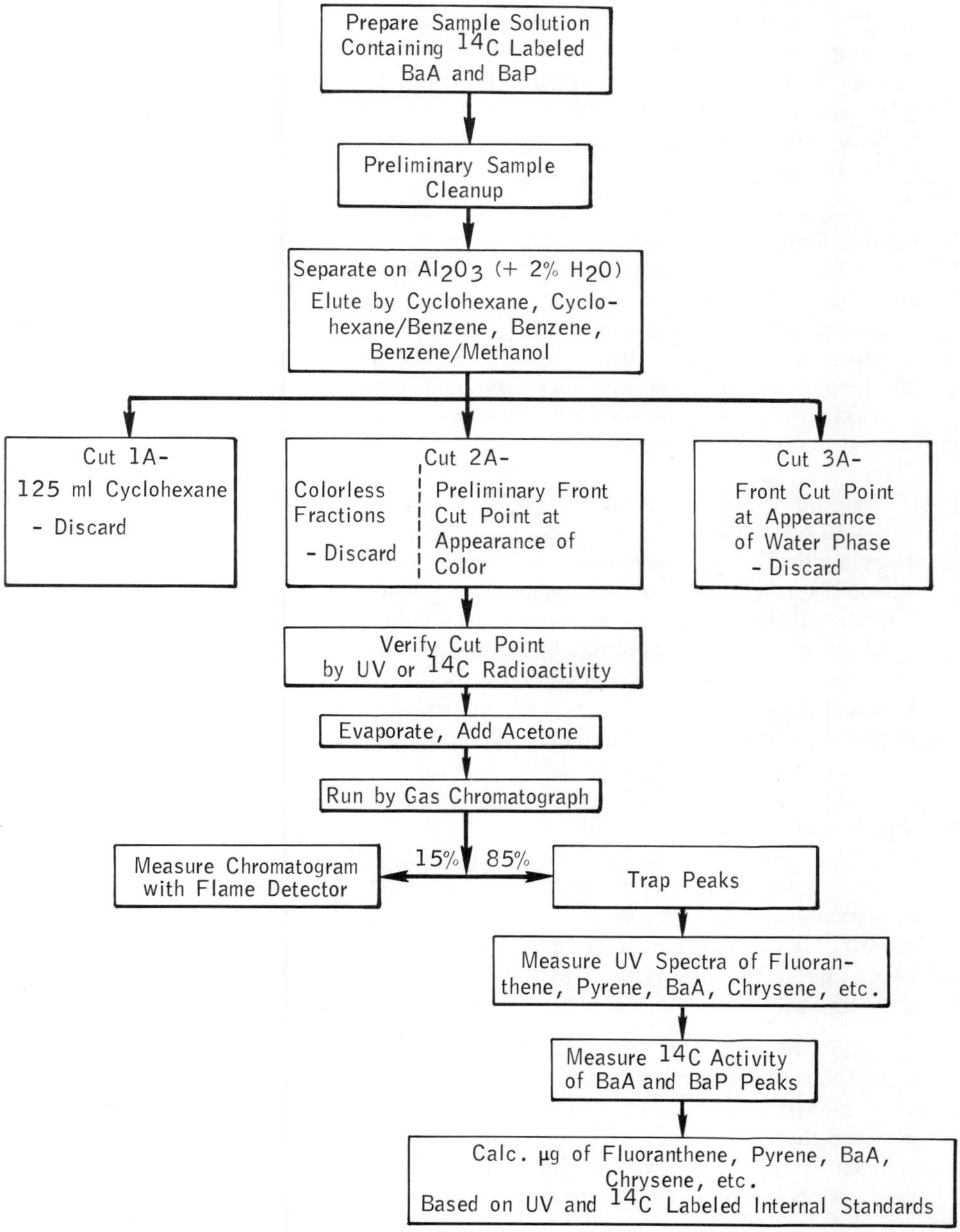

FIG. 1. Outline of method.

Recent, more extensive analysis of lube oils has resulted in our use of DMSO extraction for sample cleanup, a step which is employed in front of separation on the partially deactivated alumina column. This additional sample cleanup has been beneficial in providing cleaner uv spectra for the quantitative measurement. While the DMSO extraction has not been applied to heavy fuels and crude oils, it is felt that this procedure can also be advantageously applied to such samples.

10 g of sample containing the carbon-14 spike of BaA and BaP is placed in 150 ml of isooctane. The procedure then follows that of Howard et al. [11]. This isooctane solution is extracted with five 100 g portions of 10:1 dimethyl sulfoxide : H_3PO_4. Then 720 ml of water is added to the combined DMSO extracts and this solution is back extracted with 100 ml portions of isooctane. Each isooctane portion is washed with three 250 ml portions of water. The isooctane solutions are combined and reduced to approximately 10 ml on a steam bath under N_2. This solution is now ready for elution through the partially deactivated alumina column.

d. Heavy Fuels, Crude Oil [16]. For crude oil only, it is desirable to distill off 204C and lighter material using ASTM Method D 86. A weighed aliquot (about 12 g) is dissolved in 100 ml of n-pentane, and carbon-14 standards are added. At this point it is desirable to separate the hydrocarbons as saturates and aromatics from the polar compounds utilizing 100/200 mesh silica gel. The column, 760 mm by 22 mm, is fitted with a 500-ml reservoir and a Teflon stopcock; 180 grams of the gel is placed in the column and prewetted with 50 ml of n-pentane to remove saturated hydrocarbons. Elution by 500 ml of acetone removes the aromatics. The acetone eluate is then evaporated on a steam bath to small volumes three successive times. Each time the volume is brought up to about 25 ml with cyclohexane, to finally give a cyclohexane solution for separation on the alumina column.

e. Auto Exhaust Tar. In our work, the starting sample of exhaust tar was in solution as several gallons of cyclohexane/acetone solvent containing the tar. It was necessary to distill off solvent to a residual volume of one liter. Prior to starting the distillation, ^{14}C BaA and ^{14}C BaP are added to the sample.

Caustic extraction is needed for tar samples. 500 ml of the 1-liter sample is extracted three times with 0.5 N aqueous sodium hydroxide and then washed successively with 0.1 N HCl and water. This extraction step reduced the background of the gas chromatogram. 250 ml of the caustic treated sample (1/4 of the original sample) was charged to the alumina column. This volume usually contains 100 to 400 mg of tar.

f. Thermal Tars, Pitch. A weighed quantity (0.1 g) is dissolved in toluene and the internal standards are added. It is usually unnecessary to prepare a PNA concentrate by elution through a partially deactivated alumina column. Thus, after evaporating toluene so as to obtain a concentrate, an aliquot of the solution can be injected directly to the gas chromatograph.

g. Air Particulates, Coke Oven Effluents. Samples are collected on a silver membrane filter [18] or a Hi Vol Sampler. Soxhlet apparatus can be used for extraction but a quicker method of suitable proficiency has been employed as described by Chatot [3]. He demonstrated that an ultrasonic bath efficiently extracts PNAs from filters. To apply this technique to a Hi Vol Sampler filter, the filter is cut into small pieces and placed in a 125-ml Erlenmeyer flask with 100 ml of cyclohexane. After adding carbon-14 labeled internal standards, the sample is placed on a steam bath for 30 min and then placed in an ultrasonic bath for 2 min. Contents of the flask are filtered, using a pressure funnel. The filtrate is evaporated under flowing nitrogen on a steam bath to a residual volume of 2 ml. For air particulates the sample is then ready for separation on the partially deactivated alumina column. For a coke oven effluent, the filtrate would simply be reduced to a small residual volume prior to direct injection to the gas chromatograph.

2. PNA Fraction from Adsorption Chromatography

A PNA concentrate is prepared using the materials and conditions listed in Table 3. After the elution of 100 ml of cyclohexane from the column, 10-ml fractions are collected and examined for radioactivity. The first show of radioactivity represents the appearance of BaA in the alumina eluent. This fraction and higher ones will contain the 4-rings plus (except pyrene and fluoranthene, which appear in the fractions just preceding the ^{14}C activity).

TABLE 3. Conditions for Stepwise Elution of PNAs from Samples

Column dimensions:	50 cm length and 1.1 cm ID
Alumina:	Woelm neutral alumina dried at 160°C and then deactivated with 2 wt. % water
Elution schedule:	100 ml of cyclohexane, including starting volume of sample
	100 ml of cyclohexane/benzene (4:1)
	100 ml of benzene
	100 ml of benzene/methanol (1:1)

In an alternative technique, the 10 ml fractions are inspected by uv for the appearance of pyrenes (absorption band at 340 nm). The first fraction containing pyrene and all subsequent fractions to the appearance of a two-liquid phase boundary are combined as the PNA concentrate. The appearance of pyrene has varied from the second to the seventh 10-ml fraction of the cyclohexane-benzene elution step.

A second alternative method is also available and from experience this may frequently be the most convenient. In this method, the first colored fraction and all subsequent fractions to the appearance of a two-liquid phase boundary may be combined as the PNA fraction containing 3-ring plus hydrocarbons. For most samples, the initial colored fraction occurs near the end of the cyclohexane/benzene (4:1) solvent followed by nine to eleven additional fractions. Appearance of water from the alumina indicates the end of the colored fraction. This technique is not applicable to lighter hydrocarbon samples such as gasoline, diesel fuel, and number 2 heating oil because their fractions are colorless.

3. Gas Chromatography

The operating conditions for the gas chromatographic separation are shown in Table 4.

The influence of operating conditions on the recovery of PNAs indicated that the sample injection port temperature was very important.

Figure 2 shows the gas chromatograph trapping assembly. To trap a fraction the stainless steel tube is slid into position as shown.

TABLE 4. Operating Conditions for GC Separation of PNA Fractions

Chromatograph	Perkin-Elmer Model 900
Gas flow	Helium at 30 ml min^{-1}
Column	300 cm-0.22 cm ID stainless steel, packed with 2% SE 30 (GC grade) on Chromosorb "G" 80/100 mesh (acid washed and DMCS treated)
Injection port	305°C
Detector	FID-345°C
Program	175°C to 300°C at 4°/min, then hold

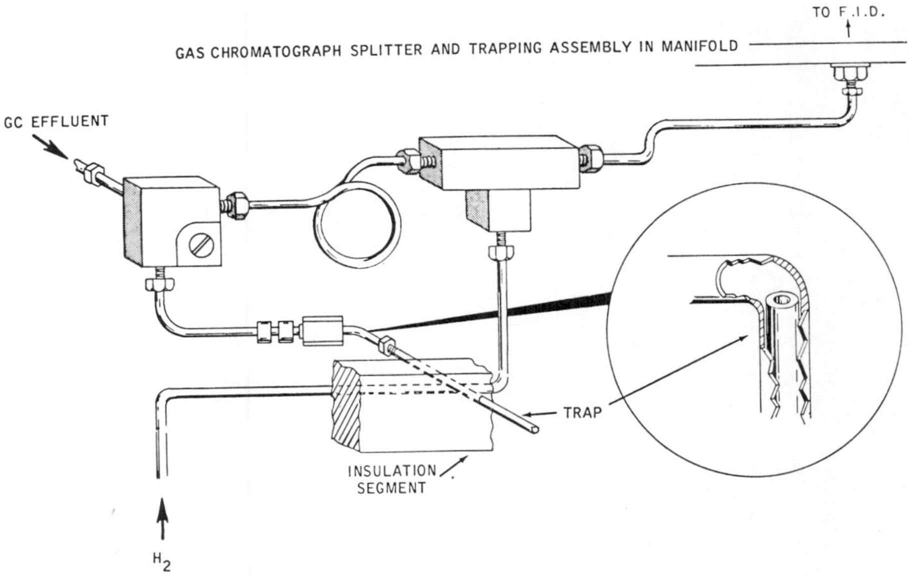

FIG. 2. Gas chromatograph splitter and trapping assembly in manifold.

As sample is run on the chromatograph, selected peaks are trapped and the trap content rinsed out with cyclohexane for uv measurement. This rinse is not 100% effective for compounds such as perylene, B(g,h,i)P, coronene, and anthanthrene. For GC peaks containing these PNAs, the 5-ml graduate containing the trap and cyclohexane is placed in an ultrasonic bath for 30 sec.

Typical retention times for 2-ring plus PNAs are shown in column 2 of Table 5. For trapping purposes, these values provide a guide to the occurrence of these compounds. Retention times vary somewhat; however, an experienced technician soon learns to recognize how a given column performs.

4. Ultraviolet Absorption Spectrophotometry

The cyclohexane rinse of the GC peak is made up to 3.8 ml and the uv absorbance measured in a 1-cm path cell. From work in our laboratory and other published data, Table 5 presents a compilation of uv absorptivity coefficients for 42 individual 3- to 7-ring PNA hydrocarbons. Slant base line measurements are used and anchor points are shown along with the analytical band. In most cases, the calibration coefficients are based upon measurements done in triplicate. They are also corrected for the observed purity of each compound for which purity was established by GC and MS.

TABLE 5. Gas Chromatographic Retention Times and UV Absorption Calibrations for Polynuclear Aromatic Hydrocarbons

GC conditions: Column, 300 cm of 0.32 cm OD, 2% SE-30 on Chromosorb G (acid washed and DMCS treated) 80/100 mesh; temperature program, 175°C to 300°C at 4°C min^{-1}

Compound	GC retention time (min)	Background wavelengths (nm)		Wavelength (nm)	Absorptivity (ml μg^{-1} cm^{-1})	Reference number or footnote
3-Ring						
9,9 Dimethylfluorene	4.4	298	307	302	0.050	1
		238	287	267	0.074	
Fluorene	4.5	297	306	301	0.044	1
		233	285	262	0.090	
9-Methylfluorene	5.0	298	307	302	0.041	1
		238	287	267	0.083	
2-Methylfluorene	6.2	302	312	306	0.034	1
		239	290	267	0.095	
1,9-Dimethylfluorene	6.2	297	307	301	0.027	1
		238	286	267	0.005	
1-Methylfluorene	6.3	297	305	300	0.040	1
		240	285	267	0.10	

4-Methylfluorene	6.4	296	301	297	0.019	1
		238	295	267	0.11	
Phenanthrene	7.2	288	298	293	0.064	1
		226	270	252	0.33	
2-Ethylfluorene	7.4	302	312	306	0.038	1
		239	290	269	0.105	
Anthracene	7.6	—	265	253	1.26	1
2,3-Dimethylfluorene	8.2	303	313	308	0.049	1
		238	291	267	0.099	
2-Methylphenanthrene	9.8	290	301	295	0.05	1
		228	273	254	0.34	
1-Methylphenanthrene	10.3	294	307	300	0.064	1
		—	273	256	0.33	
5H-Benzo(a)fluorene	15.1	262	269	263	0.63	h
11H-Benzo(b)fluorene	15.1	258	271	264	0.18	
<u>4-Ring</u>						
Fluoranthene	12.9	285	292	288	0.199	10a
Pyrene	13.6	327	343	336	0.24	2,10a,10b
	—	327	345	336	0.25	4

TABLE 5 (continued)

Compound	GC retention time (min)	Background wavelengths (nm)		Wavelength (nm)	Absorptivity (ml μg^{-1} cm^{-1})	Reference number or footnote
<u>4-Ring</u> (cont.)						
Methylpyrenes	15.6, 16.0	328-333	345-353	337-344	0.21	h
	—	—	284	276	0.18[a]	4
Dimethylpyrenes	19.2	340	356	348	0.13[b]	h
Benzo(c)phenanthrene	19.4	250	298	282	—	h
Triphenylene	20.0	252	263	259	0.480	2,10a,10b
Chrysene	20.4	263	277	269	0.464	2,10a,10b
	—	260	275	267	0.53	4
Methylchrysenes	—	262	278	271	0.38[c]	4
Dimethylchrysenes	—	263	279	272	0.38[d]	4
Benz(a)anthracene	20.1	283	295	289	0.332	2,10a,10b
	—	—	297	289	0.40	4
Methylbenz(a)anthracenes	23.3	285	300	293[e]	0.27	2,10b
Dimethyl/ethylbenz(a)-anthracenes	25.0	285	300	293[e]	0.27	2,10b
7,12-Dimethylbenz(a)-anthracenes	25.7	285	307	298	0.22	h

3-Methylcholanthrene	29.2	293	303	298	—	h
5-Ring						
Benzo(g, h, i)fluoranthene	19.2	288	300	291	0.051	9
Benzo(b)fluoranthene	25.9	297	308	302	0.120	9
Benzo(j)fluoranthene	25.8	330	337	332	0.063	9
Benzo(k)fluoranthene	25.8	393	405	398	0.085	9
Dibenz(a, h)anthracene	32.4	294	305	298	0.48	h
Benzo(a)pyrene	27.3	373	390	383	0.092	10a, 10b
Methylbenzo(a)pyrenes[f]	29,0, 29.6, 30.7	373	390	384[g]	0.075	2, 10b
Benzo(e)pyrene	27.2	325	340	333	0.135	2, 10a, 10b
Methylbenzo(e)pyrenes[f]	29,0, 29.6, 30.7	325	340	335	0.110	2, 10b
Perylene	27.8	418	460	437	0.130	9
6-Ring						
Benzo(g, h, i)perylene	33.3	370	390	382	0.074	10b
Anthanthrene	33.7	298	313	305	0.32	h
		423	440	430	0.08	

TABLE 5 (continued)

Compound	GC retention time (min)	Background wavelengths (nm)		Wavelength (nm)	Absorptivity (ml μg^{-1} cm^{-1})	Reference number or footnote
7-Ring						
Coronene	40.6	295	312	303	0.737	9
1,3,5-Triphenylene	28.5	225	290	254	0.142	

[a]1-Methylpyrene.
[b]1,9-Dimethylpyrene.
[c]3-Methylchrysene.
[d]Used 3-methylchrysene since a dimethylchrysene isomer was not available.
[e]Occurs in range 290-293 nm.
[f]Measured in three different GC peaks.
[g]Occurs in range, 383-385 nm.
[h]In laboratory measurement.

5. Calculation of PNAs in GC Peaks

Concentrations of the individual PNAs are calculated from uv spectra of GC peaks based on absorptivity coefficients as listed in Table 5. With minor exception the given analytical band is unique for its compound in terms of the mixture to be anticipated for the GC peak. In some cases, the GC peak contains only one PNA hydrocarbon so as to assure uniqueness. In a typical mixture, for example, three different benzofluoranthenes occur in the same GC peak. However, each compound can be measured independently of one another by using the analytical bands at 302 nm, 332 nm, and 398 nm for benzo(b)fluoranthene, benzo(j)fluoranthene, and benzo(k)-fluoranthene, respectively.

Of the 42 individual hydrocarbons that have been quantitatively measured by the GC/UV method, only eight of them occur as mixtures in GC peaks where there is overlap among the analytical bands. One such mixture includes BaA, chrysene, and triphenylene as depicted by a typical uv absorption spectrum in Figure 3. To calculate the concentration of each compound, a set of three simultaneous equations is solved by substitution. The starting matrix of coefficients is given in Table 6.

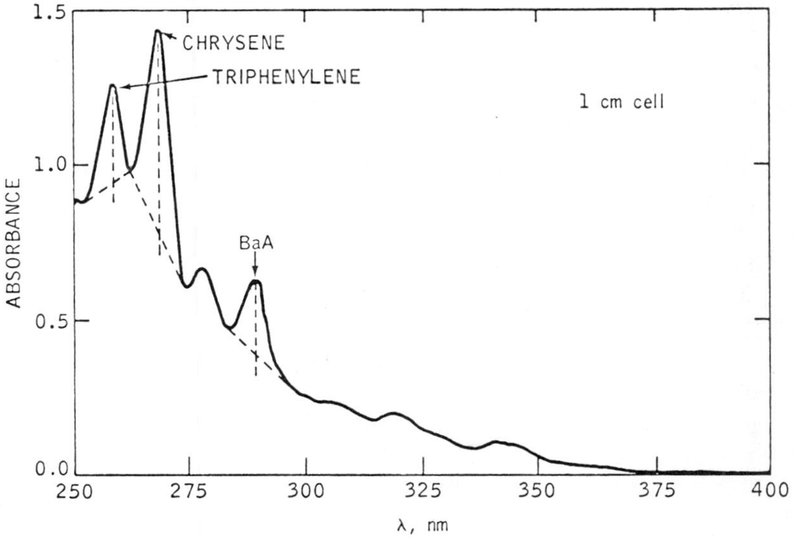

FIG. 3. Ultraviolet spectrum of GC fraction containing BaA, chrysene, and triphenylene.

TABLE 6. Calibration Matrix for Calculation of BaA, Chrysene, and Triphenylene

	BaA	Chrysene	Triphenylene	A[a]
(1)	0.332	0	0	at 289 nm
(2)	0.039	0.464	0	at 269 nm
(3)	0.015	0.107	0.480	at 259 nm

[a]Ultraviolet absorbance of GC peak in a 1 cm cell.

The equations are:

$$\text{BaA} = \frac{\text{A289}}{0.332}$$

$$\text{Chrysene} = \frac{\text{A269 - BaA} \times 0.039}{0.464}$$

$$\text{Triphenylene} = \frac{\text{A259 - BaA} \times .015 - \text{Chrysene} \times 0.107}{0.480}$$

A second example is the GC peak which consists of BaP, perylene, and benzo(e)pyrene (BeP). BeP can be directly calculated from its analytical band at 333 nm. Perylene, however, has a measurable uv absorption at the analytical band of BaP and a matrix calculation is also required in this situation.

In still another case, the two isomeric benzo(b)fluorenes have identical retention times and their uv spectra are too similar for the two hydrocarbons to be separately resolved. Thus, it is necessary to group the two based upon an average absorptivity.

In the calculation of uv spectral measurements, it should also be noted that methyl-substituted BaP and BeP isomers require special handling as they occur in three separate GC peaks. In practice, each of the three peaks is separately trapped and measured by uv. Summed values are then used to calculate the quantity of each methyl isomer. A more efficient approach would be to blend the solutions of the three fractions prior to the uv measurement. This is not done because there are variations in the wavelength position of the maximum absorption and the uv absorption of the blend would therefore be nonadditive. For methyl BaP the band position was observed to vary from 384 to 386 nm, and for methyl BeP it was 334 to 337 nm. Exact details on the calculations to a sample basis have been published in the three references previously cited [2, 10a, 10b].

C. Role of Other Techniques

Because trapped GC fractions are taken in the GC/UV method, it is convenient to examine such fractions by other techniques to seek greater sensitivity, confirm uv absorption measurements, or to obtain additional information. To this end, rather extensive use of mass spectrometry has been made to obtain greater compositional detail. Other techniques, such as uv fluorescence and high resolution gas chromatography, have also been used.

1. Mass Spectrometry

The principal value of mass spectral measurements is to provide more detail on individual GC peaks. This involves the ability to introduce into the mass spectrometer microgram quantities of material in a relatively large volume of solvent, usually cyclohexane.

A satisfactory procedure is to add a standard solution of dioctylphthalate and triphenylbenzene to the cyclohexane solution. The phthalate is relatively nonvolatile and serves as a sink for the PNA during the evaporation of the cyclohexane, whereas the triphenylbenzene is the internal standard. Its use allows a quantitative measurement to be made of any PNA in the GC fraction. Mass spectrometry is generally carried out at low ionizing voltage conditions.

This technique has proven helpful in many instances. In one auto exhaust tar sample, for example, the extent of methyl substituted BaP and BeP was measured as shown in Table 7.

Peak 20 contained BaP and BeP. Peak 22 showed only 266 as a molecular-weight ion, which indicates the presence of only methyl BaP and methyl BeP. In peak 23, however, only 58% of MW 266 was found and there was 42% of MW 280. Mol. wt. 280 is probably due to dimethyl and/or ethyl BaP and BeP. Peak 24 consists primarily of mol. wt. 280 compounds.

2. Ultraviolet Fluorescence

Although uv fluorescence is a much more sensitive technique than uv absorption it is not generally used for quantitative analysis. Reasons for this include its lower reliability and its inability to handle as complex a mixture. Because the GC/UV method frequently does provide relatively "pure" GC peaks, however, it would appear that fluorescence should be a valued supportive technique [17].

In a large number of samples, fluorescence measurement of BaP has been compared with values from the corresponding uv absorption spectra. For coke oven effluents [22] and tar from auto exhaust gas, agreement between the two techniques was within 1 μg (1 σ) at about 8 μg. Thus far,

TABLE 7. Identification of GC Fractions Containing Alkyl Substituted BaP and BeP

Peak number	BaP series nm	BeP series nm	Mol. wt.	Identification
20	383	332	252	BaP, BeP
22	384	334	266-100%	Methyl BaP, Methyl BeP
23	384 to 388[a]	334, 337	266-58%	Methyl BaP, Methyl BeP
			280-42%	Dimethyl/ethyl BaP, Dimethyl/ethyl BeP
24	385	335	266-15% 280-85%	Same as fraction number 23 above

[a]Broad peak.

fluorescence has been of value to cross-check on the measurement of BaP, BaA, chrysene, benzo(b)fluoranthene, and perylene. Calibration curves were prepared based upon the work of Sawicki et al. [20].

In general, the fluorescence technique is only applicable for a GC peak containing minimal quantities of extraneous uv absorbers. An "unclean" GC peak with a high background of continuous uv absorption is not amenable to a reliable measurement, apparently, due to quenching effects.

3. Capillary Gas Chromatography

Coated open tubular (capillary) columns are used in the measurement of PNAs. In these methods samples are prepared so that upon injection to a chromatograph individual PNAs are identified by a single peak having a predetermined retention time. Grimmer and co-workers have extensively applied such a method to a variety of sample types [5, 6, 8]. However, not all PNAs can be individually resolved by presently known GC columns.

The capillary column separation does provide a convenient way of cross-checking selected compounds from a GC/uv analysis. In the analysis of a South Louisiana crude oil [16], the capillary gas chromatogram of the GC peak containing BaP, BeP, and perylene was found to agree with a known blend whose composition was the same as initially measured by GC/uv. This is demonstrated in Figure 4.

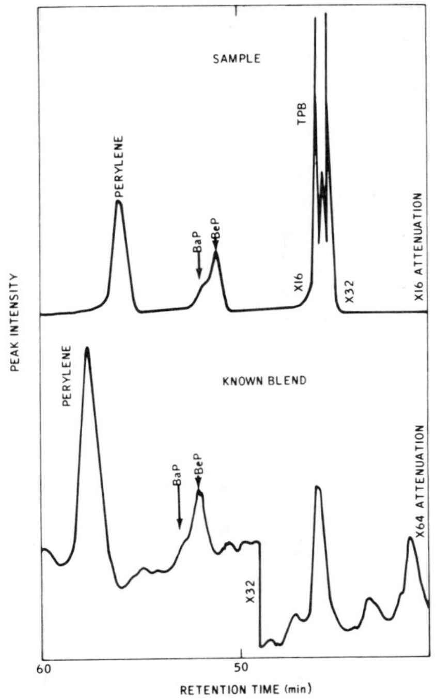

FIG. 4. Capillary gas chromatogram of fraction 7 and a known blend.

D. Performance of GC Trapping Apparatus

Based on many years of experience, this trap has been found to be both convenient and efficient. It is so easy to change traps that it is normal practice to individually trap most peaks even though this may not be necessary. By storing solvent solutions of these peaks in small vials, it is possible to examine them at a later time if additional information is required for a given sample. The normal protocol involves trapping of peaks that are only seconds apart or only even partially resolved. Sometimes large peaks precede small ones and vice versa. Regardless of the circumstances, uv and MS examination of successively trapped peaks consistently show variations in spectra which demonstrate that the trap is accurately tracking changes in GC effluent.

An example of the trap efficiency is the ability to separately resolve 2-methylphenanthrene, 1-methylphenanthrene, and an alkylfluorene, as found in petroleum [1]. The partially resolved gas chromatogram of this mixture is shown in Figure 5. The uv spectra of peaks 5 and 6 (Figure 6)

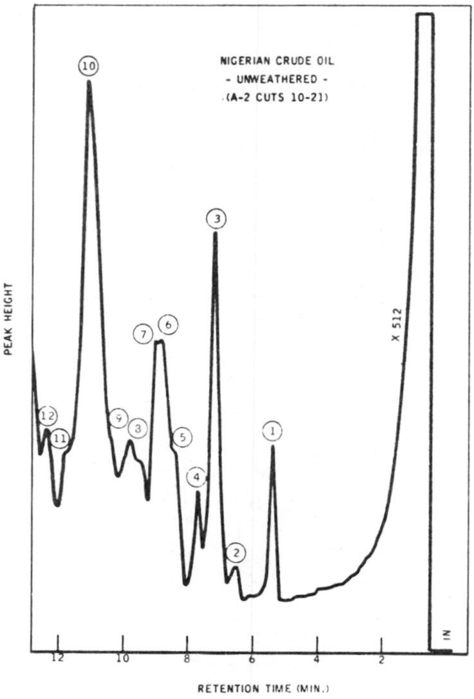

FIG. 5. Nigerian crude oil (A-2 cuts 11-21).

are seen to agree closely with the respective pure compounds, 2-methylphenanthrene and 1-methylphenanthrene. A reference spectrum is not available for peak 7 but from other data it can be reliably identified as an alkylfluorene. The spectra of the successive traps are different from one another and there appears to be little, if any, smearing of one hydrocarbon to another. This lack of smearing may be confirmed by close examination of Figure 6.

E. Examples

Figure 7 shows the chromatograms for some typical samples, including a petroleum pitch, auto exhaust gas tar, and a crude oil. The concentration or quantity of many individual PNAs is as summarized in Table 8 [9, 16]. As previously discussed, petroleum pitch is relatively simple to analyze. The sample is shaken with warm toluene in order to extract the PNAs after adding the internal standards. The toluene solution is evaporated so as to prepare a concentrate for the GC/uv steps. The sample is somewhat

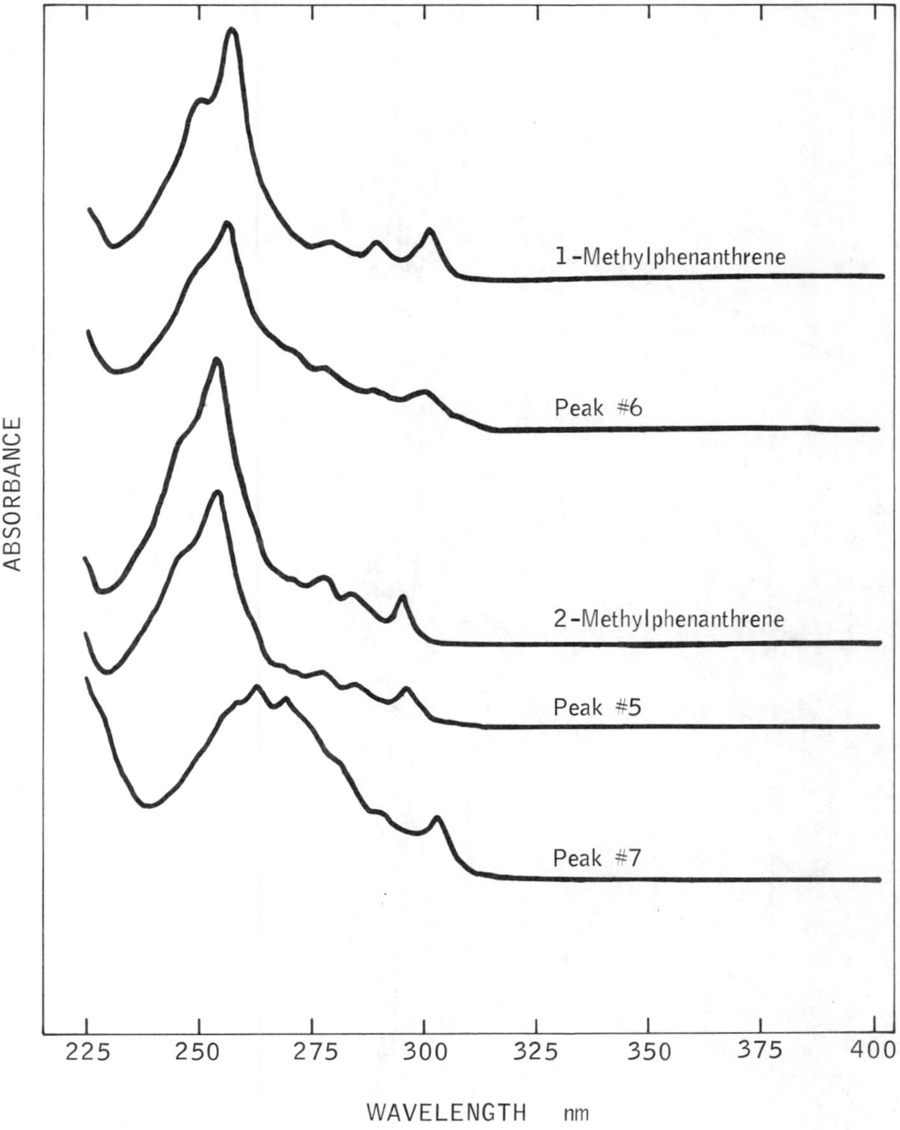

FIG. 6. Ultraviolet spectra of 1-methylphenanthrene, 2-methylphenanthrene, and GC peaks of Nigerian crude oil.

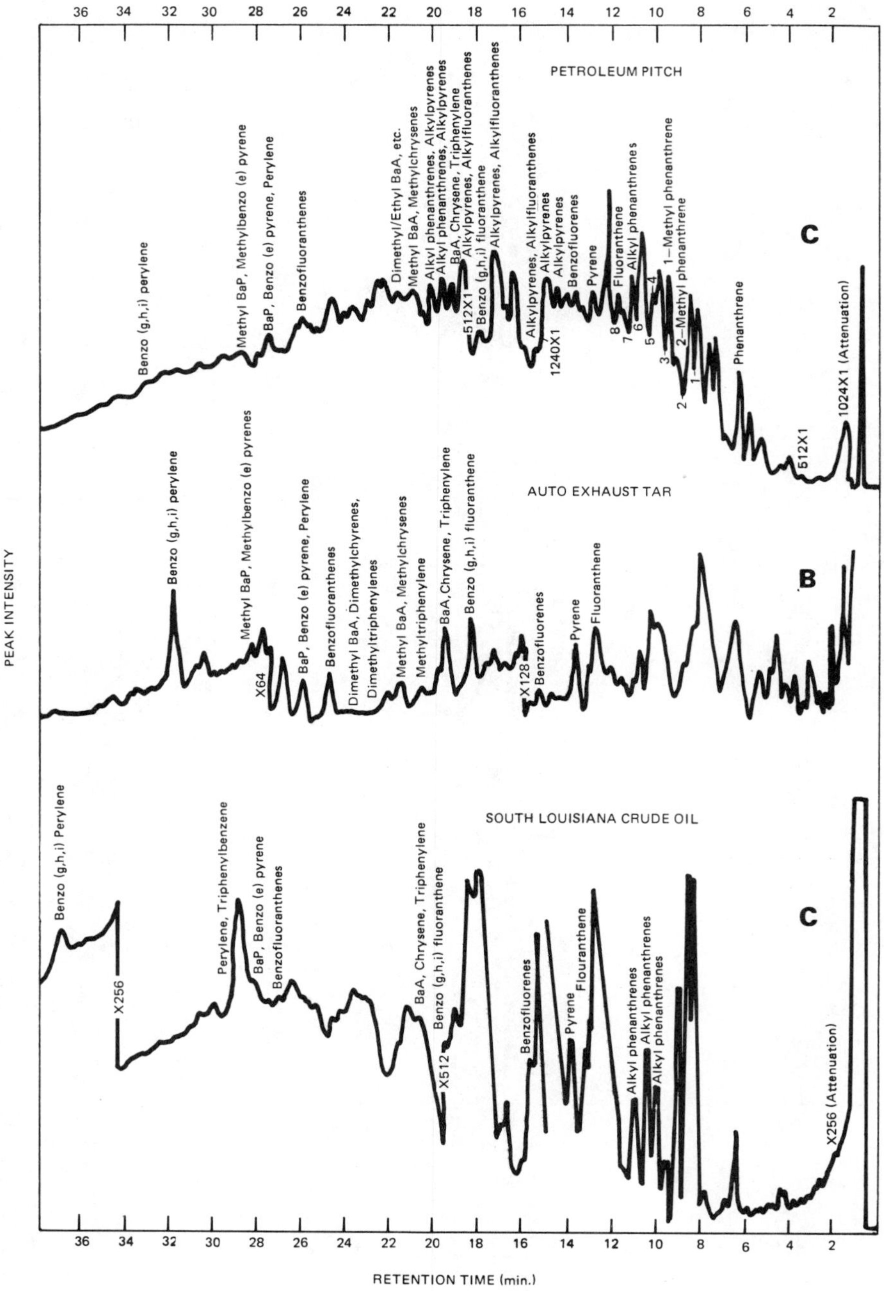

FIG. 7. Chromatograms of some typical samples.

TABLE 8. Polynuclear Aromatic Hydrocarbons in Some Typical Samples

	Petroleum pitch (ppm)	Tar from auto exhaust gas (μg)	So. Louisiana crude oil (ppm)
Phenanthrene	—	—	70
2-Methylphenanthrene	—	—	144
1-Methylphenanthrene	—	—	111
Fluoranthene	490	523	5.0
Pyrene	1180	421	3.5
Benz(a)anthracene	500	38	1.7
Chrysene	1510	77	17.6
Triphenylene	340	21	10
Methylbenz(a)anthracene	70	21	—
Dimethyl/ethylbenz(a)-anthracene	<50	3.4	—
Benzo(g,h,i)fluoranthene	460	106	1
Benzo(b)fluoranthene	300	50	<0.5
Benzo(j)fluoranthene	150	8.6	<0.9
Benzo(k)fluoranthene	190	1.2	<1.3
Benzo(a)pyrene	500	21	0.8
Methylbenzo(a)pyrene	100	5.7	—
Benzo(e)pyrene	310	75	2.5
Methylbenzo(e)pyrene	120	32	—
Perylene	100	3.9	34.8
Benzo(g,h,i)perylene	<50	78	1.6

unusual in having a range of alkyl phenanthrenes, fluoranthenes, and pyrenes that appear throughout much of the chromatogram. These compounds did not seriously interfere with the measurement of the other PNAs, however.

Tar from auto exhaust gas is reported in terms of micrograms in a total sample. Since a 1/4 aliquot was used for the analysis, the measured quantities were actually 1/4 of those reported in Table 8.

Of the three sample types discussed here, crude oil is the most difficult to analyze. As previously described in Sec. III.B.1, considerably more sample preparation is required to obtain a concentrate for the GC/uv steps. This is necessitated by the somewhat lower concentrations in crude oil and its much more complex composition. This is particularly evidenced by the presence of interfering uv absorbers.

IV. OTHER METHODS

Previous sections of the chapter refer to modifications of the GC/uv method as used by other workers in the analysis of volatiles from coal tar and petroleum pitch. Schulte et al. [21] and Seim et al. [23] analyze coke oven effluents for fluoranthene, pyrene, benz(c)acridine, chrysene, BaA, benz(a)anthrone, benzo(b)fluoranthene, benzo(k)fluoranthene, benzo(j)fluoranthene, BaP, BeP, dibenz(a,h)anthracene, benzo(g,h,i)perylene, and anthanthrene. Greinke and Lewis [4] analyzed petroleum pitch volatiles for methylphenanthrenes, pyrene, methylpyrenes, chrysene, BaA, methylchrysenes, dimethylchrysenes, BaP, and BeP. Individual components were identified by uv absorption spectroscopy and mass spectrometry.

Starting in the mid-1960s, Grimmer and co-workers have extensively published on an analytical method and its application to samples such as dust [5], auto exhaust [5, 6, 7], foodstuffs [8], and others. The precise steps vary with sample type but may include solvent extraction and liquid-solid chromatography to prepare PNA concentrates for direct measurement by capillary gas chromatography. Internal standards are added just prior to the GC measurement. In a typical case two separate fractions are examined, one identified as naphthalene-pyrene and a second as pyrene-coronene. Identification is by GC retention time and the quantitative measure is by the flame ionization detector of the chromatograph. A typical array of hydrocarbons reported in this method includes phenanthrene, anthracene, pyrene, fluoranthene, BaA, chrysene, BaP, and BeP, anthanthrene, benzo(g,h,i)perylene, dibenz(a,h)anthracene, and coronene.

In the analysis of auto exhaust tar for PNAs, Jentoft and Gouw employ tritiated BaA and BaP as internal standards [12]. The principal steps include separation by supercritical fluid chromatography.

The analysis of coke oven emissions, coal tar, and air particulates is described by Lao et al. [14]. The three-step method consists of (a) a preliminary separation of PNA by solvent and/or column chromatography, (b) identification by a combination of gas chromatography with quadrupole mass spectrometry and computer, and (c) measurement by computerized gas chromatography using internal standards. Most of the PNAs present in a sample are measured individually or in groups to provide detailed compositional detail covering 3- to 7-ring compounds. No precision or accuracy data are presented. So far, demonstration of the method has been accomplished for samples that are relatively simple to analyze. Its ability to handle more complex samples containing smaller concentrations of the PNAs will be of great interest.

ACKNOWLEDGMENT

The author gratefully acknowledges the courtesy of the American Chemical Society for permission to publish Figure 2 in this chapter.

REFERENCES

1. R. A. Brown, "Identification Based on Gas Chromatograms of Tricyclic Aromatics in Petroleum," Contract 68-01-0058, Office of Research and Monitoring, EPA, Washington, D.C., 1972.
2. R. A. Brown, T. D. Searl, W. H. King, Jr., W. A. Dietz, and J. M. Kelliher, Rapid Methods of Analysis for Trace Quantities of Polynuclear Aromatic Hydrocarbons in Automobile Exhaust, Gasoline, and Crankcase Oil, Final Report, CRC-APRAC Project CAPE-12-68, U.S. Document #PB-219-025, 1971.
3. G. Chatot, Anal. Chim. Acta 53:259 (1971).
4. R. A. Greinke and I. C. Lewis, Anal. Chem. 47:2151 (1975).
5. G. Grimmer and H. Böhnke, Z. Anal. Chem. 261:310 (1972).
6. G. Grimmer, A. Hildebrandt, and H. Böhnke, Erdöl und Kohle 25(8): 442 (1972).
7. G. Grimmer, A. Hildebrandt, and H. Böhnke, Erdöl und Kohle 25(9): 531 (1972).
8. G. Grimmer and H. Böhnke, J. Assoc. Off. Anal. Chemists 58(4): 725 (1975).
9. G. P. Gross, Gasoline Composition and Vehicle Exhaust Gas Polynuclear Aromatic Content, Final Report, CRC-APRAC Project CAPE-6-68, U.S. Document #PB-233-529, 1974.

10a. Tentative Method of Analysis for Polynuclear Aromatic Hydrocarbons in Coke Oven Effluents, International Standard No. 11104-03-73T, Health Lab. Sci. 11:218 (1974).

10b. Tentative Method of Analysis for Polynuclear Aromatic Hydrocarbons in Automobile Exhaust, International Standard No. 11104-04-73T, Health Lab. Sci. 11:228 (1974).

11. J. W. Howard, T. Fazio, R. H. White, and B. A. Klimeck, J.A.O.C.S. 51:122 (1968).
12. R. E. Jentoft and T. H. Gouw, Chemistry, Occurrence, and Measurement of Polynuclear Aromatic Hydrocarbons, Preprint, Div. Petrol. Chem., ACS, Chicago, August 1975, p. 797.

13. M. Kuratsune and T. Kirohata, Nat. Cancer Inst. Monograph 9:117 (1962).
14. R. C. Lao, R. S. Thomas, and J. L. Monkman, J. Chromatog. 112: 681 (1975).
15. A. Liberti, G. P. Cartoni, and V. Cantuti, J. Chromatog. 14:141 (1954).
16. R. J. Pancirov and R. A. Brown, Conference on the Prevention and Control of Oil Spills, San Francisco, CA, 1975, Amer. Petrol. Inst., Washington, D.C., p. 103.
17. R. J. Pancirov and R. A. Brown, Chemistry, Occurrence, and Measurement of Polynuclear Aromatic Hydrocarbons, Preprint, Div. Petrol. Chem., ACS, Chicago, August 1975, p. 812.
18. R. T. Richards, D. T. Donovan, and J. R. Hall, Amer. Ind. Hyg. Ass. J. 28:590 (1967).
19. E. Sawicki, Chemist-Analyst 53:24 (1964).
20. E. Sawicki, T. R. Hauser, and T. W. Stanley, Int. J. Air Poll. 2:253 (1960).
21. K. A. Schulte, D. J. Larson, R. W. Harnung, and J. V. Crable, Am. Ind. Hyg. Assoc. J. 36:131 (1975).
22. T. D. Searl, F. J. Cassidy, W. H. King, Jr., and R. A. Brown, Anal. Chem. 42:954 (1970).
23. H. J. Seim, W. W. Hanneman, L. R. Barsotti, and J. J. Walker, Am. Ind. Hyg. Assoc. J. 35:718 (1974).

Chapter 16

SYNTHETIC FUELS

Douglas M. Jewell
Gulf Research and Development Company
Pittsburgh, Pennsylvania

I. INTRODUCTION

With increasing concern over petroleum reserves to meet the needs of an industrialized society, serious efforts are being directed to produce alternative or "synthetic" liquid fuels. These efforts are centered primarily on coal, oil shale, and tar sands, because of their availability; a smaller effort is being placed on organic wastes of various sources. If one wishes to determine the best use of these fuels, it is necessary to (1) determine the compositional characteristics of each and (2) be able to compare the characteristics of one fuel directly with those of another and with the petroleum being replaced. In order to achieve the first, numerous chromatographic techniques must be applied, which would be true in any complex natural product problem. Several useful and current techniques are discussed below.

Experience has shown that in order to obtain meaningful information on coal liquids, shale oils, or petroleum, it is necessary to predictably separate a sample into a reasonable number of well-defined fractions and then analyze each fraction in depth. Conclusions as to the composition of the original sample are then drawn from these analyses in a manner consistent with the steps used to obtain them.

II. GAS CHROMATOGRAPHY

Chapters 4 and 5 discussed gas chromatographic techniques which are frequently used for analyzing petroleum. The same techniques can be used for synthetic fuels; however, one quickly observes the lack of suitable reference compounds for the more complex coal liquids or shale oils, e.g., cyclic structures are dominant in coal liquids and tar sands bitumen, while olefins (of all types) occur in shale oils.

Gas chromatography is most useful through the kerosene range ($\sim C_{10}$) for precise applications (component identification). However, simulated distillation by GLC is also useful in synthetic fuels up to approximately 1050°F (565°C).

With the advent of high-performance liquid chromatography, significant improvements in the isolation of saturated, olefinic, and aromatic hydrocarbons, as distinct groups, can be expected [1]. Further resolution of each of these groups by capillary gas chromatography would be made easier.

III. LIQUID-SOLID CHROMATOGRAPHY

A. Nonhydrocarbons

Chapter 9 discussed numerous separation techniques, most of which employ some form of liquid-solid chromatography. In principle, most of these techniques will work on synthetic fuels if suitable precautions or variations are made. Synthetic fuels differ from typical petroleum in that each usually contains much higher levels of nonhydrocarbons or possesses unique solubility or stability problems: coal liquids are highly aromatic and rich in oxygen compounds; shale oils are highly olefinic and rich in nitrogen compounds. The nonhydrocarbons frequently comprise $>50\%$ of synthetic fuels and these compounds must be removed before attempting to isolate the hydrocarbons. The API and SARA techniques described in Chaps. 9 and 11 are completely applicable to synthetic fuels. Figure 1 illustrates how these techniques are used on any type of sample from either petroleum or nonpetroleum origin (coal liquids, shale oils, kerogen, oil spill residues, etc.).

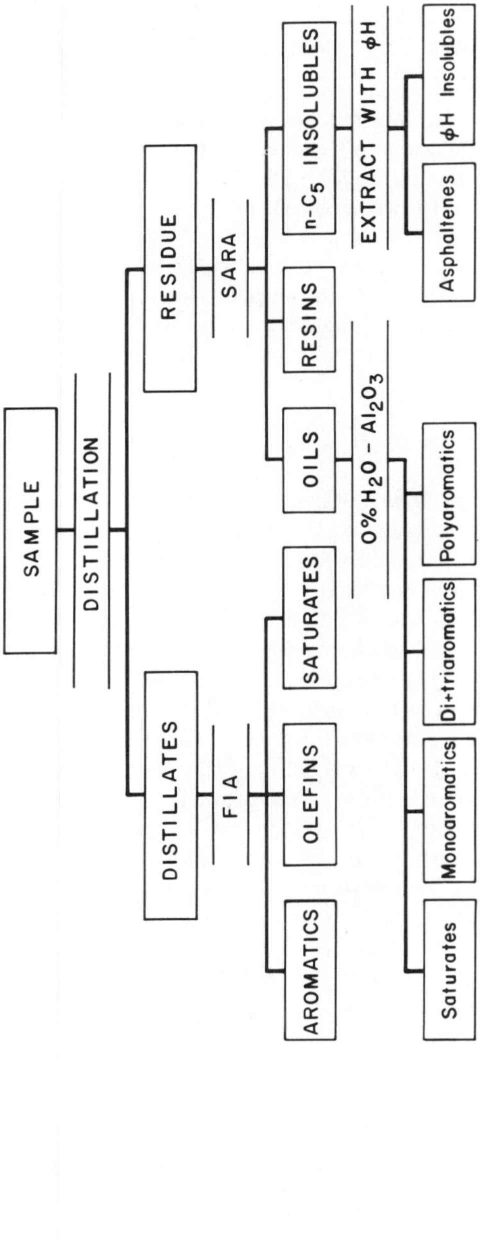

FIG. 1. General separation scheme.

If the sample is a liquid, volatile components (bp $<232°C$, $<450°F$) must be removed or accounted for. If sample quantities prohibit a distillation, the volatiles can be determined by weight-loss observed by keeping the sample under a stream of N_2 at 70°C until constant weight is reached. Components having vapor pressures less than that of dodecane ($n-C_{12}$) are lost. If distillates are collected, they may be analyzed by FIA + GLC and high-performance liquid chromatography (HPLC) as mentioned earlier.

Polar nonhydrocarbons are synonymous with resins and asphaltenes; the latter two differ in solubility, molecular weight, and multiplicity of functional groups. Resins can be separated according to major functional groups as described in Chap. 11. Table 1 demonstrates typical examples of this scheme applied to two different coal liquids [2]. The order in which the ion exchange resins are used is not as critical as maintaining consistency in a chosen sequence. The absolute quantities of bases or acids may change if the sequence is reversed due to the presence of both acidic and basic functional groups in the same molecule. Phenolic pyridines and quinolines are common examples of multifunctional resins.

B. Aromatic Hydrocarbons

The hydrocarbon constituents of synthetic fuels vary significantly, not only in quantities, but also in types. Using the 0-% water-alumina technique described in Chap. 9, one can readily obtain these data. If only total saturates and aromatics are desired, 0% water-silica gel adsorbents are very satisfactory, as well as the new high surface area commercial silica gels (5 to 10 μm particle size) designed for high-performance liquid chromatographic instruments. These new silica gels have recently been found effective for separating hydrocarbons analytically as well as preparatively [3]. When using HPLC in place of the regular column and the silica gel ordinarily employed with the SARA technique (Davison Grade 923), but still together with the precolumn for resin removal, significant improvements in speed and quantities (capacities) can be realized. With the advent of commercially available nonaqueous ion exchange resins, chemically similar to the Amberlyst 15, A-29, and IRA-904, and designed specifically for HPLC systems, one may expect further improvements in speed and efficiencies.

IV. CHARGE-TRANSFER CHROMATOGRAPHY

A. Procedures

Many studies of synthetic fuels require the actual separation of aromatic hydrocarbons into groups, usually according to their number of aromatic rings; this separation is most helpful in simplifying the interpretation of

TABLE 1. Distribution of Resins in Coal Liquids

	Big Horn coal (wt % charge)	Pitt Seam coal (wt % charge)
Very weak bases	0.9	1.5
Weak bases	1.0	2.2
Strong bases	0.5	1.3
TOTAL BASES	2.4	5.0
Very weak acids I[a]	2.3	2.5
Weak acids I	0.9	0.6
Strong acids I	0.1	0.8
Very weak acids II[a]	0.7	0.2
Weak acids II	0.3	0.4
Strong acids II	1.5	2.2
TOTAL ACIDS	5.9	6.6
Neutral resins	1.6	1.6
TOTAL RESINS	9.8	13.2
Neutral oils	84.4	53.3
n-C_5 Insolubles	2.4	31.8
Charge	100.0	100.0
Recovery	96.6	98.3

[a]Acids I are stronger than corresponding Acids II.

spectroscopic data obtained on the various fractions by reducing much of the overlap and structural redundancies that inhibit low-resolution mass spectrometry and NMR spectra of these complex mixtures. In addition to the gradient elution chromatography system using the 0% water-alumina mentioned earlier, the same basic principles can be further utilized by combining them with selective charge-transfer complexation for improved separation of aromatic ring systems [4]. Polynitroaromatic compounds and other strongly electronegative molecules frequently form useful charge-transfer complexes with aromatic hydrocarbons. When these reactions are sufficiently fast, they can be performed in situ with liquid chromatography. Figure 2 sketches a separation system employing 2,4,7-trinitrofluorenone

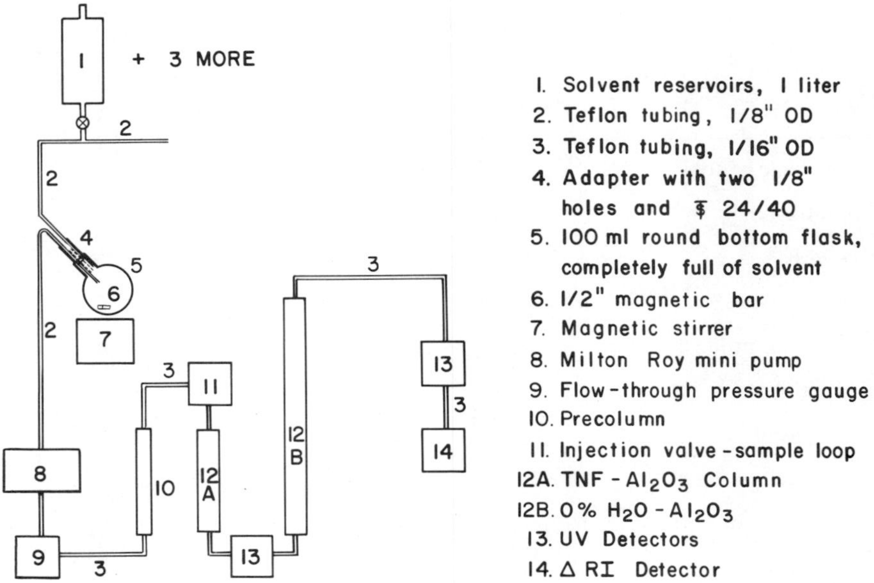

FIG. 2. Apparatus for charge transfer, GEC separations. (From Ref. 4, courtesy ACS.)

(TNF) as acceptor. TNF forms particularly strong complexes with compounds possessing 3 or more condensed aromatic rings. Weak complexes form with 1- to 3-ring systems. The overlap of 2- and 3-ring systems is not readily resolvable by normal (preparative) gradient elution chromatography. The system shown in Figure 2 is designed for the separation of 0.5 g of aromatics or "oils" and uses: (a) a zone of pure powdered TNF (1 to 2 g), as reaction site, in front of a column of 100 g hydrated basic alumina (>6% H_2O)*; (b) a column of 200 g 0 % water-alumina for separating mono- and diaromatic systems; and (c) two on-line detectors (dual-channel uv absorbance monitors) at the end of each column. (Note: the pressure of this system is limited by the particular cell design.) The remainder of the system is self-explanatory. The function of the hydrated alumina is to decompose the complexes as they slowly migrate from the reaction zone with increasing polarity of eluting solvent. Since the solubility of TNF complexes in hydrocarbon solvents is inversely proportional to their strength, the strong (three or more condensed rings) complexes will remain in the reaction zone until a polar solvent is begun. Figure 3 shows the detectors'

*Exact water content is not critical; 6 to 10% range is optimal.

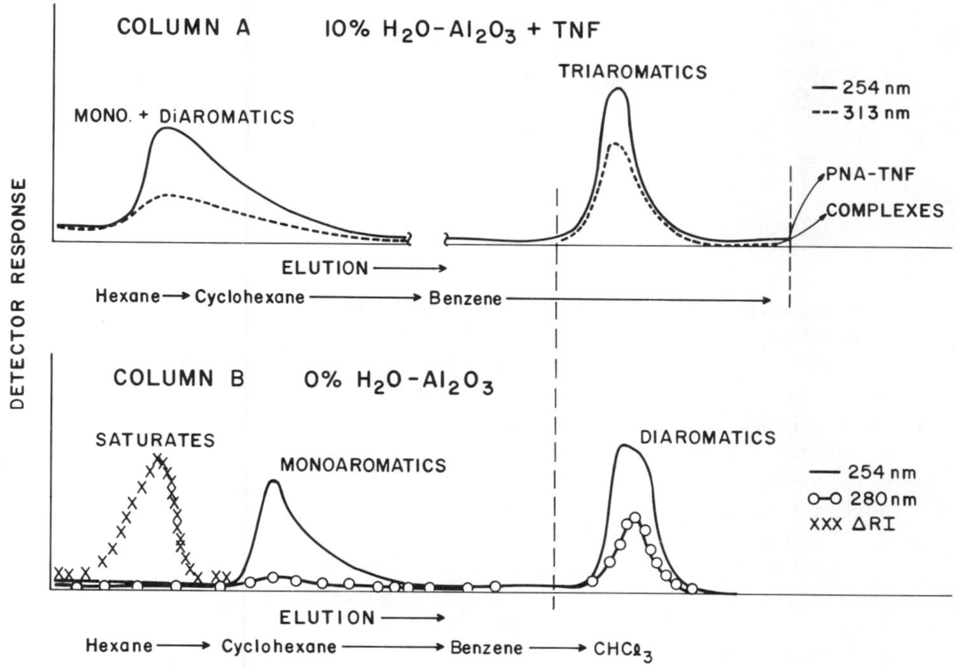

FIG. 3. Profiles of charge transfer, GEC separations. (From Ref. 4, courtesy ACS.)

responses as gradient elution proceeds (exponential gradient, hexane → cyclohexane → benzene). Since saturates (if present), monoaromatics, and diaromatics do not react with TNF, they migrate as a group through >6% H_2O-Al_2O_3, and are monitored from Column A at two different wavelengths as they proceed directly to Column B. The same solvent system (since only one pump is used initially) is continued until saturates and monoaromatics have been collected; diaromatics cannot be eluted from the 0% H_2O-Al_2O_3 by cyclohexane and this class remains at the bottom of Column B. When the detector shows that the monoaromatics are completely eluted, the detector for Column A will also have returned to baseline. The "tail" from Column A is completely due to diaromatics. The two columns are then independently eluted with benzene → chloroform gradient to desorb diaromatics from Column B and triaromatics from Column A. Many polyaromatic TNF complexes (>4 condensed rings) are very stable and resist cleavage on basic alumina and are eluted as a highly colored complex together with free TNF. These polyaromatics can best be recovered from TNF by performing a mild catalytic reduction of the nitro- to amino-groups in the presence of a macroreticular cation exchange resin. The resin retains the basic amino intermediates and removes them from solution.

Synthetic fuels and coal liquids in particular always contain more polyaromatics than petroleum. The polyaromatics are rarely larger than 4 rings, from any source. They will contain the traces of physiologically important structures (e.g., benzo(a)pyrene), if they were originally present, and care should be exercised in their handling.

B. Applications

The combination illustrated in Figure 2 is primarily used in preparative rather than analytical applications because quantitative measurements cannot easily be made. Not only would an accurate mass detector be required, but the polyaromatics would have to be separated from the simultaneously eluting TNF. The value of rapidly separating saturates and the different aromatics lies in the fact that each group is more amenable to study by spectroscopic and/or high-resolution GLC, TLC, and HPLC techniques. Figure 4 shows the NMR spectrum of a discrete concentrate of fluorenes which were preparatively isolated as triaromatics by the TNF technique. Spectroscopic studies of the triaromatics and polyaromatics frequently indicate that they are not as highly condensed as expected; the presence of single aromatic rings is indicated. This may be caused by (a) the exertion of additive influence by heteroatoms on the adsorptivity and strength of TNF complexes; (b) the presence of noncondensed systems (e.g., biphenyl type) in which individual aromatic moieties show additive effects in determining the strength of their complexes; and (c) both (a) and (b) in the same molecule. "Oils" from tar sands frequently contain large quantities of sulfur compounds which are completely thioether structures. These ether and thioether groups are mostly but not always in aromatic furan or thiophene ring systems. Selective chemical reactions are needed and must be used to supplement chromatographic techniques if these compounds are to be separated from aromatic hydrocarbons.

One example of combining selective chemical reactions and chromatography lies in the area of sulfur constituents of petroleum and tar sands; the level of organic sulfur compounds in coal liquids and shale oils is usually quite low. Concentrates of "pure" sulfur compounds can be prepared via their sulfoxide-sulfone intermediates by the method of Drushel [5] and further separated by a scheme involving the charge-transfer (TNF) technique, as outlined in Figure 5. The various groups of sulfur compounds are individually characterized by desulfurization reactions and spectroscopic techniques. With this approach, one can readily observe the true complexity of the mixtures. These reactions and separations can be done on a microscale, in conjunction with molecular size (GPC) separations to provide useful compositional information on high-molecular-weight residuals from either natural or synthetic origin. The knowledge of the number and functionality of sulfur atoms and the type of ring structure can prove extremely useful to modern process developments.

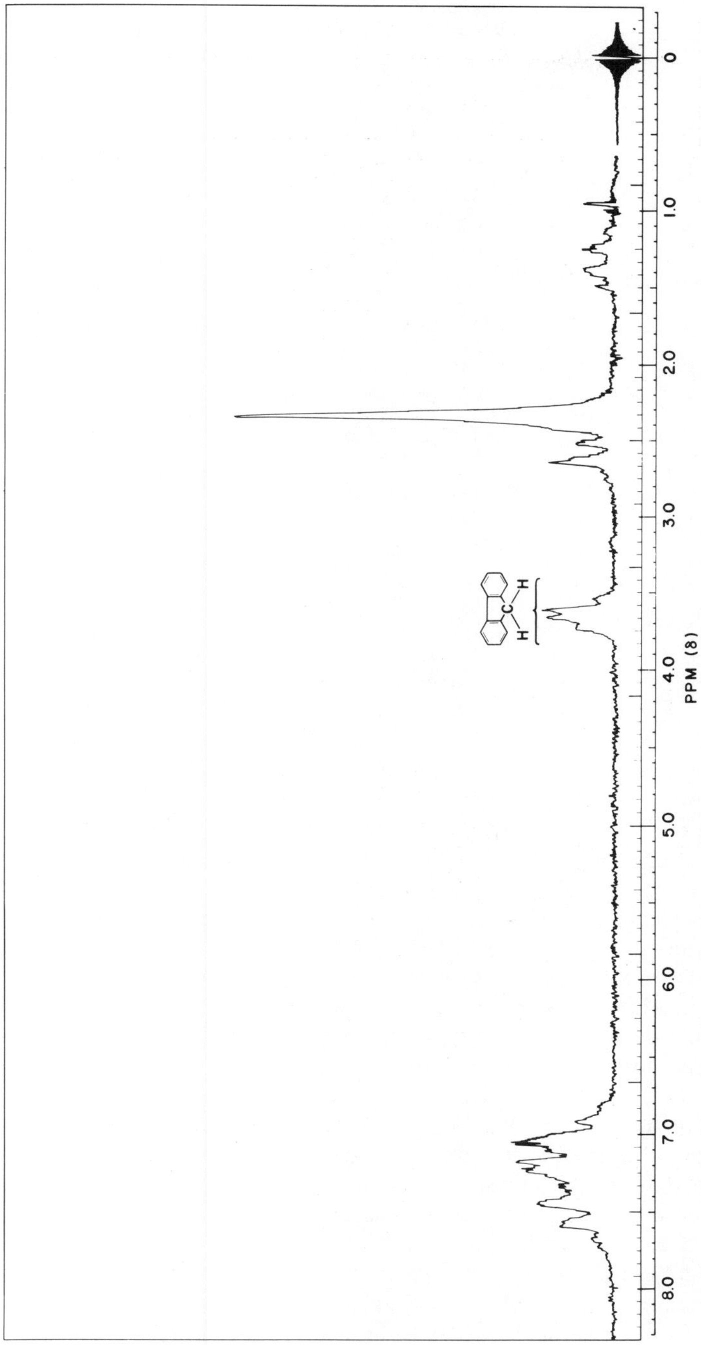

FIG. 4. ^{1}H NMR spectrum of fluorenes isolated by charge transfer chromatography. (From Ref. 4, courtesy ACS.)

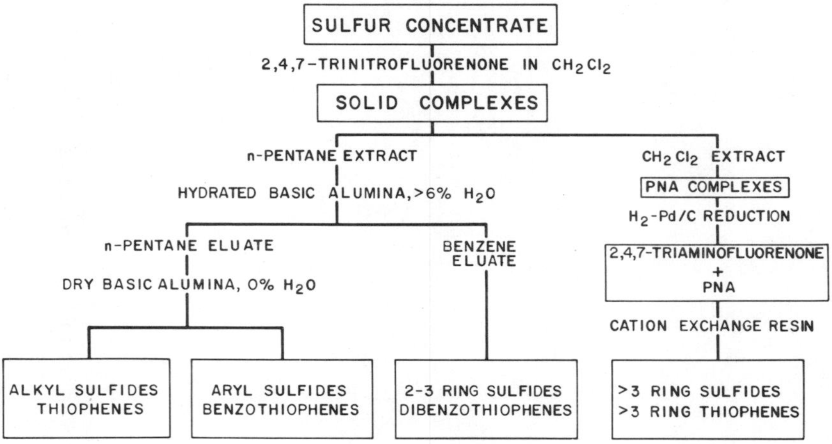

FIG. 5. Application of charge transfer, GEC to sulfur concentrates. [Courtesy Preprints, ACS, Div. Petrol Chem., 20(1):19 (1975)].

V. SATURATED HYDROCARBONS

Saturated hydrocarbons are probably the most valuable structural class of compounds in petroleum and synthetic fuels and have been studied more extensively than any other type. The isolation of saturates, as a group, is straightforward as shown in earlier sections. Although GLC is a valuable tool for studying saturates, its value is enhanced if used discreetly in conjunction with other chromatographic techniques which may selectively isolate structural subgroups. The normal paraffins are readily removed by their selective retention by 5A Molecular Sieves. Low-molecular-weight ($<C_{20}$) n-paraffins will be retained on a sieve precolumn in a gas chromatograph, but high-molecular-weight homologs must be removed in the liquid phase. A relatively simple method of achieving this separation employs a small 5A Molecular Sieve column in a conventional liquid chromatograph equipped with a differential refractometer detector. Equipment described earlier for the isolation of total saturates (Figure 5) will also suffice if the columns and detectors can be thermally controlled up to 90°C. The block diagram shown in Figure 6 illustrates the few components needed for this LC separation. In this application, the alumina column need not be highly activated since its sole function is to provide sufficient "band spreading" for integration purposes. The 5A Molecular Sieve (60-80 mesh) is freshly activated by calcining at 400°C in a vacuum desiccator just prior to filling the column; further activation and regeneration is achieved by purging with N_2 or He while heating at 400°C. If diagram A is followed, care must be taken to avoid excessive pressures on the cells of the detector; the advantage

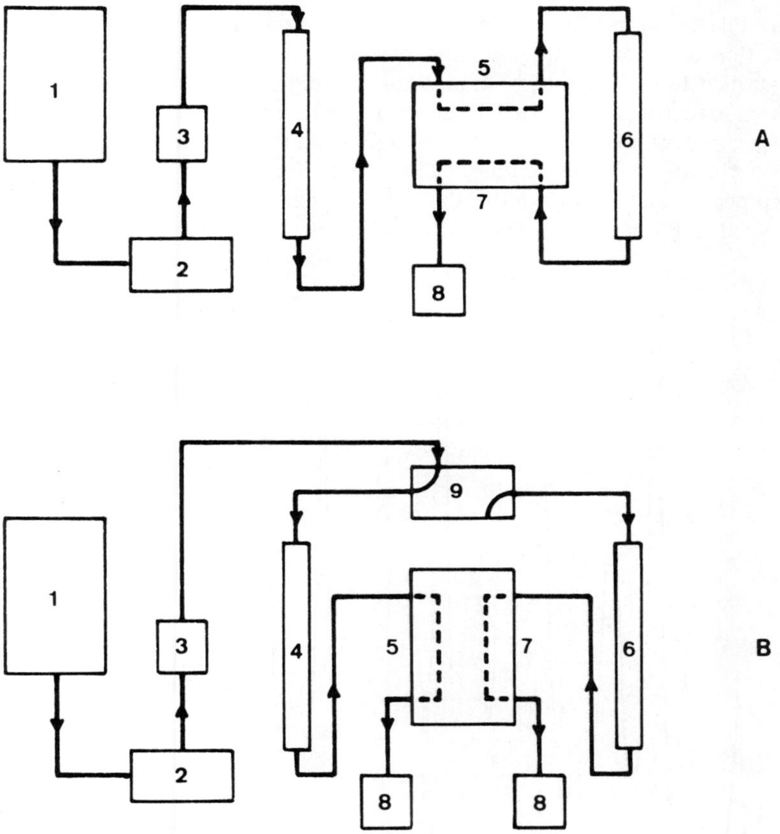

FIG. 6. Apparatus for liquid chromatographic separation of n-paraffins. Key to figure is as follows: 1, iso-octane reservoir; 2, pump; 3, injection valve; 4, alumina column; 5, reference side of RI detector; 6, molecular sieve column; 7, sample side of RI detector; 8, collectors; 9, switching valve. Source: D. M. Jewell and R. G. Ruberto, unpublished.

of this technique is that with only one injection both elution patterns (+, -) can be obtained on the recorder and immediately evaluated with an integrator.

Diagram B requires two identical injections in succession: one for the purpose of generating a response signal for the total saturates and the second for nonnormal paraffins only. Implicit in either technique is the assumption that no significant change in response factor results from the selective removal of one structural type. This assumption is well justified for the larger paraffins since differences in refractive index among structural types decrease as the molecular weight increases. If mass detectors are used, this assumption is valid for all types of saturate mixtures. When using this technique with saturates above C_{20}, the sieve column should be maintained in the 85 to 90°C temperature range to insure optimum retention of the n-paraffins. Figure 7 is an example of a typical GLC chromatogram of low-molecular-weight saturates, before and after the removal of n-paraffins by the above LC technique. The n-paraffins exhibit characteristic sharp peaks (and R_V values) in the top GLC chromatogram which have disappeared in the bottom curve. The weight percent n-paraffins determined

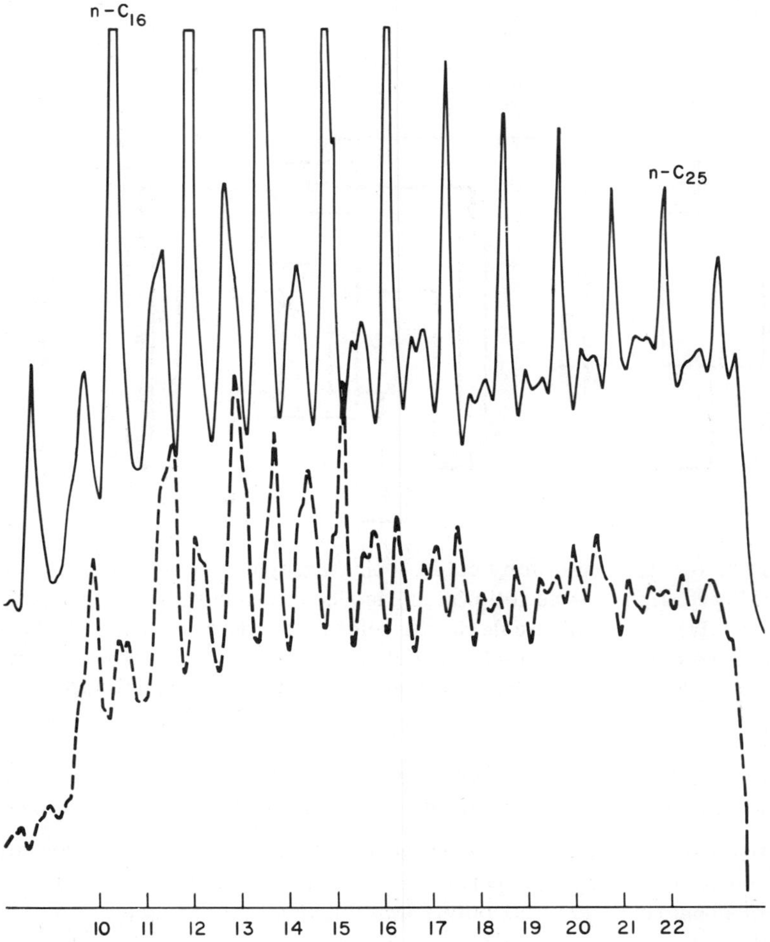

FIG. 7. GLC tracings of saturates before (-) and after (---) removal of n-paraffins. (Unpublished observations of D. M. Jewell and R. G. Ruberto.)

with the LC apparatus by area differences (Δ refractometer detector) are in excellent agreement with conventional batch-wise experiments; the former are determined in a fraction of the time (minutes) usually required for the latter (hours). The n-paraffin values are particularly important parameters for shale oils and their refined (hydrogenated) products.

The iso-paraffins are also important structures in synthetic fuels. Special 7 A Molecular Sieves have been shown to selectively separate certain isoparaffins from cycloparaffins [6], but no reliable commercial supply of these sieves exist yet. If these sieves should become available, the same LC procedure shown above for n-paraffins could be applied to the separation of isoparaffins from cycloparaffins.

Iso-methyl paraffins and certain singly substituted cycloparaffins can be isolated from non-n-paraffin concentrates by using urea or thiourea clathration techniques; however, these are difficult to perform routinely and do not adapt readily to chromatographic systems. Non-n-paraffins can be partially separated by size on columns of Sephadex LH-20 [7, 8]. The first structures to elute are chain structures (largest hydrodynamic volume). As structures become more condensed, their elution time (volume) also increases. Frequently, the final structures to elute are completely condensed cycloparaffins. This technique could provide the rapid preparative isolation of isoprenoid- or terpenoid-derived systems for further structural analysis.

REFERENCES

1. J. C. Suatoni, H. R. Garber, and B. E. Davis, J. Chromatog. Sci. 13:367 (1975).
2. R. G. Ruberto and D. M. Jewell, Report to the symposium Supplemental Fuels from Coal: New Analytical Needs, 26th Pittsburgh Conference on Analytical Chemistry and Applied Spectroscopy, March 1975.
3. J. C. Suatoni and R. E. Swab, J. Chromatog. Sci. 13:361 (1975).
4. D. M. Jewell, Anal. Chem. 47:2048 (1975).
5. H. V. Drushel and A. L. Sommers, Anal. Chem. 39:1819 (1967).
6. R. Curran, G. Eglinton, I. Maclean, A. G. Douglas, and G. Dungworth, Tetrahedron Letters 14:1969 (1968).
7. B. J. Mair, P. T. R. Hwang, and R. G. Ruberto, Anal. Chem. 39:839 (1967).
8. P. C. Talarico, E. W. Albaugh, and R. E. Snyder, Anal. Chem. 40:2192 (1968).

Chapter 17

THE DETERMINATION OF ADDITIVES IN LUBRICANTS

R. Amos
Analytical Department
Esso Petroleum Company Ltd.
Esso Research Centre
Abingdon, England

Edgar W. Albaugh
Chemicals and Minerals
Gulf Research and Development Company
Pittsburgh, Pennsylvania

I. INTRODUCTION

Synthetic chemicals are added to lubricants to improve their properties within the hostile environment of the engine. Prior to the 1930s a well-refined mineral oil was sufficient, but as engine demands increased, antioxidants were added to diesel engine lubricants to prevent oxidation of the base oil with consequent formation of corrosive acids and rapid increase in viscosity. Subsequently, metal soaps were added as detergents to reduce engine deposits, particularly in the piston ring zone.

Since about 1950, the introduction of increasingly powerful gasoline engines has brought about the development of highly sophisticated motor oils which can contain up to 10% of a complex additive mixture conferring properties such as detergency, dispersancy, antioxidant, antiwear, rust inhibition, viscosity index improvement, and pour-point depressancy upon the finished oil. The present day aviation gas-turbine engine presents such extreme demands that mineral oils can no longer cope, especially on the grounds of oxidative stability at high temperature and inadequate viscosity index. For these reasons, aviation lubricants are based either on diesters or neopentyl polyol esters with hindered phenolic or amine antioxidants and antiwear additives such as aryl phosphates. By the petroleum chemist such products can be considered to consist of 100% additive blends.

The determination of additives in lubricating oils therefore presents the analyst with a challenging problem. In the past, elemental analyses for, say, calcium, barium, zinc, and phosphorus have given much useful information, but if the present trend towards low-ash or ashless oils continues, such estimations would become largely redundant. Happily, the emerging techniques of high-performance liquid chromatography (HPLC) and gel permeation chromatography (GPC) seem to be the ideal tools to monitor the synthesis of new additive preparations, to use as a routine technique in process and quality control and to follow the fate of additives in service.

The successful identification of additives in unknown blends by spectroscopic methods, e.g., by infrared, nuclear magnetic resonance, ultraviolet, and mass spectroscopy depends, in the first instance, upon the isolation of sufficiently pure material to enable it to be characterized. As a further complication, additives are rarely pure chemicals. For example, the reader may be left to speculate on the meaning of "sulfurized vegetable oils" or phosphosulfurized animal oils"!

In this chapter we shall describe and discuss methods of isolation, i.e., dialysis, classical adsorption chromatography on silica and ion exchange resins, and thin layer chromatography (TLC) as well as the analytical scale and quantitative techniques of high performance liquid and gel permeation chromatography for the routine determination of additives in lubricants. Before we do so, we must first consider in more detail the nature of the additives themselves.

II. ADDITIVES AND THEIR FUNCTIONS

It may be somewhat artificial to classify additives by chemical type since many of them are multifunctional, e.g., zinc dialkyl dithiophosphates contain both oxidation and corrosion inhibition properties as well as being antiwear agents. Furthermore, synergism is widespread where, for example, certain blends of metal sulfonates and phenates are more effective as detergent inhibitors in certain applications, than when only one of the constituents is used alone.

Although most additive packages are proprietary blends, certain basic types have now been widely recognized.

A. Viscosity Index Improvers

These are long-chain polymers, usually with a broad molecular weight range somewhere within the region of 10,000 to 1,000,000. They are normally present in crankcase oils to promote easier cold starting. At ambient temperatures the molecules assume a tightly coiled configuration, but as the engine warms up they uncurl. As a result, the oil exhibits a flat viscosity versus temperature curve and oil consumption is kept to a minimum. Examples of VI improvers include polyacrylates and polymethacrylates (I), vinyl acetate-alkyl fumarate copolymers and polyolefins such as poly-isobutylene. Many of these additives also act as pour depressants, i.e., they interact with wax molecules thereby depressing the solidification point of an oil to lower temperatures.

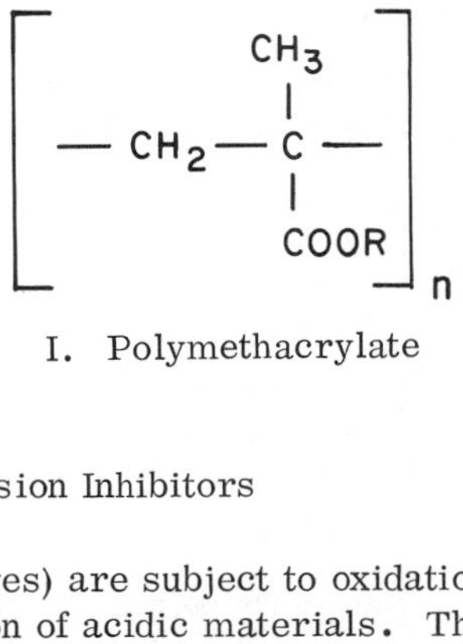

I. Polymethacrylate

B. Oxidation and Corrosion Inhibitors

Mineral oils (and additives) are subject to oxidation at elevated temperatures resulting in the formation of acidic materials. These lead to metal corrosion and an increase in oil viscosity due to the presence of insoluble oxidation and corrosion products. The two names therefore cover a single function and several chemical compound types can be recognized.

1. Hindered phenols and amines which act through a chain-breaking mechanism or sulfurized materials which act as peroxide decomposers.

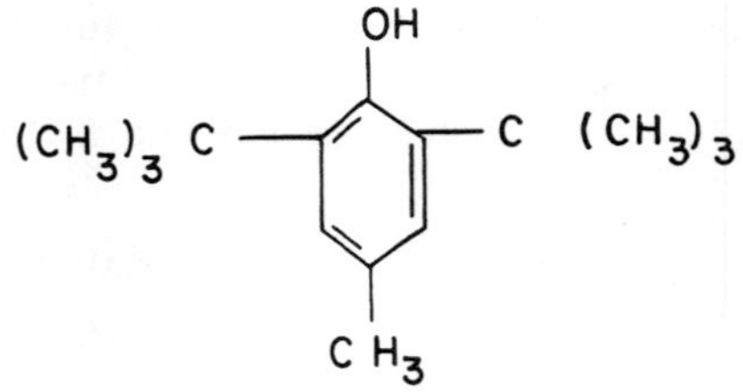

II. 2,6-Di-tertiarybutyl, 4-methyl phenol

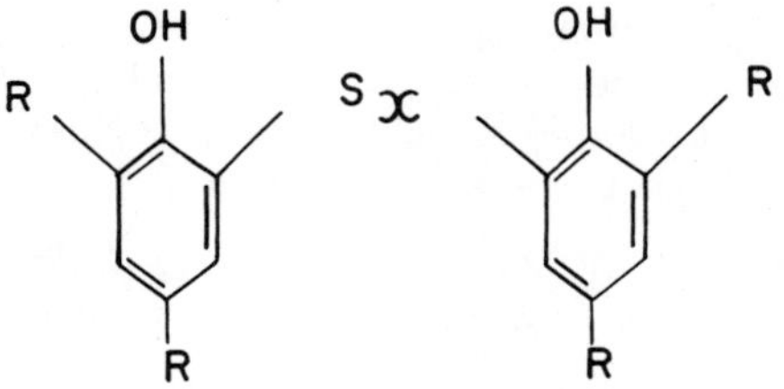

III. Thiobisphenols

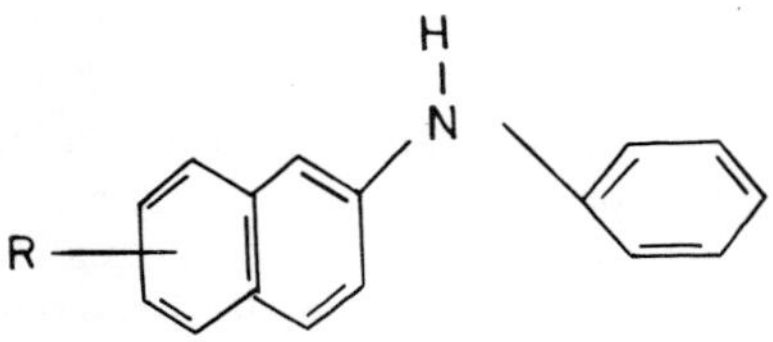

IV. Alkyl phenyl-β-naphthylamine

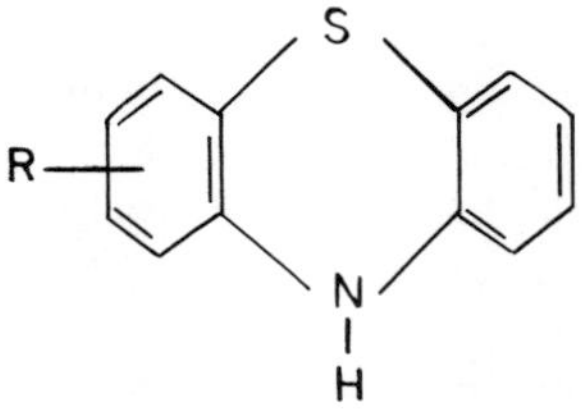

V. Alkyl phenothiazine

2. Zinc dialkyldithiophosphates and carbamates which also act as peroxide decomposers. These are further described below.

3. Metal deactivators such as benztriazole (VI) or mercaptobenztri azole derivatives (VII) which are used as copper corrosion inhibitors.

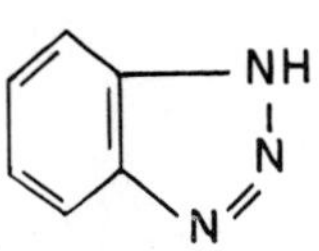

VI. Benztriazole

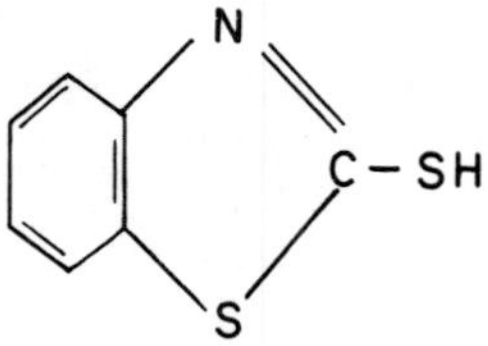

VII. Mercaptobenztriazole

4. Overbased detergent inhibitors which contain alkaline earth metal hydroxides or carbonates in colloidal form and which neutralize acidic products: These are described further below.

5. Miscellaneous bases such as amines, ethanolamines, or Schiff's bases, which neutralize acids.

C. Antiwear and Extreme Pressure Additives

These are used to maintain a protective film of lubricating oil between moving parts under conditions of high temperature and stress, so as to reduce frictional wear and prevent seizure. The most widely used in crankcase oils are the zinc dialkyl (aryl) dithiophosphates (VIII) and zinc dialkyl dithio carbamates (IX).

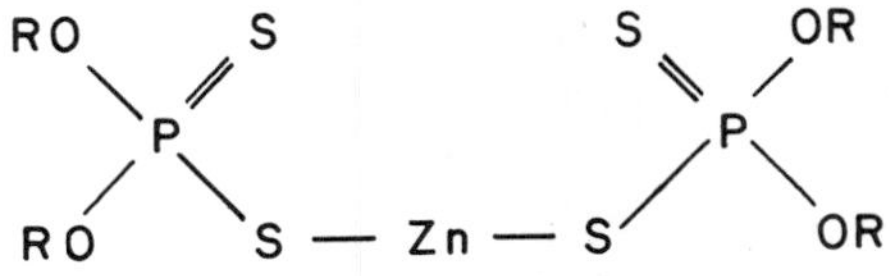

VIII. Zinc dialkyldithiophosphate

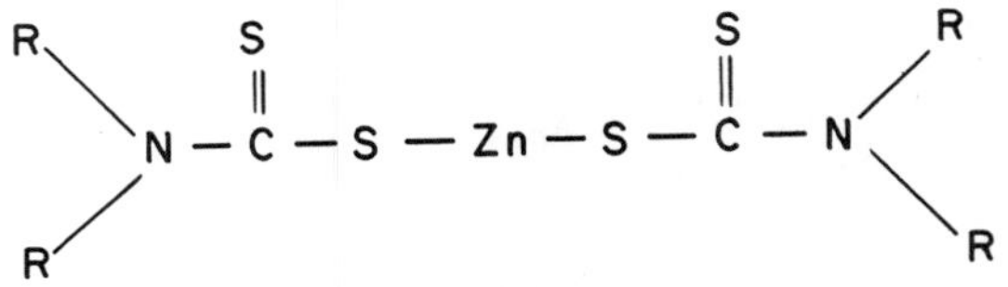

IX. Zinc dialkyldithiocarbamate

Many other additives containing chlorine, sulfur, or phosphorus have been used to combat wear, especially in gear trains and bearings in which very high loads can occur. Examples are mixed alkyl or aryl phosphates such as tritolyl phosphate, chlorinated hydrocarbons, and sulfurized or phosphosulfurized terpenes and esters. Some of the latter products are of animal or vegetable origin, e.g., sulfurized sperm oil.

D. Detergents

These are used to reduce formation of high-temperature deposits in an engine and to keep them suspended in the oil. They are usually calcium, barium, or magnesium salts of (sulfurized) phenols (X), salicylic acids (XI), sulfonic acids (XII), carboxylic acids, or phosphosulfurized alkenes in the equivalent weight range of 400 to 800, i.e., the molecules contain alkyl chains of sufficient length to impart oil solubility. They generally include the corresponding metal hydroxide or carbonate in the form of micelles such that the base number of the additive can range from about 10 (i.e., neutral) to 400 mg (i.e., strongly overbased) KOH per g.

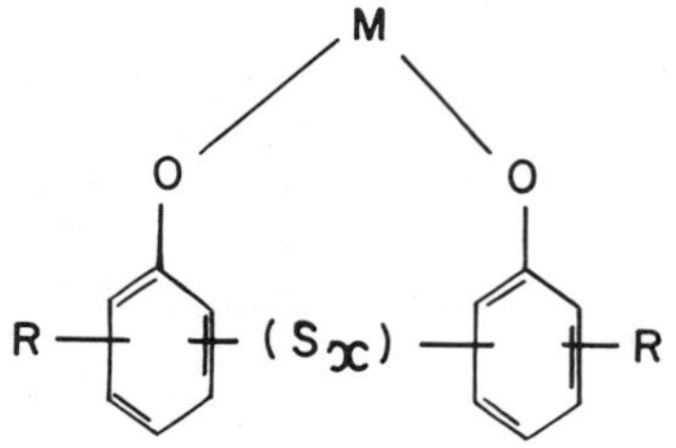

X. (Sulfurized) phenate

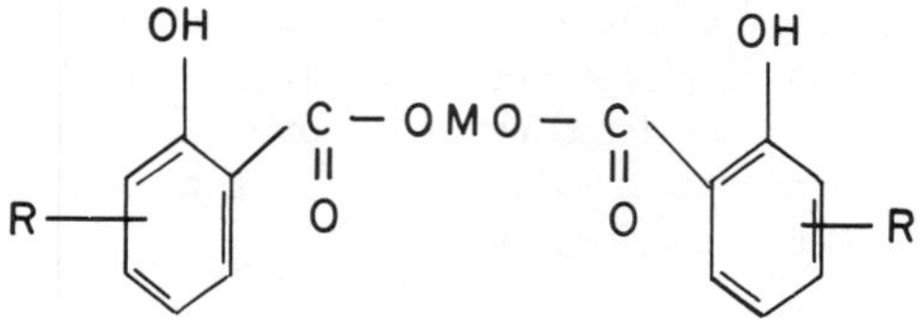

XI. Salicylate

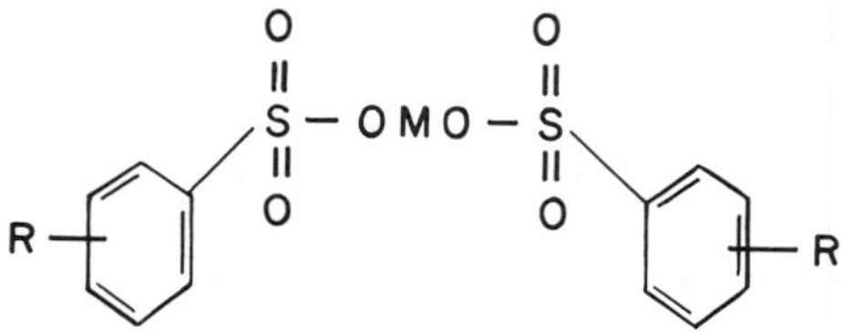

XII. Sulfonate

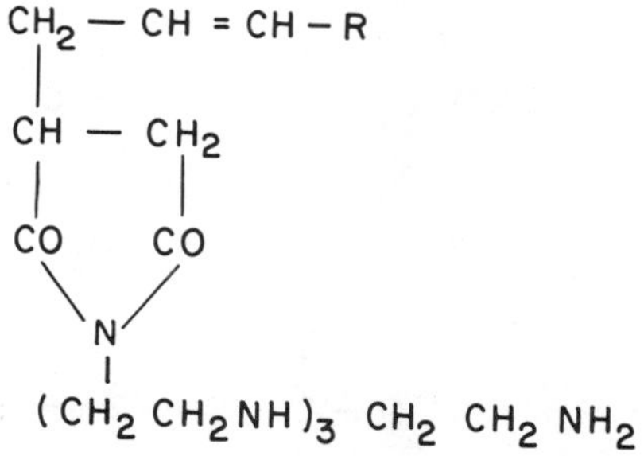

XIII.

E. Dispersants

These have a similar function to detergents but are more active in reducing sludge formation under low temperature operating conditions and preventing agglomeration of particles produced by degradation. The more common types are polyisoalkenyl polyamides (XIII) in the molecular weight range of 1000 to 50,000. Some VI improvers containing polar groups can behave as dispersant VI improvers.

The above descriptions can give no more than a brief outline of the most common types of additives in use at the present time. The patent literature is replete with compounds claimed to have beneficial effects, the vast majority of which have never been used, and never will be used, in practice. Nevertheless, it is safe to assume that new additives with different novel compositions will continue to be introduced into commercial service.

It can be seen that the petroleum analyst may be confronted with an oil containing an array of metallo-organic additives which may themselves be complex mixtures.

The types of analyses that are most frequently needed, in which chromatography makes a significant contribution, can be classified as follows:

1. To monitor the synthesis and determine the composition and structure of a new additive.
2. For process and blending control.
3. To detect and determine the additive content in lubricants of unknown formulation.
4. To follow additive depletion in used oils.

In applications 1, 2, and 4 it is inferred that there is considerable knowledge of the likely composition of the additive or lubricant and the appropriate method of analysis can be more easily chosen.

It will be therefore more rewarding for us to consider the various chromatographic methods in relation to the problem of determining the composition of an unknown oil. This can be done most conveniently by describing a general separation scheme and discussing in detail each of the individual separation techniques in turn.

III. GENERAL SEPARATION SCHEME

The successful identification of an unknown lubricant by spectroscopic methods depends, in the first instance, on the successful isolation of fractions sufficiently pure to enable their spectra to be interpreted. The separation procedures used fall into two categories: first, the preparative but low-efficiency techniques of dialysis and classical liquid chromatography using adsorbents or ion exchange resins and second, the analytical scale but more efficient techniques of thin layer (TLC), high-performance column (HPLC), and gel permeation chromatography (GPC). HPLC and GPC can be expected to be increasingly used for routine quantitative analysis in process and quality control, while TLC in particular is a useful aid to spectroscopic identification.

The way all these techniques interact with each other is summarized in Figure 1.

A preliminary infrared spectrum and thin layer chromatogram of the original sample will give some idea as to its complexity. They will indicate, for example, if it is a hydrocarbon or ester-based lubricant, whether it is strongly overbased with carbonate, and whether it is rich or lean in additives. A second preliminary experiment that needs to be carried out is to test, qualitatively at first, for metals, phosphorus, chlorine, silicon, and boron. This should then be followed up by quantitative determination of those elements found. Unless the lubricant is wholly ester based, the sulfur and nitrogen contents of the original sample may not be very meaningful because these elements will be naturally present in the hydrocarbon

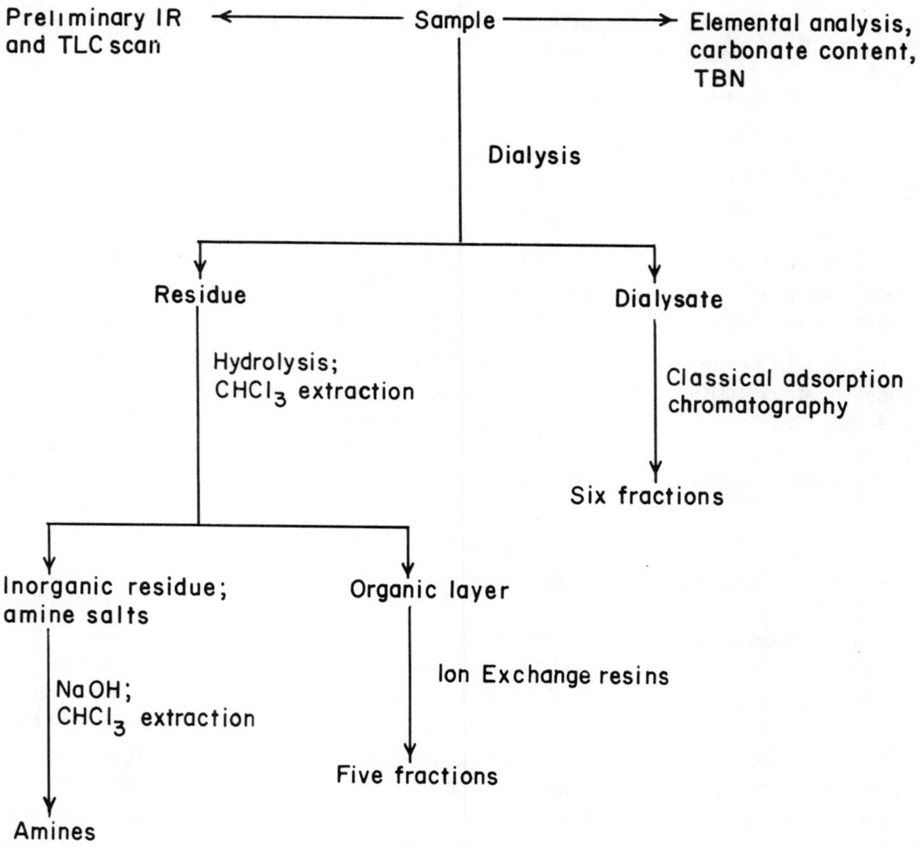

FIG. 1. General analysis scheme for a lubricant of unknown composition.

base oil. These figures become more significant when determined on the dialysis residue or other individual fractions of any significant size. Two other determinations which are of help in ascertaining the composition of an unknown lubricant are measurements of the basicity of the sample, i.e., the total base number (TBN) by ASTM D 664 and the carbonate content [14].

The first stage in the separation scheme itself is to dialyse the sample. This separates high-molecular-weight polymers and colloidal soaps (the dialysis residue) from base oil and low-molecular-weight additives (the dialysate). Polymers in the 2000-10,000 molecular weight range (e.g., some ashless dispersants) dialyse slowly and consequently some will appear in both fractions.

Low-molecular-weight additives in the dialysate can be separated from oil by classical column chromatography to yield fractions which are usually pure enough for spectroscopic identification, with the aid of TLC using specific spray reagents. Examples of additives appearing in these fractions are antioxidants of the hindered phenol and amine type, esters, zinc dialkyldithio phosphates, free phenols, and carboxylic acids. HPLC is being used increasingly for quantitative analysis and isolation of specific components of the dialysate.

The soaps in the dialysis residue are best converted by hydrolysis with hydrochloric acid (20% in methanol) to form free acids, i.e., sulfonic, carboxylic, and phenols, which can be separated from each other and from neutral polymers using ion exchange resins prior to their spectroscopic identification. GPC is the best technique to isolate and characterize VI improvers and ashless dispersants, whereas HPLC, again, is being used increasingly to handle polar detergent additives.

It should be emphasized that additives in a lubricating oil can form a complex colloidal surfactant system which must first be destroyed either by dialysis or hydrolysis before they can effectively be chromatographed. Passage of an untreated premium diesel or motor lubricant down a chromatographic column packed with adsorbent or ion exchange resins produces very little separation because the colloid, protected by its film of oil, behaves as a hydrocarbon and passes straight through the stationary phase unadsorbed.

If the composition of a lubricant is known, the general scheme we have just described can be modified to suit the type of analysis required. Frequently, information on only one particular additive component is required and here rapid analyses by HPLC and GPC are particularly attractive. When prior destruction of the colloid is necessary, this can be achieved by evaporating to "dryness" on a water bath a solution of 1 g oil in 20 ml of a 1:1 chloroform-isopropanol mixture, to which 1 ml concentrated HCl has been added. This is often an effective alternative to the lengthy dialysis procedure.

We shall now describe and discuss applications of the various separation techniques in greater detail.

A. Dialysis

Dialysis is a very useful initial separation procedure since it provides a very clean separation between high-molecular-weight polymers (e.g., molecular weight greater than 10,000) and low-molecular-weight additives (e.g., metal sulfonates, phenates, carboxylates of molecular weight less than 1000). Other additives, such as some ashless dispersants, in the molecular weight range 2000 to 10,000 dialyse slowly and, under the test conditions which will shortly be described, portions will appear in both fractions, i.e., the dialysate and the residue. In addition to achieving this separation, the process of dialysis destroys the colloidal structure of the additive package and yields two fractions which, in contrast to the colloidal mixtures, are amenable to chromatographic analysis.

The procedure is based on the work of Hill and Munsell [9], who studied the differential diffusion of colloidal components of crude oil, asphalt, and products containing certain petroleum additives through a thin, gum-rubber membrane using low-boiling paraffinic hydrocarbons as solvent.

The apparatus now used is shown in Figure 2. About 5 to 10 g oil is weighed into a surgeon's rubber fingercot, which has previously been washed with isopropanol to remove excess talc and allowed to dry in a stream of nitrogen. The open end of the fingercot is tied with glass tape over the flared end of the fingercot holder (50 cm by 6 mm ID glass tubing with the lower end flared to a diameter of 10 mm). This is placed in the Soxhlet extractor as shown and supported with a glass stopper to prevent the fingercot from blocking the entrance of the sidearm. The sample in the fingercot is then extracted with boiling pure n-heptane which allows low-molecular-weight materials to diffuse through the rubber membrane and find their way into the boiler. Meanwhile, polymers and colloids are retained within the fingercot.

The rate of dialysis is given by the expression

$$\ln\,[1 - (\text{wt of fraction dialysed in time t})]^{-1} = \frac{P \cdot A \cdot t}{\ell \cdot V}$$

where

ℓ = thickness of fingercot
P = permeability coefficient
A = surface area of fingercot
V = total volume of liquid in the fingercot

that is, for a given membrane a plot of $\ln\,[1 - (\text{wt fraction dialysed in time t})]^{-1}$ against time, for a particular substance, gives a straight line that goes through the origin. Some plots for polypropylene glycols are given in Figure 3, which illustrates dramatically diffusion rate differences between low and even only moderately high molecular weight materials. Figure 4 shows the relationship between the amount dialysed after 24 hr with molecular weight (of polypropylene glycols). From this, it can be seen that, for practical purposes, substances with molecular weights greater than 10,000 will remain exclusively within the fingercot, while those with molecular weights of 2000 or less will dialyse through the membrane within 24 hr and accumulate in the flask of boiling heptane. It is important to ensure the solvent is pure since the presence of a small concentration of aromatic impurities will

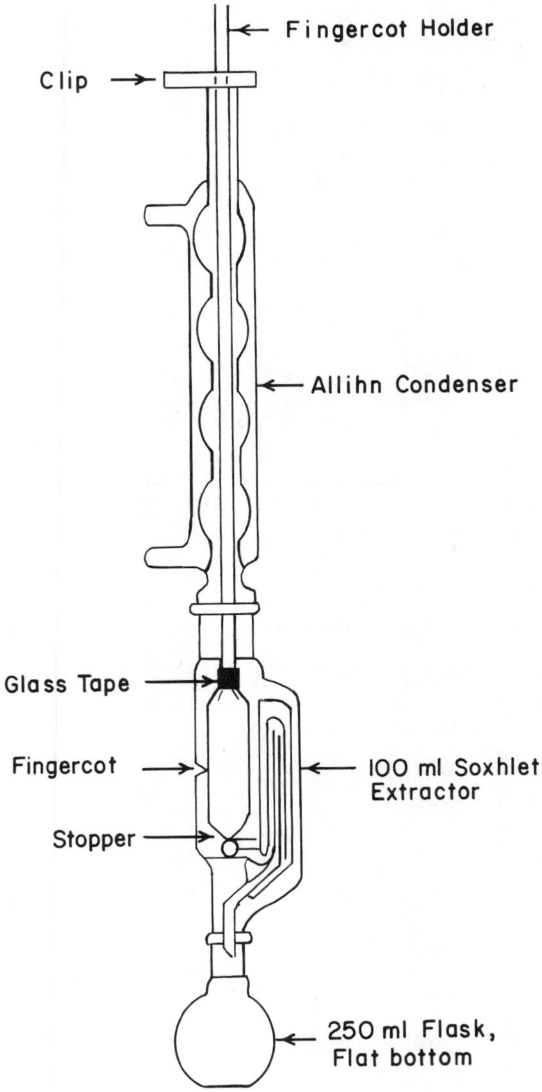

FIG. 2. Equipment for dialysis

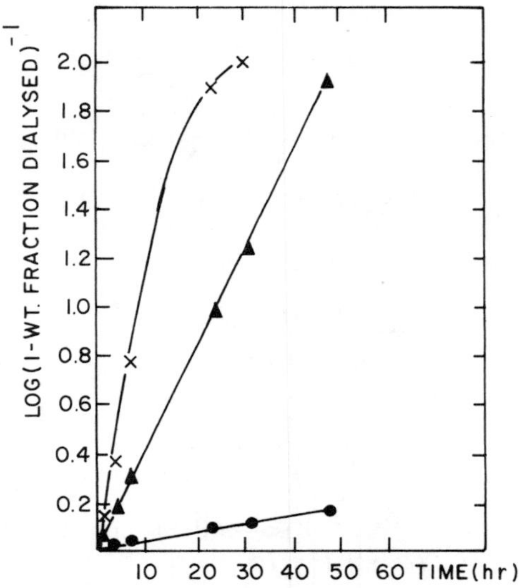

FIG. 3. Variation of diffusion rates through rubber membrane. Polypropylene glycols in n-heptane. × = 10,000 molecular weight (MW); ▲ = 3000 MW; ● = 2000 MW.

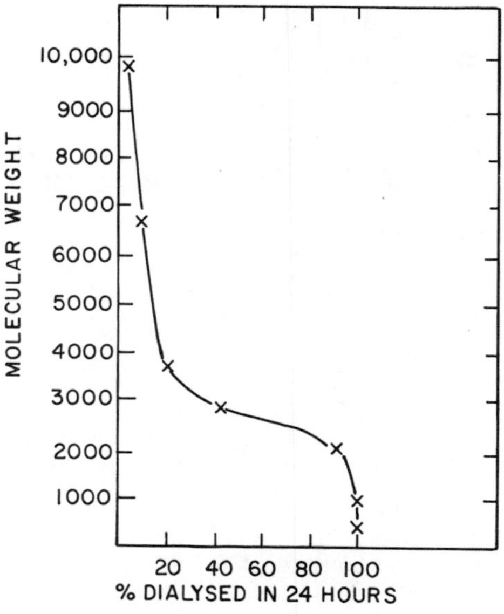

FIG. 4. Variation of amount dialysed with molecular weight.

TABLE 1. Behavior of Different Additive Types on Dialysis

Additive type	Dialysate	Residue
Base oil (hydrocarbon or ester)	Yes	None
VI improvers	Solvent oil	Polymer
Hindered phenolic and amine antioxidants	Yes	None
Zinc dialkyl-dithio-phosphates and dithiocarbamates	Yes	Small amounts of inorganic residues
Metal phenates	Solvent oil + unreacted phenols	Metal soaps
Metal sulfonates	Solvent oil	Metal sulfonates
Colloidal metal carbonates, hydroxides, carboxylates	None	Yes
Phosphate esters	Yes	None
Ashless dispersants	Solvent + low molecular weight polymer	Higher molecular weight polymer

cause the rubber membrane to swell and significantly affects its separation characteristics.

Table 1 shows the behavior of various additive types on dialysis and, with the exception of ashless dispersants, it can be seen that clear-cut separations into two fractions are obtained. The dialysate is recovered from the boiler after removal of heptane by evaporation, and the residue is similarly recovered from the contents of the fingercot.

Despite the lengthy duration of the test, it is an invaluable first step when examining an unknown oil for the first time since it produces two sizable fractions on which further exploratory work can be carried out by a wide variety of analytical techniques.

One or two further practical points may be mentioned. Some hindered phenols, e.g., 2,6-ditertiary butyl, 4-methyl phenol, which are commonly found in industrial oils, are rather volatile and may be lost from the dialysate while evaporating off the heptane solvent. For this reason, it is necessary to test for the presence of such additives, e.g., by TLC, before dialysis. Similarly, some zinc dialkyl dithiophosphates may undergo thermal decomposition during dialysis. Again, such compounds will have been detected during a preliminary TLC examination and from the elemental analysis figures of the original sample.

B. Classical Column Adsorption Chromatography

Separation by displacement chromatography in a column packed with silica or Florisil using a series of eluents of gradually increasing polarity is a useful way of separating polar additives from nonpolar hydrocarbon oil in macro quantities, i.e., on a scale such that further substantial analytical work can be carried out on the fractions. It is more usual to carry out this technique on the sample, provided carbonate is absent, or on the dialysate. It can also be applied to the hydrolyzed dialysis residue, but stronger acids, i.e., carboxylic or sulfonic may be esterified or saponified, so it is preferable to separate the hydrolyzed dialysis residue in columns packed with ion-exchange resins (see Sec. C).

In the procedure described by Berthold [4], a glass column, 50 cm long by 1.1 cm ID, equipped with a stopcock and 250 ml capacity solvent reservoir, is filled with either Davison Silica gel Code 62* or Florisil,† 30-60 mesh. Unless the object of the separation is to recover the saturated and aromatic portions of the hydrocarbon oil as separate fractions, activation is unnecessary. About 250 mg sample, dissolved in 5-ml n-heptane is charged to the top of the column and 200 ml portions of a series of solvents of gradually increasing eluent strength (Chap. 14, Table 1) is passed, in succession, down the column at a flow rate of about 1.5 ml min^{-1}. A suitable series of solvents has been found to be n-heptane, carbon tetrachloride, benzene, diethyl ether, acetone, and methanol. As a result, six fractions are collected and Table 2 indicates in which fractions certain additive types may be expected to appear. In comparison with HPLC, this is a very low efficiency separation, and some additives may appear in more than one fraction. Nevertheless examination by TLC, infrared, and mass spectrometry is usually sufficient to completely identify each fraction.

C. Ion Exchange Resins

Ion exchange resins are very useful for isolating basic, acidic, and some polar components of additives from lubricating oils. They are particularly

---

*W. R. Grace and Co., Davison Chemical Division, Baltimore, Maryland 21203.

†Floridin Co., Pittsburgh, Pennsylvania 15235.

TABLE 2. Separation by Classical Adsorption Chromatography

Eluent	Dialysate	Hydrolyzed residue
n-Heptane	Saturated hydrocarbons	Polyolefins
Carbon tetrachloride	Aromatic hydrocarbons, hindered phenols	
Benzene	Hindered amines, alkyl phenols	Alkyl phenols
Ether	Zinc dialkyldithiophosphates	Polymethacrylates, sulfurized phenols
	Carboxylate esters, sulfurized phenols, phosphate esters	
Acetone	Carboxylate esters, zinc dialkyldithiophosphates	
Methanol	Naphthenic acids	Sulfonic acids, carboxylic acids, phosphonic acids

effective in separating the components of the hydrolyzed dialysis residue. The "macroreticular" ion exchange resins [11] have a rigid macroporous structure, with pore diameters up to several hundred Angstrom units, superimposed upon the normal gel structure. Their physical structure is therefore similar to that of conventional adsorbents such as silica or alumina. Otherwise, they are chemically similar to conventional polystyrene-divinyl benzene based resins. As a consequence of their macroporous structure they can be used in solvents in which resins do not swell (i.e., nonpolar organic solvents), and for the adsorption of large organic molecules or ions. Furthermore, they show no marked deterioration when subjected to repeated changes from an aqueous to a nonaqueous medium.

A number of macroreticular resins are now commercially available and the two most used in petroleum additive analysis are Amberlyst 15,* a sulfonic acid cation exchanger and Amberlyst 29,* a quaternary ammonium anion exchanger. Webster and Wilson et al. [15, 16] used macroreticular resins to extract and recover acids, phenols, and bases from petroleum products and demonstrated the superiority of these resins over those originally used by Munday and Eaves [12]. They also demonstrated the use

*Rohm and Haas Co., Philadelphia, Pennsylvania 19105.

of dissolved gases in polar solvents as eluents and showed that in its transitional metal form a macroreticular cation exchanger could extract ligands from nonaqueous systems.

This work has now been confirmed and extended by Amos [2] and by Crump and Lynes [7]. We shall now describe three examples in the use of macroreticular ion exchange resins in petroleum additive analysis, viz., the further separation of the hydrolyzed dialysis residue, the recovery of basic components and the determination of specific additives via ligand formation.

1. Separation of Hydrolyzed Dialysis Residue

The hydrolyzed dialysis residue may contain VI improver and ashless dispersant, phenols, carboxylic, and sulfonic acids. These can easily be separated by passing down a column packed with a macroreticular anion exchange resin to provide pure fractions for spectroscopic analysis. The use of ion exchange resins has advantages over other methods of separation. Extraction with bases, e.g., NaOH leads to the formation of troublesome emulsions, whereas attempted separations on columns packed with conventional adsorbents, e.g., silica, magnesium silicate, or alumina are hampered by chemisorption effects causing saponification, esterification, and incomplete recoveries.

Amberlyst 29 can be used in the chloride form in which it is supplied or, better, after conversion to the hydroxide, carbonate, or acetate forms. In general it can be said that the resin in the chloride or hydroxide form acts as a strong adsorbent which chemisorbs even weakly acidic materials. The carbonate and acetate forms, on the other hand, offer more selectivity because phenols, for example, are physically adsorbed from hydrocarbon solutions, and can be recovered as a separate fraction by elution with polar solvents such as methanol.

We shall now describe the procedure we employ for the separation of hydrolyzed dialysis residues with the acetate form of the resin.

The sample is prepared by treating the dialysis residue with 50 ml 20% HCl in methanol and allowing the mixture to evaporate to dryness on a boiling-water bath under a stream of nitrogen. The residue is extracted with 50 ml toluene and filtered. The filtrate is applied directly to the prepared ion exchange column.

The resin is conditioned by soaking the beads in deionized water for 24 hr and pouring the slurry into a glass chromatographic column, 30 cm by 1 cm, with sintered disc support, stopcock, and solvent reservoir (250 ml capacity). Before use on actual samples, the resins must undergo three or four exhaustion and regeneration cycles to eliminate organic impurities. The resin is first converted to the acetate form, via the hydroxide, by passing through, in succession, 500-ml 4% aqueous NaOH, 200 ml deionized water, 500 ml 4% aqueous sodium acetate and 200 ml deionized water. Nonaqueous conditions are then established by passing 100 ml methanol and 50 ml toluene down the column at which stage the sample solution (in toluene) is added. Four fractions are then collected: unadsorbed material eluted with toluene followed by fractions eluted with methanol, 20% acetic acid in toluene, and 20% HCl in methanol. Flow rates of 1 to 2 ml min^{-1} are maintained throughout. The separations thus achieved are summarized in Figure 5.

If strong acids are known to be absent, elution with HCl-methanol can be omitted and the resin, which is left in the acetate form, is available for immediate reuse. Otherwise, the resin finishes up in the chloride form and the whole regeneration cycle has to be repeated. It should be noted that the resin slowly decomposes with loss of amine if allowed to stand in the hydroxide form for any appreciable length of time. It should therefore be stored in either the chloride or acetate forms. Provided these simple precautions are observed, the same resin may be recycled for up to 3 months before replacement becomes necessary.

Similar results may be obtained with the carbonate form of the resin, as with the acetate form. In this case weak acids are displaced with methanol saturated with CO_2. The danger here is that not all acidic materials are soluble in methanol and poor recoveries may occur. An attraction of

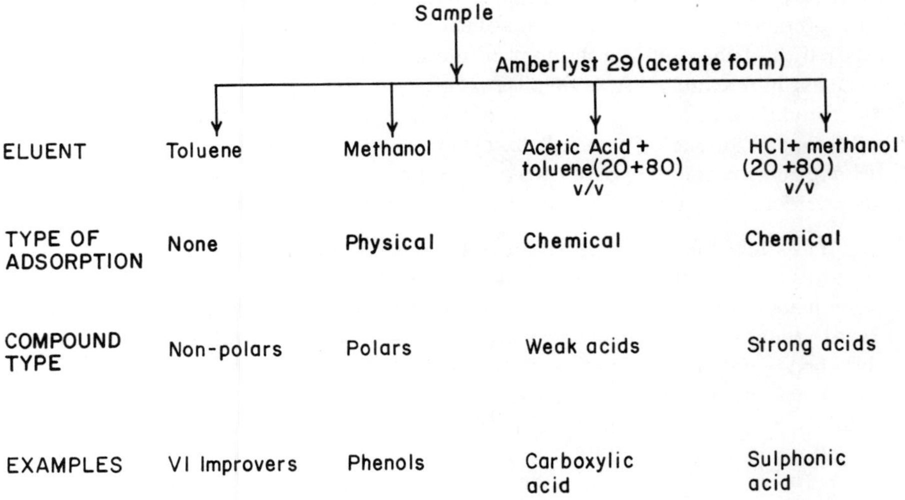

FIG. 5. Separation of hydrolyzed dialysis residue on Amberlyst 29 (acetate form).

the carbonate form of the resin, especially in monotonous routine work, appears to be the spectacular evolution of CO_2 from within the bowels of the column. Purists will agree that this is not sound chromatographic practice!

2. Recovery of Basic Components

Amines or amine salts can be recovered in a similar fashion to acids by using the cation exchange resin, Amberlyst 15. The column is packed in the same way as the Amberlyst 29 column and the resin is used directly in the free sulfonic acid form in which it is supplied. After establishing nonaqueous conditions with methanol and toluene, a toluene solution of the sample, containing not more than 0.5 g of basic material is passed down the column at a flow rate of 1 to 2 ml min^{-1}. Acidic and neutral substances pass through unadsorbed while bases are adsorbed. Amine salts are decomposed, with the acidic portion passing through the resin bed unhindered. Adsorbed bases are eluted by passing 200 ml 5% 0.880 ammonia in methanol down the column. In the case of long chain amines which are not very soluble in ammonia, an eluent consisting of chloroform + isopropanol + ammonia (35 + 60 + 5 [vol %]) is more effective.

3. Determination of Specific Additives via Ligand Formation

A feature of cation exchangers in their transitional metal form is their ability to adsorb complexing agents and to undergo "ligand exchange" with other complexing agents in solution [8]. Webster et al. [15] have studied the behavior of a number of metal deactivators commonly found in aviation lubricants, e.g., benztriazole, quinizarin, and derivatives of 1-salicylal-amino-guanidine and found they produce colored complexes with copper, nickel, or cobalt forms of Amberlyst 15. The most successful application they reported was the quantitative determination of bis-salicylidene-ethylene-diimine, a copper deactivator in kerosene.

The nickel form of the resin is formed by passing down 500 ml 4% aqueous nickel sulfate. Excess salts are removed with deionized water. Nonaqueous conditions are established with methanol and toluene in the usual manner. The kerosene sample is then passed down the column where the copper deactivator reacts with the resin to form a colored complex which can be recovered quantitatively with methanol. Its concentration in a solution of known volume can be measured colorimetrically at 405 nm.

D. Thin Layer Chromatography

TLC is a valuable technique for the identification and estimation of a wide range of additives in lubricating oils [3]. It can rapidly provide information on the additives in a drop of oil, which may be time consuming or even impossible to get in any other way. For example, while the infrared spectrum of an oil containing an overbased detergent-additive may be completely dominated by carbonate absorption bands, a rapid TLC test will establish whether the additive is of sulfonate or phenate type. Similarly, the infrared spectrum of an oil containing several additives may be too complex to interpret, whereas TLC of the same oil rapidly separates the additive components and, with the help of specific spray reagents, provides valuable information on the type and quantity of specific additives present.

Current methods are based on the original work of Killer and Amos [10], who developed procedures for a wide range of additive types in both new and used oils. These procedures have now been consolidated by a number of workers, including Amos [1], Coates [6], and Brook et al. [5].

The apparatus and technique of TLC has already been described in Chap. 14. The method can be applied to the original sample or to any fraction derived from the general separation scheme we have described in previous pages. A 5-μl portion of a 50% solution of the oil in heptane, or a 10% solution of any additive fraction in heptane or chloroform, is applied to a silica-coated plate. Nonpolar additives are developed in a single pass with the appropriate mobile phase, but polar additives such as dialkyldithio phosphates and detergent inhibitors (e.g., phenates and sulfonates) need a double development. The first step is to develop the full length of the plate with toluene. This removes hydrocarbon base oil to near the top of the plate, leaving polar additives at or near the start. A second development with a polar mobile phase to halfway up the plate then separates the more polar additives.

TLC is applicable to all the low-molecular-weight additives, including oxidation and corrosion inhibitors, antiwear and extreme pressure additives, and metal detergent inhibitors. TLC is not applicable to polymeric additives, such as ashless dispersants and VI improvers.

Some typical mobile phases and chromogenic reagents that have been reported in the literature are summarized in Table 3. Recipes for chromogenic reagents are given in Appendix 1.

E. Gel Permeation Chromatography

Gel permeation chromatography (GPC), or liquid exclusion chromatography, separates primarily by size (Chap. 12) and is an excellent technique for analyzing viscosity index improvers and many dispersants. The effectiveness and shear stability of viscosity improvers are influenced by their molecular weight distributions. Figure 6 shows the chromatogram of an experimental viscosity index improver. Here, the experimental data from

TABLE 3. TLC Systems for Lubricant Additives

Compound type	Mobile phase	Revealing reagent	Color reaction
Hindered phenols and amines	Carbon tetrachloride; n-hexane + di-isopropyl ether + methanol (70 + 30 + 5) vol %	III Gibb's reagent	Various characteristic colors
	Iso-octane + ethyl acetate (97 + 3) vol %	IV antimony pentachloride	Various characteristic colors
	Iso-octane + di-isopropyl ether (97 + 3) vol %	V diazotised p-nitroaniline	Orange-yellow
		VI tetra-azotised o-dianisidine	Various characteristic colors
Phenothiazines and naphthylamines	Carbon tetrachloride; benzene	III Gibb's reagent; V diazotised p-nitroaniline	Various characteristic colors
	Benzene + ethyl acetate (90 + 10) vol %	VI tetra-azotised o-dianisidine	Various characteristic colors
	Iso-octane + ethyl acetate (97 + 3) vol %	VII benzoyl peroxide	Various characteristic colors
		VIII chloramine T	Various characteristic colors
		IX palladium chloride	Strong blue with phenothiazines
Benztriazoles and mercaptobenzthiazoles	Benzene + ethyl acetate + acetone (100 + 5 + 1) vol %	III Gibb's reagent	Benztriazole yields a grey-brown spot. Mercaptobenzthiazole yields an orange spot.
Phosphate esters	Dichloromethane + benzene (50 + 50) vol %	X sodium molybdate + hydrazine sulfate	Blue spots on white background
	Dichloromethane + methyl ethyl ketone (94 + 6) vol %	XI KOH + tetrazotised o-dianisidine	Orange-brown spots on light mauve background

	Benzene + acetone (90 + 10) vol %		
	Carbon tetrachloride + acetone (70 + 30) vol %	XII fluorescein + lead acetate	White spots on green background
	n-Hexane + diethyl ether + methanol (70 + 30 + 5) vol %		
Zinc dialkyl dithiophosphate	n-Heptane + acetone (80 + 20) vol %	XIII dithizone	Zinc and other heavy metals yield pink spots on green background[a]
	n-Heptane + acetic acid (95 +5) vol %		
	Toluene + methyl ethyl ketone + acetic acid + pyridine (60 + 20 + 10 + 10) vol %	XIV Rhodamine B	Zinc yields blue spots on pink purple background[b]
	Acetic acid + ethyl acetate + iso-octane (15 + 25 + 60) vol %	IX palladium chloride	Yellow-brown spot on pale buff background for sulfur compounds
Alkyl phenols and sulfurized alkyl phenols	Toluene; chloroform	V diazotised p-nitroaniline	Orange-brown yellow spots
	Benzene + ethyl acetate (90 + 10) vol %	IX palladium chloride	Brown spots on pale buff background for sulfurized phenols
Carboxylates, phenates, salicylates, and sulfonates	Propanol-2 + 0.88 ammonia (80 + 20) vol %	XV Pinacryptol yellow	Natural sulfonates: dull orange-brown fluorescent spots; synthetic sulfonates: bright yellow-orange fluorescent spots; phenates and salicylates: black absorbing spots; carboxylates: pale blue-green spots
	Methyl acetate + ethanol + 0.88 ammonia (70 + 20 + 10) vol % Chloroform + 2-propanol + 0.88 ammonia (25 + 50 + 25) vol %		

[a]Adsorbents containing zinc phosphors will also react.
[b]Zinc dialkyldithiophosphates interfere. Confirm by hydrolysis and TLC examination of the resulting phenols.

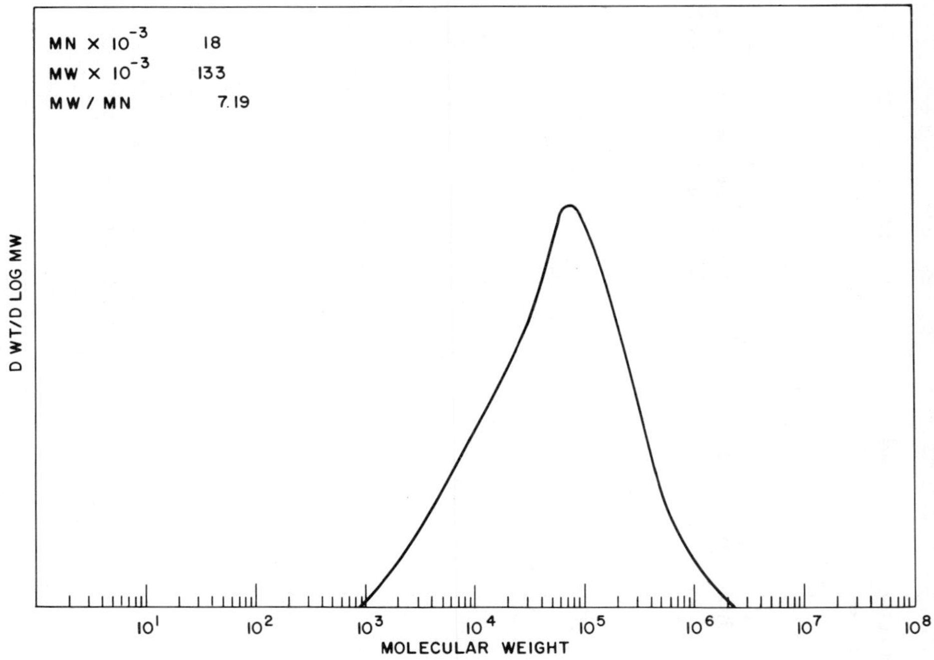

FIG. 6. Determination of molecular weight distribution, $\overline{Mw}$ and $\overline{Mn}$ of VI improver by gel permeation chromatography. Experimental conditions, see Table 4.

the chromatograph were logged with an automatic data acquisition system and a computer was used to calculate the number average molecular weight, weight average molecular weight, and plot the differential distribution. As reactor conditions are changed, the influence on distribution can be rapidly determined and the effect on performance can often be predicted. In this type of application, mainly for synthesis purposes, 1 to 2 mg of polymer is normally used. Many types of polymers have been analyzed in this way, as is well documented in the polymer literature.* Viscosity index improvers are usually supplied as a concentrate in a diluent oil or with other additives in a "package." In Figure 7 the separation of a polymethacrylate VI improver from the diluent oil is shown. Here, a larger sample (50 mg) was

*Bibliography on Liquid Exclusion Chromatography (gel permeation chromatography) AMD 40, American Society for Testing and Materials, 1916 Race Street, Philadelphia, Pennsylvania 19103.

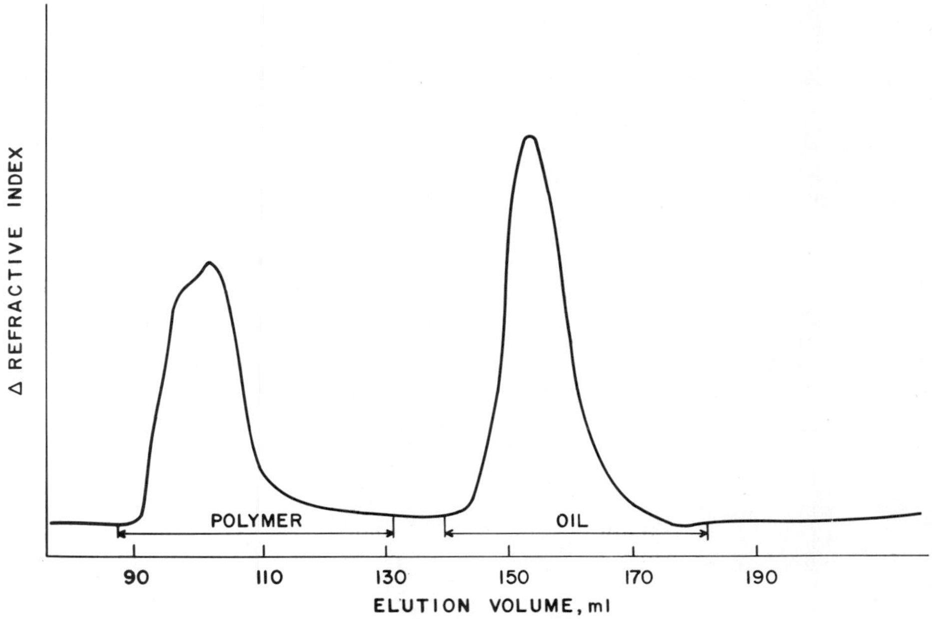

FIG. 7. Gel permeation chromatographic separation of polymethacrylate Viscosity Index improver. Experimental conditions, see Table 5.

taken. With proper calibration, the amount of polymer and oil can be determined from the chromatogram, and lot to lot variations are easily monitored.

Dispersants, in general, are lower in molecular weight than VI improvers but some do impart viscosity properties to a base oil in addition to dispersant action. In Figure 8 the chromatograms of dispersants from different vendors are shown. It is very evident that vendor B's dispersant will have a higher average molecular weight than vendor A's, and that their molecular weight distributions are quite different. One of the difficult problems in analyzing the polymeric additives has been the separation of different polymeric species. Gel permeation chromatography can be most helpful if the molecular weight ranges are somewhat different, as they usually are with viscosity index improvers and dispersants. In Figure 9 the results from the exclusion chromatographic fractionation of a fully blended oil are shown. In this fractionation, cuts were taken from the center of the distributions of the polymeric additives and then rechromatographed. The chromatograms show that these fractions of the two polymeric species had been completely isolated. Infrared analysis indicated that they were free from other additives and base oil.

TABLE 5. Experimental Conditions for the Separation of Polymeric Additives by Gel Permeation Chromatography (see Figures 7-13)

Columns	Styragel: 2×10^4 A
	Styragel: 2000-700 A
	Styragel: 350-100 A
	Styragel: 350-100 A
Mobile phases	Tetrahydrofuran with differential refractometer; chloroform with infrared detector
Oven temperature	25°C
Flow rate	1.5 ml min^{-1}
Sample size	100 mg dissolved in 2 ml mobile phase

TABLE 4. Experimental Conditions for the Determination of Molecular Weight Distribution of VI Improvers by Gel Permeation Chromatography (see Figure 6)

Columns	Styragel: 7×10^5 - 5×10^6 A
	Styragel: 5×10^4 - 1.5×10^5 A
	Styragel: 2000 - 500 A
	Styragel: 200 - 700 A
Mobile phase	1,2,4-trichlorobenzene (TCB)
Oven temperature	100°C
Syphon temperature	110°C
Detector temperature	110°C
Flow rate	1 ml min^{-1}
Sample size	2 mg ml^{-1} TCB

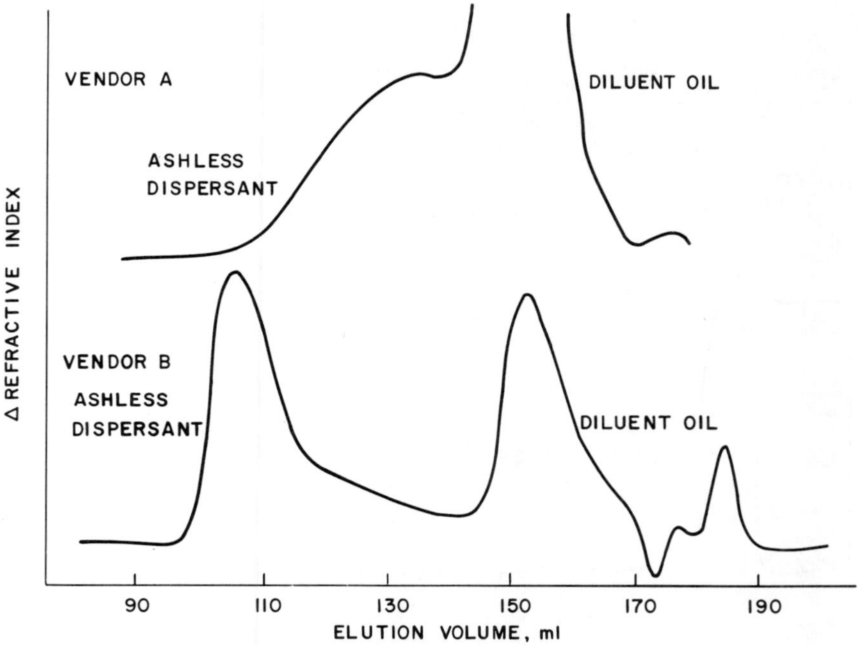

FIG. 8. Gel permeation chromatographic separation of ashless dispersants. Experimental conditions: see Table 5 .

None of the nonpolymeric additives normally found in lubricating oils appeared in the polymeric elution volume range of the exclusion system used in Figures 7 to 9. The nonpolymeric additives investigated included zinc dialkyldithiophosphates, calcium and magnesium sulfonates, calcium phenate, dioctyldiphenylamine, and phenol sulfide. All of these were eluted with the base oil. The recovery of the polymeric additives from the exclusion column was excellent except for one highly polar dispersant. The recovery of the nonpolymeric additives ranged from 50% for the sulfonates and phenates to over 90% for the zinc dialkyldithiophosphates.

Gel permeation chromatography is an excellent tool to follow the fate of polymeric additives during use. If identical weight samples are taken and chromatographed under identical conditions, one chromatogram can be directly compared with another. Figure 10 shows the chromatograms of a new oil and samples taken after 2000, 4000, and 6000 miles from a road test program. This is a multiviscosity oil containing a styrene-acrylate viscosity index improver. From comparison of the chromatograms, it is seen that with use (a) the higher molecular weight portion (leading edge) of the viscosity index improver is decomposing and forming lower molecular

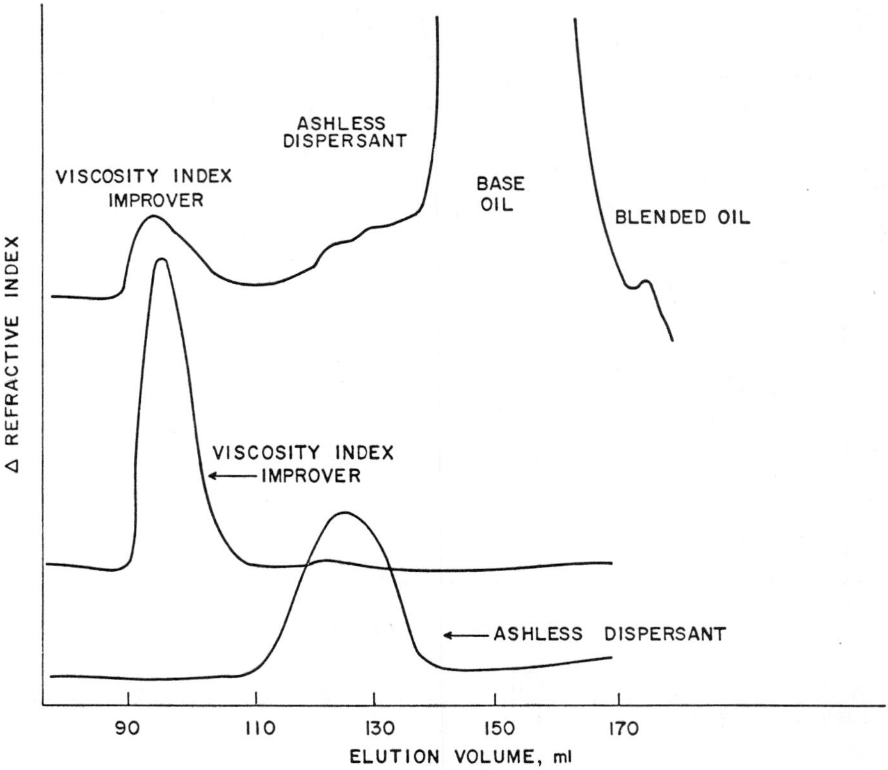

FIG. 9. Separation of viscosity index improver from ashless dispersant by gel permeation chromatography. Experimental conditions: see Table 5 .

weight fragments; (b) some of these fragments are accumulating on the lower molecular weight portion (trailing edge) of the viscosity index improver distribution; (c) the polymeric portion of the dispersant disappears with use and none is evident at 4000 miles; and (d) there is a buildup of a new material (decomposition product) in the 170 to 190 ml elution volume range.

Infrared analysis indicates that fragments from the viscosity index improver are distributed throughout the entire sample and that material building up in the elution volume range (179 to 190 ml) consists of carbonyl compounds different from the additives, presumably petroleum-derived acids. Also, none of the ashless dispersant was detected in any of the used oil fractions.

The fate of the dispersant can be better seen from the analysis of a single-weight blended oil (Figure 11). In this formulation there is only a

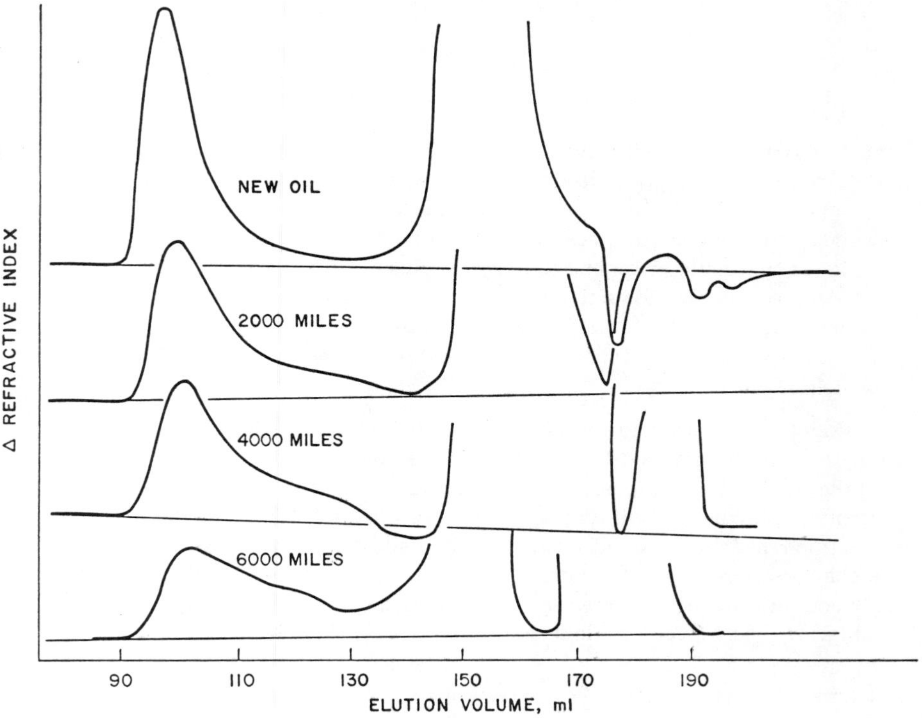

FIG. 10. Gel permeation chromatographic separation of used oils monitored with refractive index detector. Experimental conditions: see Table 5 .

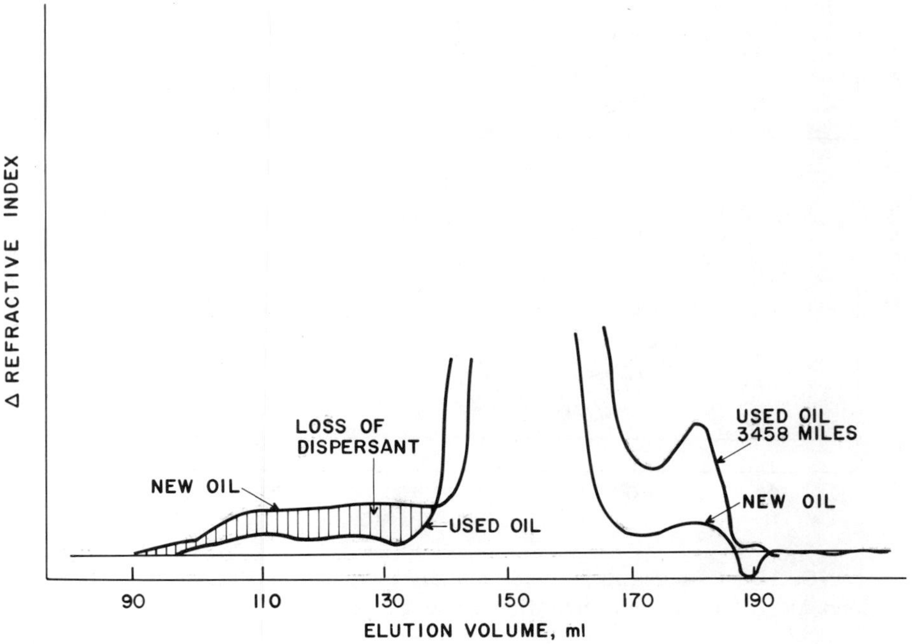

FIG. 11. Gel permeation chromatographic separation of single weight oil. Experimental conditions: see Table 5 .

small amount of viscosity index improver; thus, the changes in the polymeric region can be attributed to the dispersant. As seen in Figure 11, the dispersant does disappear with use and in the manner indicated in Figure 10. It seems that the dispersant disappears as an entity rather than that the higher molecular weight species are cleaved first. The single-weight oil in Figure 11 too shows the accumulation of material in the 170 to 190 ml elution volume range. This further indicates that this carbonyl containing material is not derived from viscosity index fragments, but rather from petroleum.

Quantitative information concerning the amount of the viscosity index improver can be gleaned from these chromatograms and also some insight into the mode of decomposition. Figure 12 shows comparisons of the chromatograms of the new oil and samples taken after 2000 and 6000 miles. By superimposing the chromatograms of the used oils on that of the new oil, the changes become very evident. If the areas representing decomposition and accumulation are divided by the area of the viscosity index improver in the new oil, numerical evaluation is possible. Figure 13 shows change

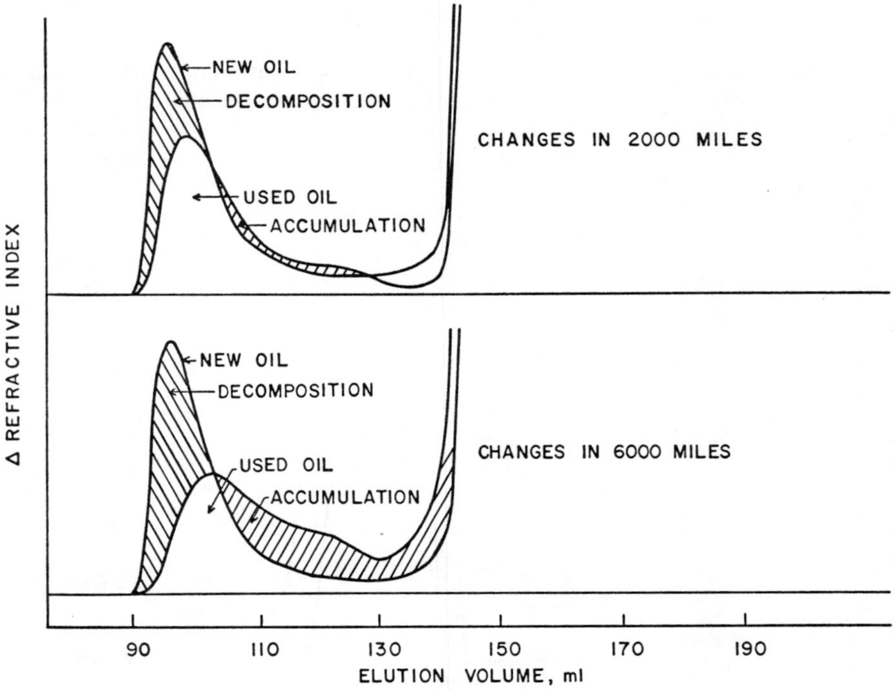

FIG. 12. Changes in polymeric additives with use in road test program (Blend A). Experimental conditions: see Table 5 .

plots derived in this manner. Let us first look at Blend A. In the first 2000 miles decomposition is rapid (30%). Then it continues more slowly until, at 6000 miles, it has reached 50%. The accumulation of fragments follows a different path; it increases slowly at first and then rises rapidly after 4000 miles.

Another view of the changes that occur during use can be obtained by monitoring the GPC separation with an infrared detector at a wavelength of 5.80 μm. The viscosity index improver can be seen by its carbonyl band, but the ashless dispersant is not detected. Figure 14 shows the chromatograms of the same new and used oils as Figure 10 in which the refractive index detector was used. In the new oil, the detector signal of the first peak (90 to 130 ml) is due to the carbonyl group in the styrene-acrylate viscosity index improver. The second peak (135 ml to 160 ml) is likely caused by traces of carbonyl compounds in the base oil. Its increase with

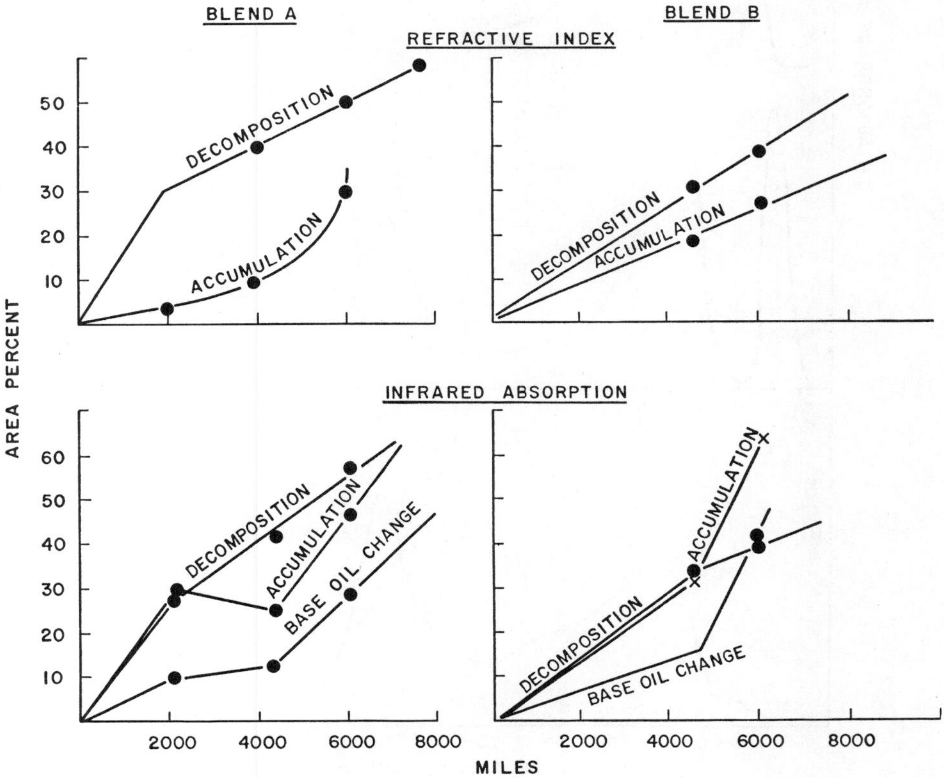

FIG. 13. Amount and mode of decomposition of two oil blends.

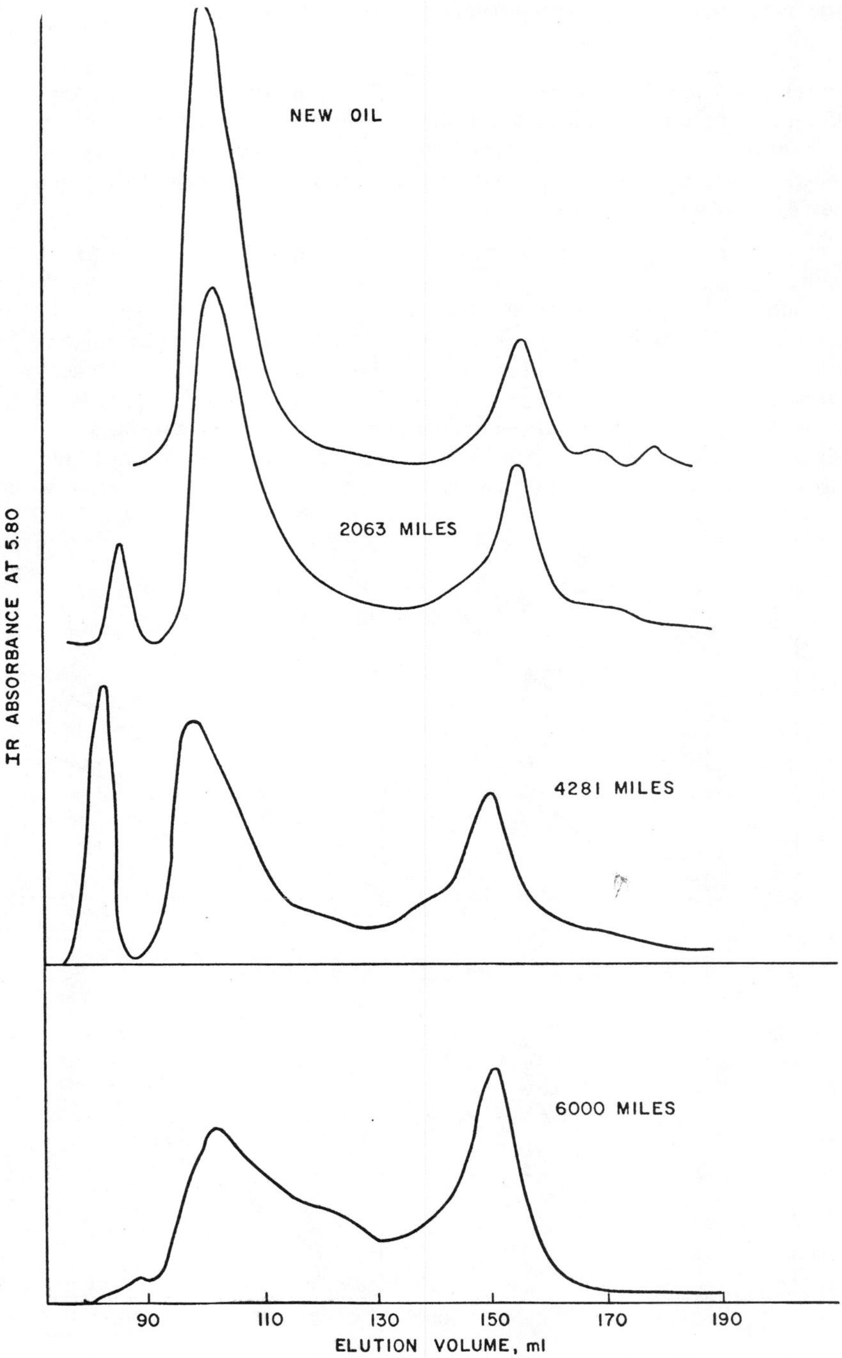

FIG. 14. Gel permeation chromatographic separation of used oils monitored with infrared detector. Road test program (Blend A). Experimental conditions: see Table 5 .

use then reflects the accumulation of more carbonyl-containing compounds from oil oxidation. In contrast, the peaks from 160 to approximately 190 ml are due to carbonyl impurities from additives. By infrared, the viscosity index improver is seen to degrade with use in the same manner as shown by the refractive index detector, i.e., the high molecular weight portion disappears and there is a buildup of intermediate molecular weight species (elution volume 110 to 135 ml).

Using the same scheme as before (Figure 12), relative changes of peak areas were calculated and plotted in Figure 13. The decomposition follows approximately the same trend as indicated by the refractive index detector, i.e., an initial rapid decomposition (28%) in 2000 miles and then a constant rate to 6000 miles with a little over 50% decomposition. However, the apparent accumulation of fragments is different from that shown by the refractive index detector, rising rapidly in the beginning, dropping between 2000 and 4000 miles, and increasing again at 6000 miles. The first large response is believed due to oxidation of certain unstable portions in the viscosity index improver. This reaction initially increases the carbonyl content of the accumulation products. Then, as decomposition proceeds, the unstable moieties are degraded, which action results in the observed plots.

Another oil, containing a straight polymethacrylate viscosity index improver from a road test program, Blend B, shows the same general trends. The high-molecular-weight portion of the viscosity index improver disappears, medium-molecular-weight fragments accumulate, the polymeric portion of the ashless dispersant has disappeared by 4000 miles, and carbonyl compounds in the base oil build up. However, mode, rate, and extent of viscosity index improver decomposition and also of fragment accumulation are different (Figure 13). Blend B does not show an initially high decomposition at 2000 miles but rather a constant rate for the full 6000 miles with an overall lower decomposition. The decomposition curve based on the infrared detector is very similar to that obtained with the refractive index detector; also, the percentage values agree very well (30 versus 33 at 4000 miles and 38 versus 38 at 6000 miles). However, the infrared accumulation curve increases more rapidly than that shown by the refractive index detector. Presumably, the ir detector sees the change in chemical structure (carbonyl content) during decomposition while the other detector sees only the mass. These observations, while not understood in full detail, are indicative of the type of information available from this technique.

F. High-Performance Liquid Chromatography

The technique of high-performance liquid chromatography (HPLC) and its relationship with thin layer chromatography (TLC) has already been discussed in Chap. 14. We have seen that TLC has the ability to separate low-molecular-weight additives and, with the help of specific spray reagents, gives some valuable information on the identity of the components.

HPLC can give fast and accurate quantitative results, provided that reference compounds can be made available. Furthermore, it is a simple and convenient technique to use for the isolation of pure fractions for further spectroscopic analysis.

In general, adsorption chromatography using silica gel or alumina is best for the separation of nonpolar additives, i.e., hindered phenolic and amine antioxidants, alkyl phenols, and esters, whereas reversed phase chromatography is best suited for the analysis of polar additives such as zinc dialkyldithiophosphates.

Figure 15 shows a chromatogram of an aviation lubricant from which accurate, quantitative determinations of the two antioxidants, dioctyldiphenylamine and phenyl-α-naphthylamine, as well as tricresyl phosphate, can be made. It can be seen that so-called tricresyl phosphate is a complex mixture and different fingerprint chromatograms are obtained in samples from different sources. "Peaks" have been trapped out and identified, by mass spectrometry, as compounds containing phenyl or xylenyl phosphates. Similarly, Figure 16 is a fingerprint chromatogram of partially sulfurized phenols obtained by hydrolysis of a phenate isolated from a dialysis residue. With this type of information the quality of the additives can be judged and the reliability of their manufacture established.

The most promising separations of zinc dialkyldithiophosphates have been achieved by reversed phase partition chromatography and Figure 17 shows such a separation from a fully blended oil. Analysis of peaks 1 and 2 by infrared spectroscopy and x-ray fluorescence indicated these peaks were alkyl isomers of zinc dialkyldithiophosphate. Mass spectrometric analysis of the same additive separated on silica gel indicated these to be the C_4 and C_6 isomers. The infrared spectrum of peak 3 indicated it also is an alkyl phosphate. Under the conditions of this separation, all other components of the oil remain on the column. This separation can serve as the basis for an analytical method to monitor vendors' lots and blended oils for the individual species of the additive.

The final example, Figure 18,illustrates the combination of exclusion with adsorption chromatography to isolate a basic magnesium sulfonate from a finished oil using a chloroform-methanol gradient and a controlled porosity silica adsorbent. The first peak is due to the polymeric additives. The separation here is believed to be a result of total exclusion from the column, while the other additives penetrate the pores. The base oil, zinc dialkyldithiophosphate and basic magnesium sulfonate, separate by virtue of their different polarities as the solvent strength is increased.

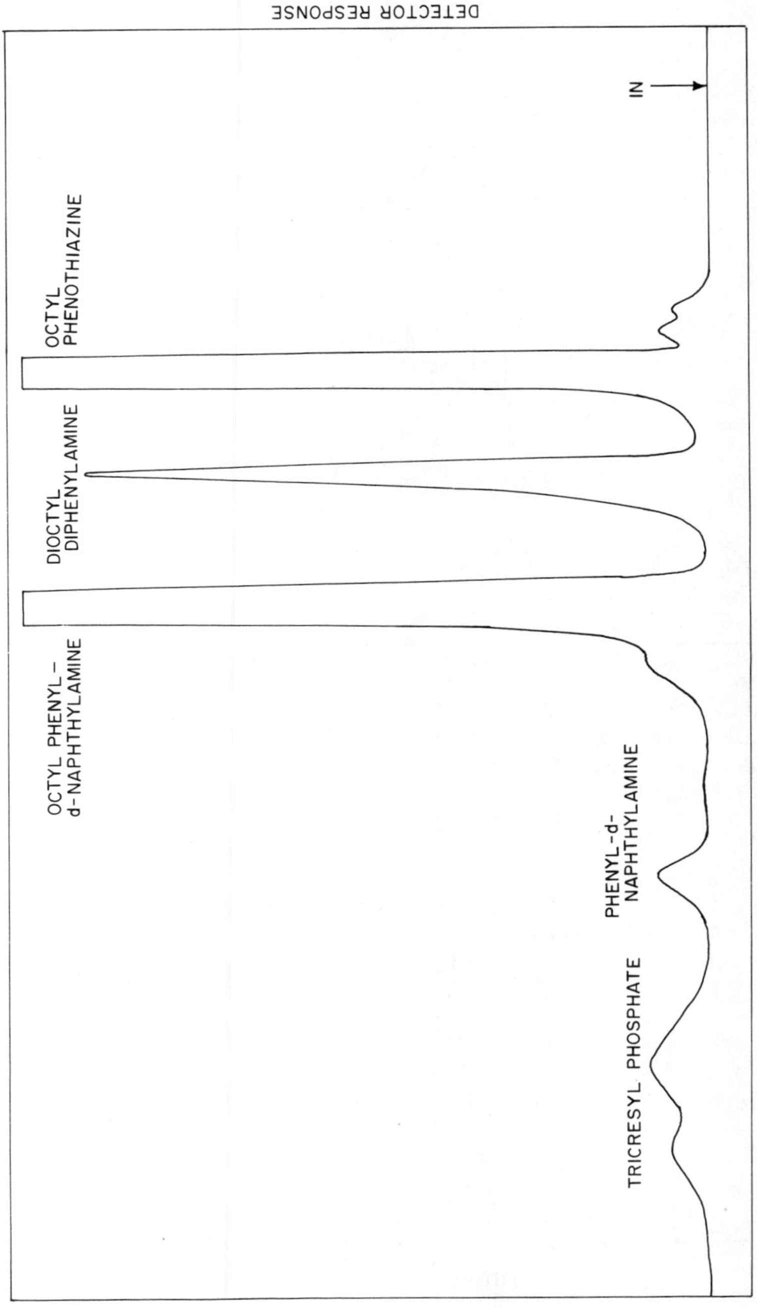

FIG. 15. Separation of aviation lubricant. Column: 25 cm by 4.6 mm ID; Partisil 10; mobile phase: heptane + ethyl acetate (99 + 1) vol %; flow rate: 1.0 ml min^{-1}; detection: uv at 254 nm.

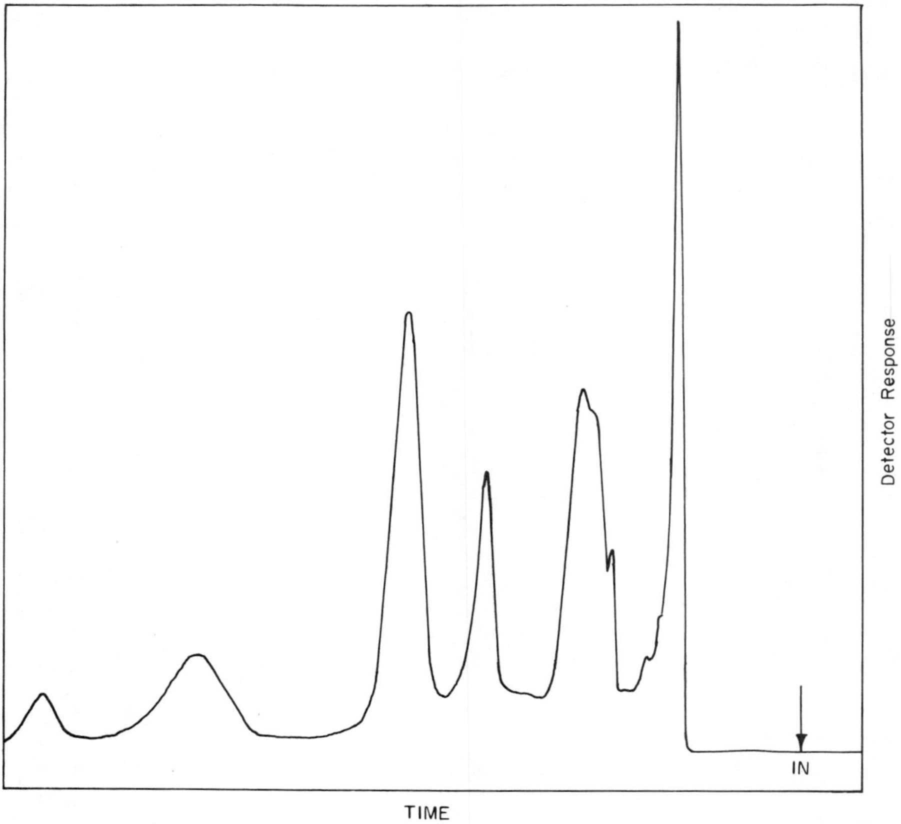

FIG. 16. Separation of phenols: column: 25 cm by 4.6 mm ID; Partisil 10; mobile phase: heptane + ethyl acetate (95 + 5) vol %; flow rate: 1.0 ml min^{-1}; detection: uv at 280 nm.

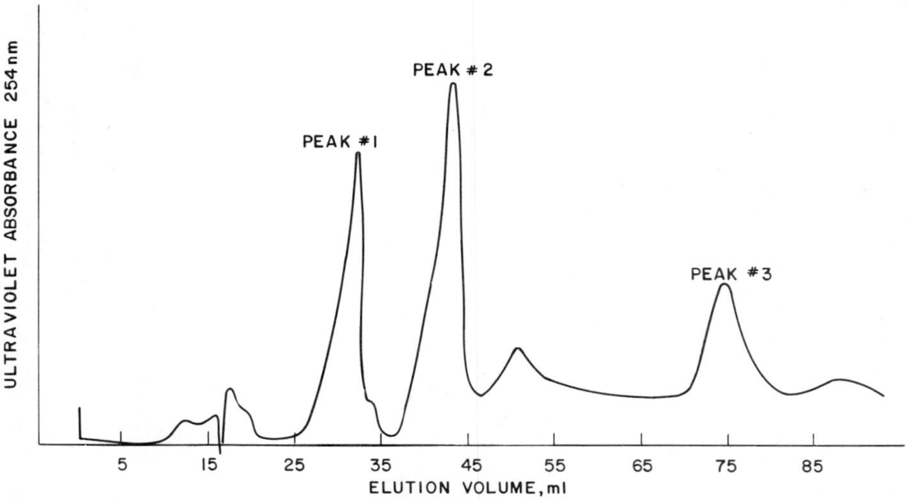

FIG. 17. Reversed phase partition separation of zinc dialkyldithiophosphate from blended oil. Column: 4 ft by 4 mm C_{18}/Porasil B (Waters Associates); mobile phase: (linear gradient) 20% to 50% tetrahydrofuran in water; temperature: 25°C; flow rate: 1 ml min^{-1}.

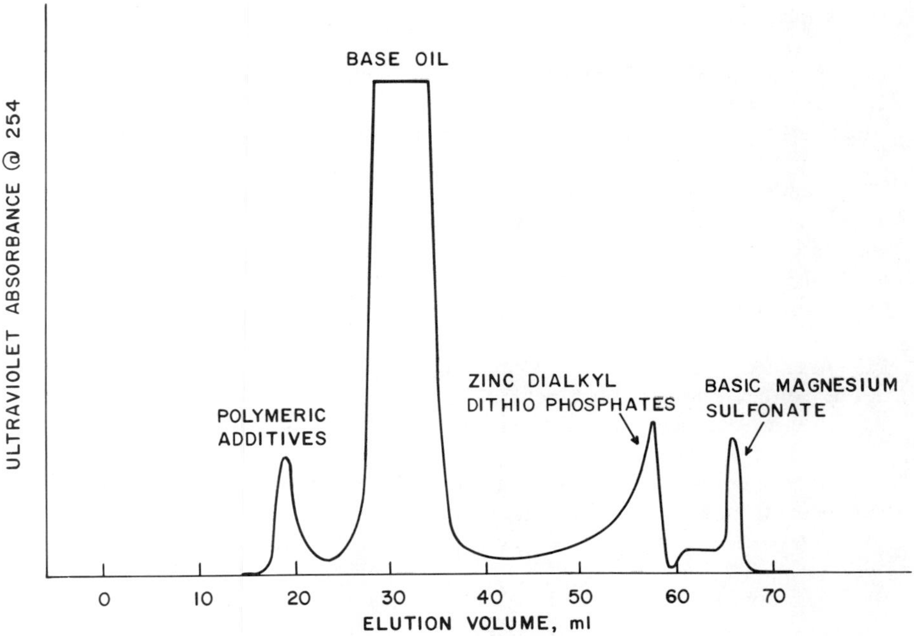

FIG. 18. Separation of blended oil with 60 A controlled porosity silica. Column: 8 ft by 4 mm Porasil A (Waters Associates); mobile phase: chloroform + methanol (linear gradient) from 90% $CHCl_3$ + 10% methanol to 100% methanol.

It is hoped that, from the examples we have chosen* and in which the chromatographic conditions were probably far removed from optimum, we have given some indication of the potential power of the technique, to obtain information that has hitherto been either time consuming or impossible to get in any other way. To substantiate the implications of this work and to exploit its benefits, data must be accumulated on additive packages and their behavior with use followed. This type of information should aid in selecting future additives and optimizing concentrations.

APPENDIX —REVEALING REAGENTS FOR LUBRICANT ADDITIVES

Vapor

I. Iodine Vapor. Place the plate in a closed jar containing a few crystals of iodine. The iodine vapor dissolves in most organic compounds which are revealed as brown spots on a pale yellow background.

*A few more applications are mentioned in Chap. 10, Sec. VIII.

Spray Reagents

II. Phosphomolybdic Acid. Dissolve 5 g phosphomolybdic acid in propanol-2. After spraying, heat the plate in an oven for 5 to 10 min. Most organic substances produce blue-black spots on a yellow background.

III. Gibb's Reagent. Dissolve 1 g dichloro-quinone-4-chlorimine (or corresponding dibromo compound) in 100 ml ethanol. Amines, phenols, and antioxidants, in general, give a wide variety of characteristically colored spots. Color differences are enhanced, if, after 15 min, the plates are resprayed with 2% borax in water-ethanol (4 + 96) vol %.

IV. Antimony Trichloride or Pentachloride. Prepare 10 to 20% solutions in chloroform or carbon tetrachloride. Heat the sprayed plate at 110°C for 5 to 10 min. A wide variety of characteristic colors are produced.

V. Diazotised p-Nitroaniline. Dissolve 0.3 g p-nitrobenzene diazonium fluoroborate in 35 ml dioxane and 65 ml water. Aromatic amines and phenols give yellow-orange-red-brown spots on a pale yellow background.

VI. Tetraazotised o-Dianisidine. Dissolve 1 g in acetone-water (40 + 60) vol %. Antioxidants give a wide range of characteristic colors.

VII. Benzoyl Peroxide. Dissolve 4 g in 100 ml benzene. Antioxidants give a wide range of characteristic colors.

VIII. Chloramine-T. 1% solution in methanol gives characteristically colored spots for some amine antioxidants, especially phenothiazines.

IX. Palladium Chloride. Dissolve 0.5 g in water + acetone (50 + 50) vol %. Add two drops of concentrated HCl. Allow 24 hr for complete dissolution. Phenothiazines give rich-blue spots. Sulfurized compounds give yellow-brown spots on pale buff background.

X. Sodium Molybdate. Dissolve 0.7 g sodium molybdate and 0.05 g hydrazine sulfate in 50% H_2SO_4 + H_2O (20 + 80) vol %. Phosphate esters give blue spots on a white background. Thiophosphates give intense magenta-colored spots.

XI. KOH + Tetraazotised o-Dianisidine. Spray with 0.5 M alcoholic potassium hydroxide and heat in oven at 110°C for 5 to 10 min. After cooling, respray with 0.5% aqueous tetraazotised-o-dianisidine. Phosphate esters give orange brown spots on light mauve background.

XII. Fluorescein-Lead Acetate. Spray with 0.5% fluorescein in water + ethanol (50 + 50) vol %. Respray with 1% aqueous lead acetate. Organic phosphates give white spots on a green background.

XIII. Dithizone. Dissolve 0.05 g diphenyl-thiocarbazone in carbon tetrachloride. Zinc-containing compounds give purple-red spots on a green background. Other heavy metals, e.g., Fe, Hg, and Mn, give similar color reactions. Beware of plates containing inorganic uv fluorescent indicators such as zinc sulfide!

XIV. Rhodamine B. Dissolve 50 mg tetraethylrhodamine in 100 ml ethanol. A wide range of organic compounds give mauve spots on a pink background. Exposure to bromine vapor decolorizes the background. Inspection under long wave uv light often shows orange fluorescent spots on a dark background.

XV. Pinacryptol Yellow. Dissolve 0.05 g in 100 ml water. After thoroughly wetting the plate, view under long-wave uv light. Sulfonates give bright yellow-orange fluorescent spots. Carboxylates give weaker green-blue fluorescent spots.

For a wider range of specific spray reagent see Perry et al. [13].

REFERENCES

1. R. Amos, Z. Anal. Chem. 236:350 (1968).
2. R. Amos, The Application of High Performance Liquid Chromatography to the Analysis of Petroleum Products, in "Gas Chromatography, 1972" (S. G. Perry, ed.), Applied Science Publishers, London, 1973, p. 235.
3. R. Amos, Talanta 20:1231 (1973).
4. I. Berthold, Erdol and Kohle 21:614 (1968).
5. A. J. W. Brook, J. E. Davies, and B. M. J. King, The Determination of Additives in Used Automotive Engine Oils by Thin Layer Chromatography and Quantitative Densitometry, in "Recent Analytical Developments in the Petroleum Industry" (D. R. Hodges, ed.), Applied Sciences Publishers, London, 1974, pp. 97-105.
6. J. P. Coates, J. Inst. Petrol. 57:209 (1971).
7. G. B. Crump and A. Lynes, Applications of Macroreticular Ion Exchange Resins to Petroleum Product Analysis, in "Recent Analytical Developments in the Petroleum Industry" (D. R. Hodges, ed.), Applied Sciences Publishers, London, 1974, pp. 109-118.
8. F. Helfferich, Nature 189:1001 (1961).
9. M. W. Hill and M. W. Munsell, Preprints, Div. Petrol Chem., ACS, 5(3):115 (1960).
10. F. C. A. Killer and R. Amos, J. Inst. Petrol. 52:315 (1966).
11. R. Kunin, E. F. Meitzner, and N. Bortnick, J. Am. Chem. Soc. 84:305 (1962).
12. W. A. Munday and A. Eaves, Proc. Fifth World Petrol. Congr., Sec. V, Paper 9, 1959.

13. S. G. Perry, R. Amos, and P. I. Brewer, "Practical Liquid Chromatography," Plenum, New York and London, 1972.
14. A. Rimmer, J. Inst. Petrol. 51:308 (1965).
15. P. V. Webster, J. N. Wilson, and M. C. Franks, Anal. Chim. Acta 38: 193 (1967).
16. P. V. Webster, J. N. Wilson, and M. C. Franks, J. Inst. Petrol. 56:30 (1970).

Chapter 18

PROCESS CHROMATOGRAPHY

M. P. Timothy Bradley*
Research and Engineering
The Standard Oil Company of Ohio
Warrensville Heights, Ohio

---

*Present affiliation: Autolab Division, Spectra-Physics, Inc., Santa Clara, California.

I. INTRODUCTION

Process chromatography is a technique that rapidly found its way into the petroleum field. Indeed, process gas chromatographs have been commercially available for 20 years, a remarkable achievement considering that the pioneering work in gas chromatography was first reported only 25 years ago [1]. Most process chromatographs are based on the use of a gas as the mobile fluid. At the time this chapter was prepared, process liquid chromatographs were being used in a limited number of applications. The explosive growth of HPLC techniques would indicate, however, that a rapid increase in process liquid chromatographs can also be expected. The current state of these chromatographs is discussed at the end of this chapter.

An American Petroleum Institute survey recently showed that gas chromatographs comprised 30% of all of the stream-compositional analyzers used at that date [2]. At the present time, it is most likely that this percentage has increased.

The reason for this dominance of a single analyzer type is that it has important characteristics particularly suitable for process analysis. Chief among these is flexibility. Column changes are readily accomplished and the system can hence be tailored to a thousand applications. By utilizing modern column technology and electronic gating—both of which will be discussed later—the system can analyze and display one or a selected number of components from a complex mixture. These instruments are sensitive in that minute components of interest can be analyzed or monitored.

Obviously, many of these characteristics are inherent to gas chromatography and are also shared by laboratory gas chromatographs. But process gas chromatographs have a number of additional attributes that have been developed over the years which make them uniquely suited for their role as a premier process control instrument.

Process gas chromatographs are used when analytical results are to be obtained with a speed comparable to or better than that of the process dynamics. They should also be used when the process stream has characteristics which make it difficult or impossible either to transport a representative sample to the laboratory or to get a representative sample out of the sample container. For instance, when the sample contains permanent gases and light hydrocarbons in the presence of heavier hydrocarbons, the former will remain in the vapor state when transported to the laboratory. The latter, however, will condense. This gives rise to a two-phase mixture in the sample bomb. There are many examples of analyses of this type. Figure 1 shows one. This rich gas from a catalytic cracker can be analyzed quite easily by a process analyzer but presents a very difficult sampling problem if the analysis is to be performed on a laboratory instrument.

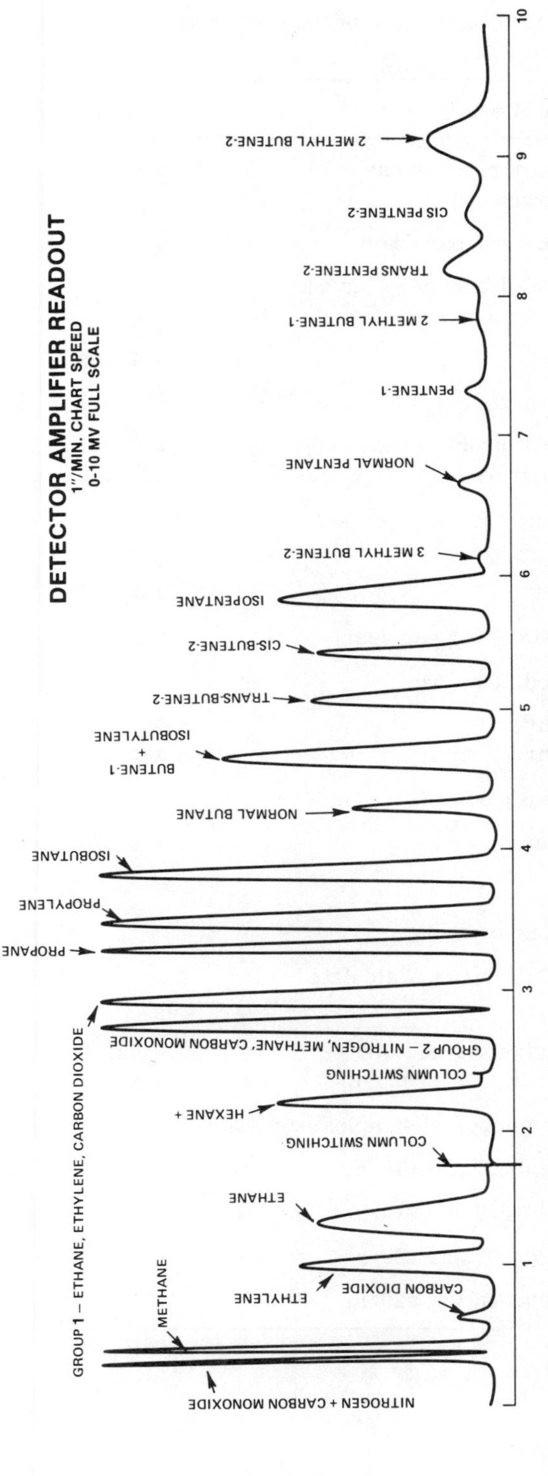

FIG. 1. Rich gas from catalytic cracker.

TABLE 1. Comparison of the Cost of Process Gas Chromatographs with Laboratory Analyzers

Item	Estimated range of cost		
Installed cost of process gas chromatograph	\$20,000	to	\$ 30,000
5-Year life (cost per year)	\$ 4,000		\$ 6,000
Maintenance/Calibration, 4 hr/wk			
at \$10/hr	\$ 2,080		\$ 2,080
(at \$20 hr)	\$ (4,160)		\$ (4,160)
Replacement parts	\$ 500		\$ 500
Cost/Year of process gas chromatograph	\$ 6,580 \$ (8,660)		\$ 8,580 \$(10,660)
Laboratory gas chromatographic analysis:			
Cost of unit	\$ 7,500		
5-Year life (cost per year)	\$ 1,500		
Maintenance, 1 hr/wk			
at \$10/hr	\$ 520		
(at \$20/hr)	\$ (1,040)		
Replacement parts	\$ 250		
Cost/Year	\$ 2,270 (2,790)		
Sampling		15 min/sample	
Sample transportation		15 min/sample	
Analysis/Calibration/Calculate report		45 min/sample	
Cost/Analysis at \$10/hr	\$ 12.50		
(at \$20/hr)	\$ (25.00)		
Breakeven number of samples per year:			
1. At \$20,000 and \$10/hr		345 samples	
2. At \$30,000 and \$10/hr		505 samples	
3. At \$20,000 and \$20/hr		235 samples	
4. At \$30,000 and \$20/hr		315 samples	

Another advantage of the process over the laboratory gas chromatograph is in the area of cost. The cost of an analysis in a laboratory varies according to the location, as does the cost of installing and maintaining a process chromatograph. An absolute break-even point cannot be defined. However, when one considers the cost to include not only the analytical technician's time, but also the cost of sampling, sample transportation, and sample disposal—using typical U.S. conditions—a typical process gas chromatograph installation is economically preferable when the desired analysis frequency is higher than 1 per day.

In fact, from Table 1 it can be seen that for a $20,000 process gas chromatograph and $10 per hr labor costs, a once/shift analysis by a laboratory gas chromatograph must cost less than $4 per sample to compete economically against a process analyzer. For a $30,000 installation, the cost must be less than $5.50 per sample.

II. SAMPLING SYSTEMS

Probably the most critical item for realizing the full potential of the on-stream analyzer is the design and construction of the sampling system. In a process gas chromatograph, it must perform the following functions:

1. Remove a sample from the process steam.
2. Transport this sample to the analyzer.
3. Condition the sample for the analyzer.
4. Inject a representative aliquot of the sample into the analyzer.
5. Dispose of the unused sample.

A. Sample Removal

Obtaining a representative sample in the petroleum industry is generally quite easy. The sampled stream usually consists of a mixture of nonreactive hydrocarbons. If reactive gases are present, they are ordinarily only there in low concentration. And often they need not be analyzed quantitatively. One of the few notable exceptions is when hydrogen sulfide is involved.

The composition of the sample stream obviously determines the degree of sophistication in sampling system design. An excellent monograph by Houser [3] discusses these problems in depth. In this chapter only the general principles of sample removal and handling will be discussed.

Sample removal from process streams in the petroleum industry is generally much less of a problem than in other industries. There, multiphase streams and highly corrosive mixtures are often the norm rather than the exception.

With hydrocarbon streams the major problems are associated with the presence of particulates and/or gas bubbles in the stream. One has to avoid or minimize the chance of allowing particulates to enter the sample probe. Also to be avoided is the intermittent removal of gas bubbles when sampling a liquid, or of water when sampling either a gas or a liquid hydrocarbon stream.

All of these problems can be avoided by correct installation and orientation of the sample probe. Figure 2 shows some typical mounting arrangements. Note that the flush-mounted probes are highly susceptible to erroneous sample removal and to contamination. On the other hand, the probe

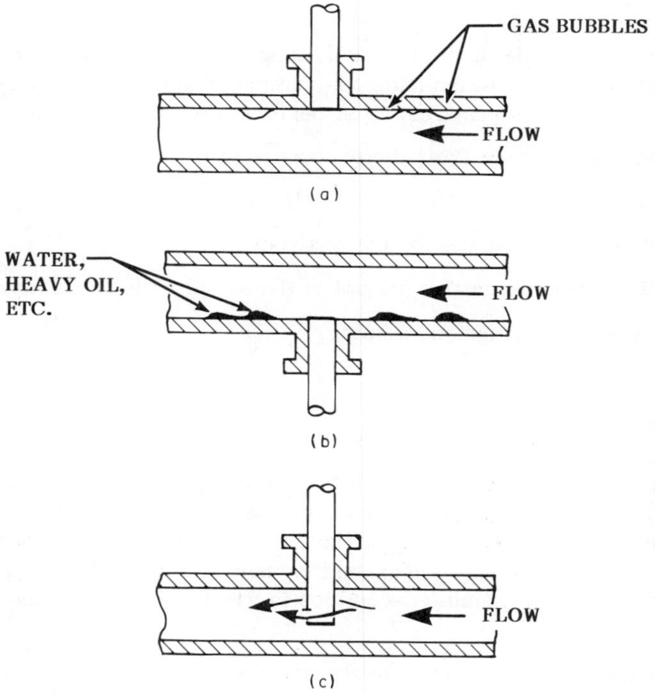

FIG. 2. Sample probe mounting arrangements: flush mounted probe, top of pipe; (b) flush mounted probe, bottom of pipe; (c) probe in center of pipe with inlet of downstream side.

that is inserted into the center of the sample stream is relatively immune to these problems. Note also (Figure 2c) that the entrance on the downstream side of the probe minimizes the amount of particulates drawn in. The sizing of the sample loop and transport system is also such that the flow velocity into the probe is significantly less than the velocity in the process pipe. Under these conditions, the probe will be highly nonisokinetic. This ensures that suspended particles are swept around the probe and are not drawn into the sampling system.

In contrast, when taking a sample from a multiphase stream, isokinetic sampling is necessary (see Figure 3). Nominally isokinetic sampling will remove a representative sample from the stream but in practice few streams are flowing under the turbulent conditions necessary to give good mixing. Stratification often takes place. The closest approximation to the true stream composition is obtained when multiple samples are taken across the cross section of the pipe.

B. Sample Transport

Transport of the sample to the analyzer should preferably take place without a change of state. The specific design of the system is quite complicated and detailed information of the physical characteristics of the stream, the required response time, etc. are needed for the correct sizing of the lines.

Transport lines should be as short and contain as few bends as possible. The greater the length of the line the greater the pressure drop and the volume of the transport system; this increases the difficulty in obtaining short response times.

The other aspects of the transport line pertain to the physical characteristics of the sample. If the stream is in the gas phase, then the entire transport line must be maintained at above the highest expected dew point temperature. If sample should condense in the transport line, it will be swept along and eventually be revolatilized at a hot spot. If this happens, the sample is no longer representative of the process stream because there is a change in the concentration of the heavier components. With liquid streams, a similar error can be induced if the temperature is too high or if the pressure is too low, such that the bubble point of the stream is exceeded.

C. Sample Conditioning

The design of sample conditioning systems is a complicated art. In many cases it is particularly difficult if only one analyzer is available to process multiple sample streams. The basic functions of the sample conditioning system are:

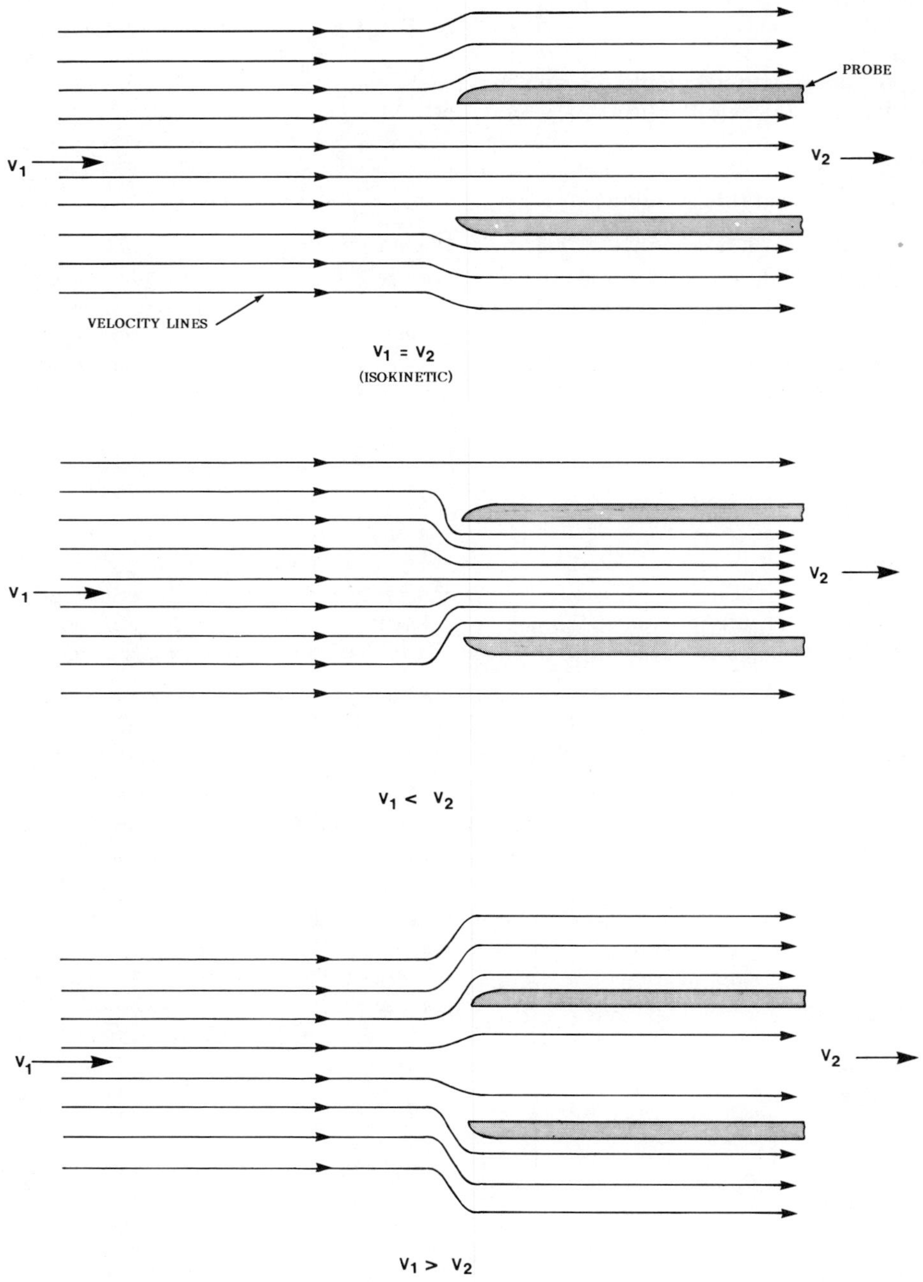

FIG. 3. Flow pattern, isokinetic sampling.

1. To remove particulates of droplets from the sample.
2. To change its physical state to that required for injection.
3. To dispose of the rest of the sample safely.
4. To select the correct stream when multiple streams are running into the analyzer.

To clean the sample, one can choose from a wide variety of filters. We can use micron filters to remove inorganic particulates, coalescing filters to remove entrained liquids (Figure 4), and cyclonic filters to remove

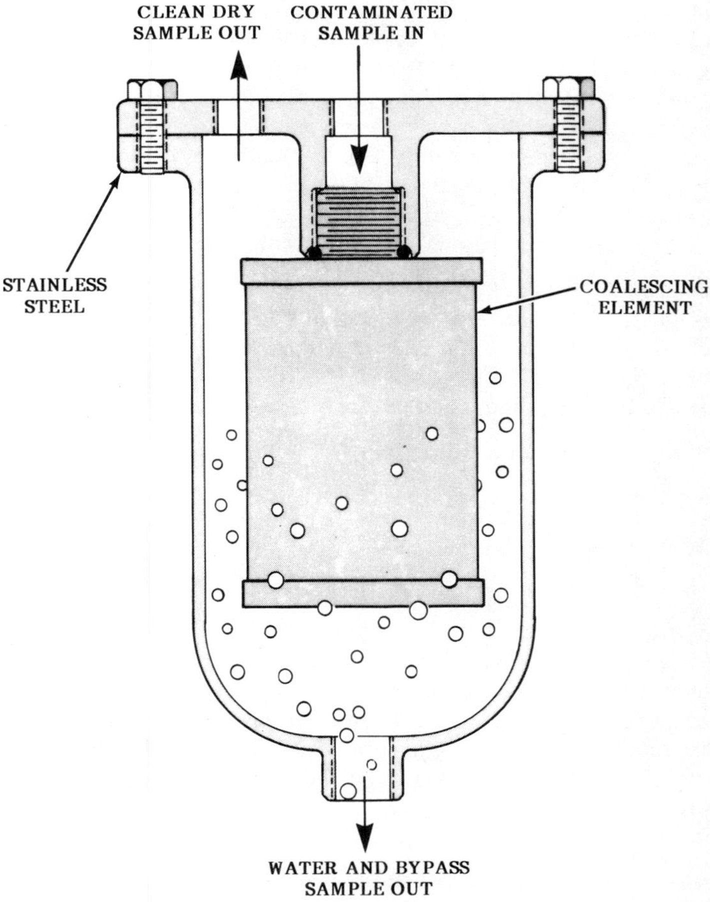

FIG. 4. Coalescing filter.

large amounts of water from hydrocarbon streams. The actual selection depends, of course, on the materials which have to be eliminated.

D. Sample Injection

To introduce the sample into the analyzer, it must be injected either as a liquid or as a gas. Sometimes it is obvious which approach should be used. In analyzing platformate for benzene, the relatively involatile sample should be injected as a liquid. On the other hand, a stream containing mostly "inerts" must be injected as a gas. Where there is a choice, for instance in analyzing C_4 streams, the advantages and disadvantages must be weighed against each other.

The advantage of liquid injection is that the physical state of the sample need not be changed. This simplifies the introduction of a representative sample because the danger of vaporizing only part of the sample or leaving the "heavies" is avoided. On the other hand, the volume of liquid that is appropriate is small—approximately 1 to 25 μl. There is danger of plugging the necessarily small passageways of the sample valve because of the presence of entrained particles. In addition, gas bubbles can cause the sample size to vary markedly, causing problems in quantitation.

The sample size for gas injection is 0.5 to 1.5 ml or approximately 250 times that of the corresponding liquid volume. The passageways in the valve and the sample loop diameter can be greater and thus will be less prone to plugging. On the other hand, as mentioned earlier, there are problems associated with vaporizing the sample in a representative manner.

Three devices are commonly used:

1. The vaporizing regulator
2. The spray vaporizer
3. The heated capillary

The vaporizing regulator is the device most commonly used for converting a liquid sample into a gas. A major advantage is that, in addition to vaporizing the sample, it also controls the pressure of the vapor sample. Figure 5 shows a typical unit. As can be seen, it is basically a regular pressure regulator. There is an extended base section to accommodate either a steam coil or an electric heater. In operation, the stream should remain liquid up to the poppet valve. As the regulator calls for more feed, the poppet valve opens and the sample flashes into the vapor space.

These devices work best when the pressure drop across the poppet valve is large and the boiling range of the sample is narrow. The upstream pressure should be high enough to maintain even the lightest component in

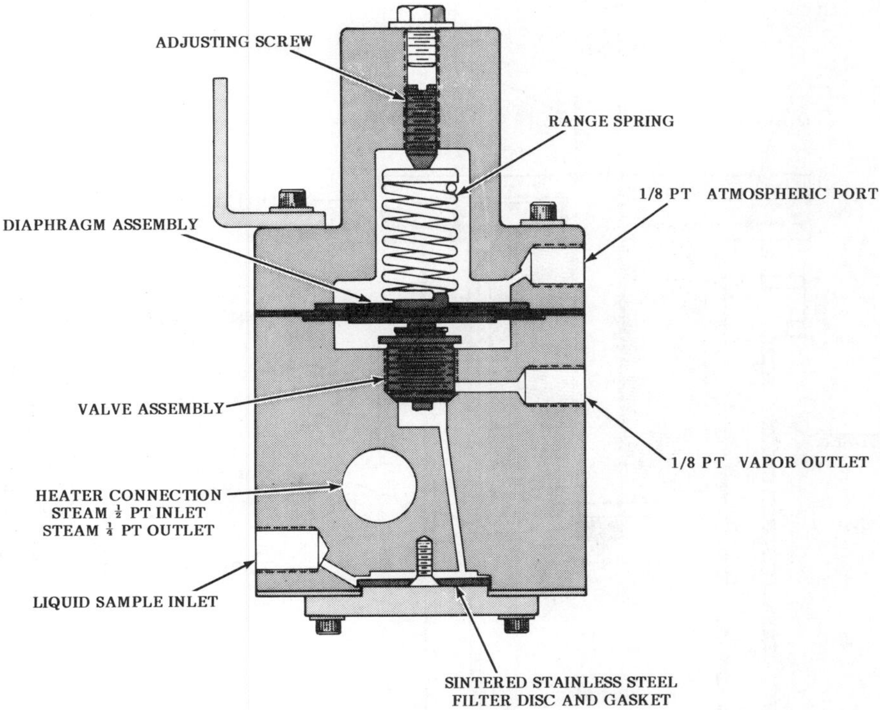

FIG. 5. Vaporizing regulator.

the liquid state, whereas the heaviest component must be converted completely into the vapor state at the lower pressure.

When the sample is likely to be contaminated with nonvolatile materials—for example, during an upset condition—the spray vaporizer is more desirable. In this unit (Figure 6), the sample is sprayed in a continuous stream from a capillary tube into a heated column. The light materials vaporize and flow to the analyzer. The heavy contaminants, on the other hand, are either deposited on the walls of the vessel or are blown out through the bypass.

The heated capillary is preferable to the vaporizing regulator when the sample pressure is relatively constant and the available pressure drop is low. The flow rate is dependent on the pressure drop and, therefore, will fluctuate with the upstream pressure. However, the unit will work with a pressure drop of as little as 1 psig.

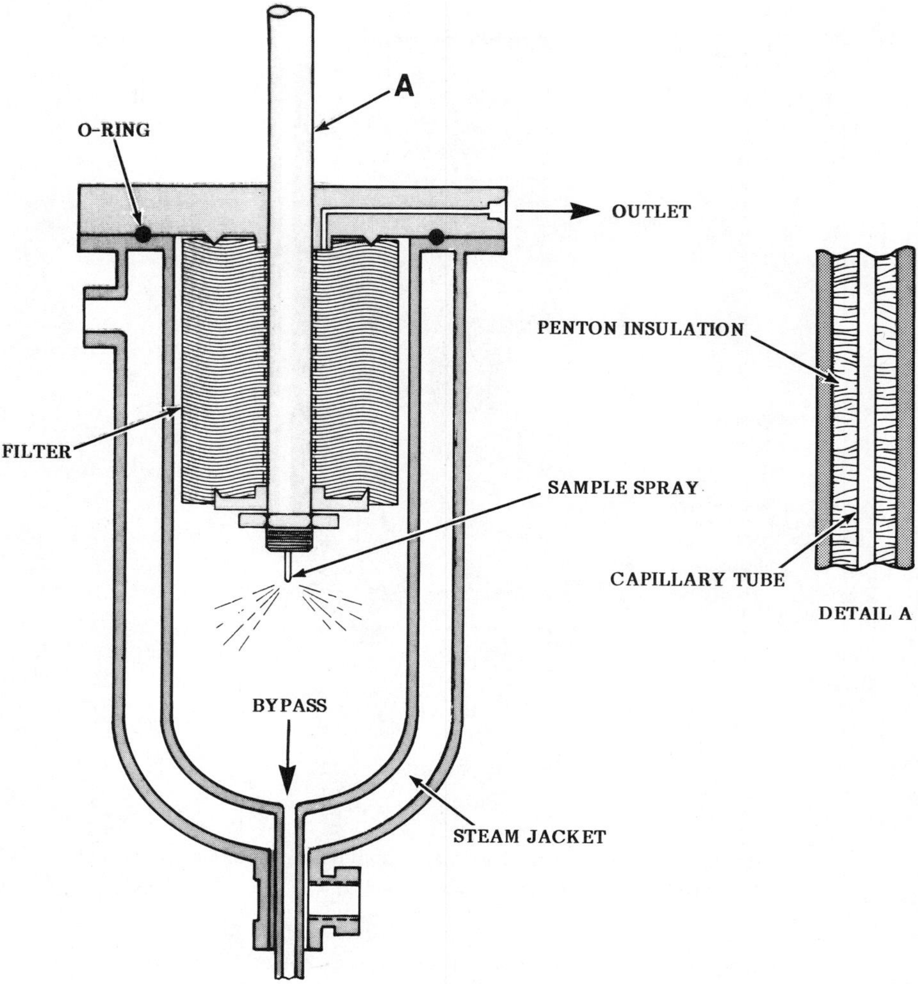

FIG. 6. Spray vaporizer.

The final device in any sample conditioning system is the filter. Filters may have been used previously in conditioning the stream, but it is important that the sample be filtered again immediately before it reaches the sample valve. For a liquid stream, a small micron particulate filter is used; for a gas stream a coalescing filter is used.

To cope with multiple sample streams and only one analyzer, the sample conditioning section must be equipped with a valve-switching system that allows any particular stream to be selected and directed to the analyzer. Ensuring the absence of cross-contamination is the most difficult aspect of stream selector design. However, with proper care and attention this can be accomplished.

The usual method of stream selection involves the use of a double block and bleed system (Figure 7). Here, when a stream is selected, the common lines of the system are purged with the sample being analyzed prior to injection.

Figure 8 shows a typical complete system. It is designed for sampling and conditioning volatile liquid and heavier streams for liquid injection.

E. Sample Disposal

Disposal of the unused sample is important in the design of a sample handling system. This is not only because of the increasingly stringent regulations dealing with the discharge of hydrocarbons to the atmosphere. Efficient, fast-responding process analyzers require high flow velocities in the sample transport and sample handling systems to minimize the lag time between the process and the analysis. Therefore, considerable amounts of sample need to be disposed of.

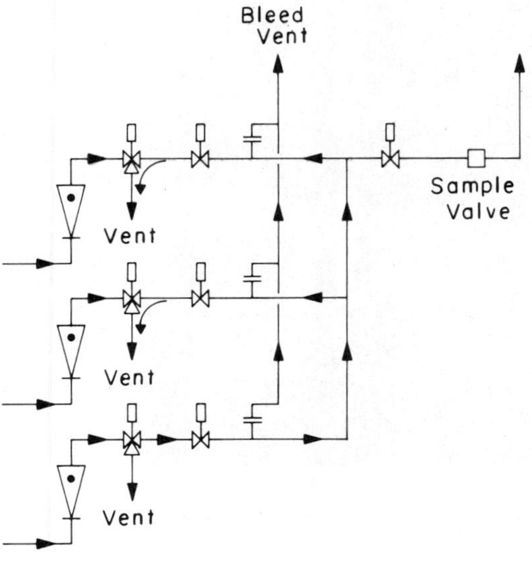

FIG. 7. Double block and bleed.

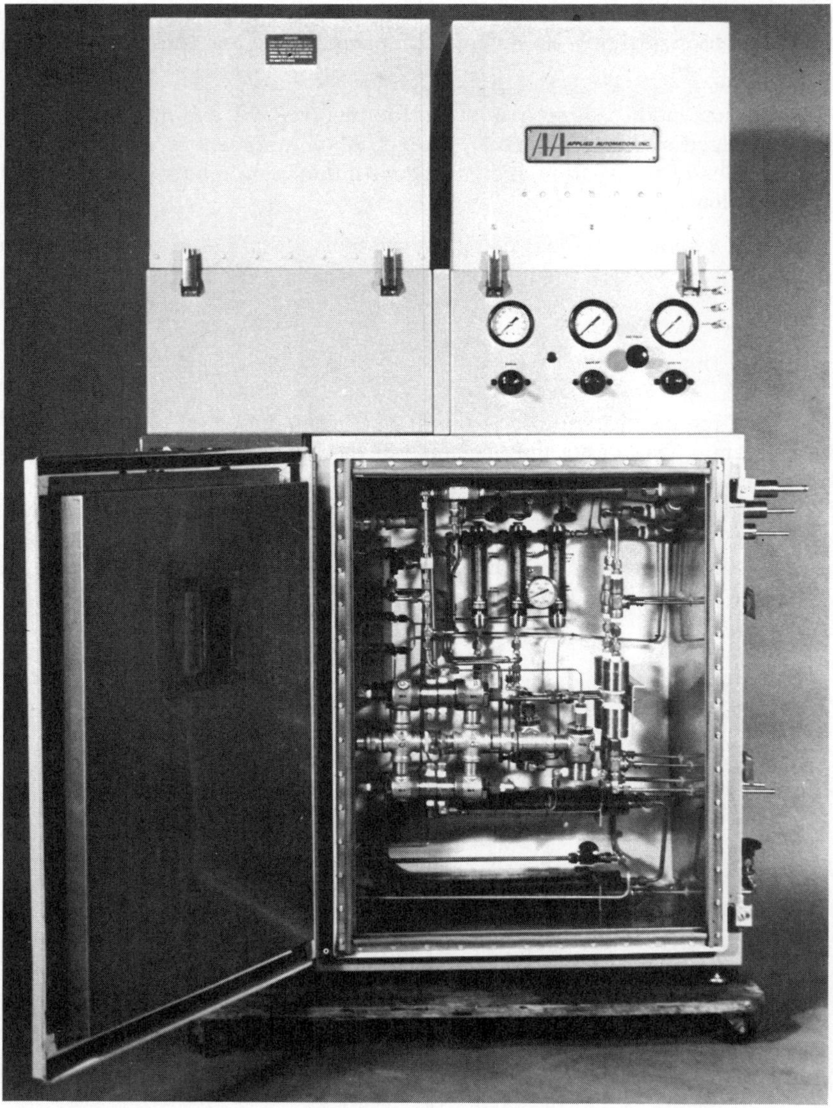

FIG. 8. Complete sample conditioning system (heated).

The preferable way to dispose of the unused sample and the sample bypass flow is by plumbing these flows back into the process. Usually one would connect to the suction side of a pump. If this is not feasible one would connect into a vent line to the flare stack. From the point of view of cleanliness and safety the former solution is preferred.

III. INJECTION SYSTEMS

A key element in any process chromatograph is the valve used to inject the sample into the analysis column. Modern valves show a high degree of sophistication and reliability. The basic types used are the rotary, slide, diaphragm, and O-ring seal valve. The last-mentioned one is not as popular as it used to be. Figure 9 shows the most common types used. All these valves have good reliability and are used not only for sample injection, but also for column switching.

IV. COLUMN SYSTEMS

As in all chromatographic systems, the heart of the unit is the chromatographic column. With process analyzers, it is usually necessary to compromise on the ideal system. The conflict often resolves itself into a trade-off of resolution versus shorter analysis time. In doing this, it is essential to consider the ultimate use of the analytical result. For instance, if the output of the analyzer is to be used directly as the input to a controller, then the analysis time should be considerably shorter than the response time of the process. However, when the response time of the process is long—on the order of hours or even days—a very rapid analysis of just a few minutes duration can lead to awkward problems. This is caused by the analyzer following the process noise. It may cause uncalled for adjustments to the process itself which will result in process oscillations.

Among process analyzer companies, there is a tendency for developing the fastest analysis procedure for a particular stream. While a rapid identification of the process state can be of immense value, the continuous update at a rapid rate may sometimes be less helpful than the recording of intermittent data.

The column system technology used in process analyzers has progressed rapidly over the past 10 years. New column packing materials such as the porous polymer materials [4, 5] have enabled systems to be designed with better peak resolution, lower stationary phase bleed, longer life, and—importantly—much greater resistance to upset conditions. The latter has greatly reduced the maintenance requirements of these systems.

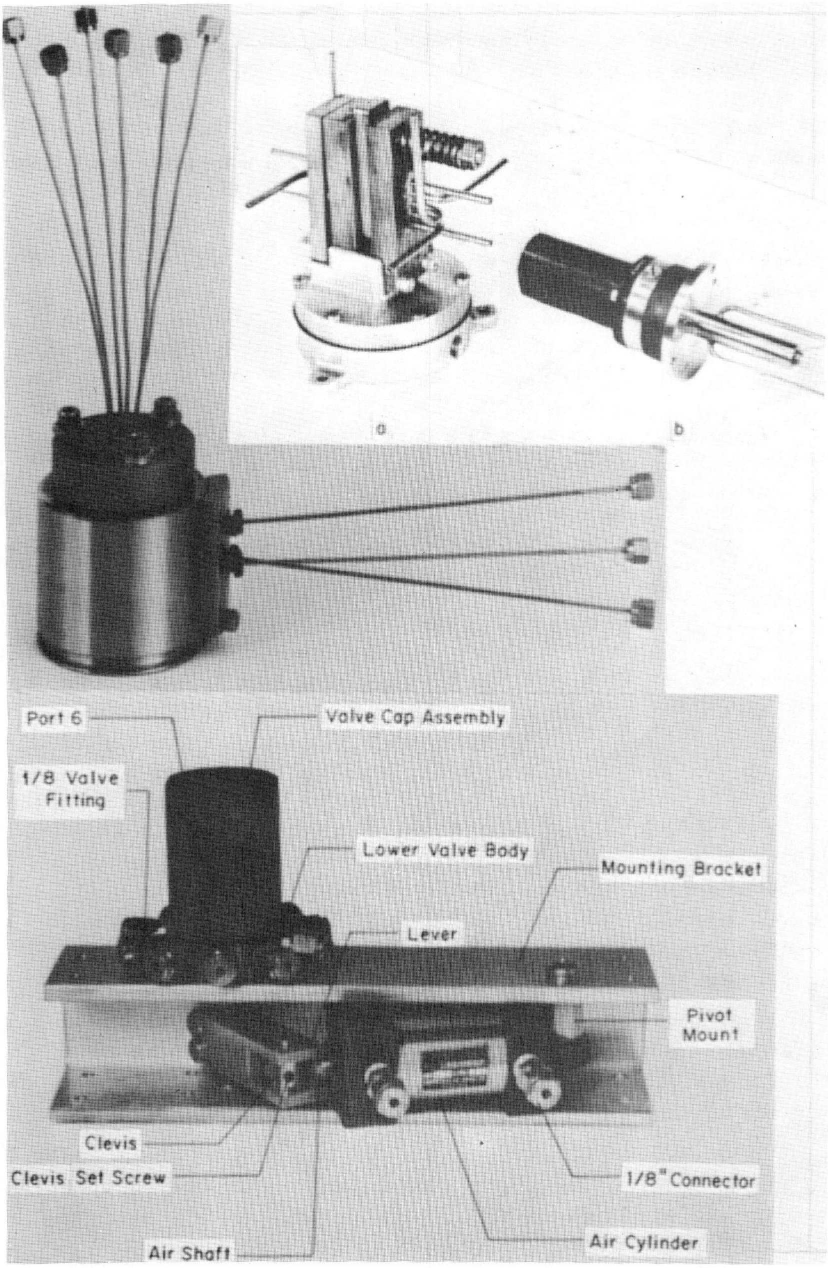

FIG. 9. Process gas chromatograph valves: a. Slide valve (Beckman Instruments); b. Diaphragm valve (Applied Automation); c. Cavity volume slide valve (Bendix); d. Rotary valve (Analytical Instruments Corporation).

Typical advanced column systems are those reported by Villalobos [6] and the micropacked porous polymer column systems frequently used by Honeywell [7].

V. DETECTORS

The workhorse detector for process gas chromatographs has been—and still is—the thermal conductivity (TC) detector. The simple, rugged, relatively inexpensive design and universal response make it ideal for process conditions. However, the ones commonly used in process analyzers do not have the sophistication or the sensitivity of the detectors found in modern laboratory chromatographs.

In the area of detectors the state of the art in laboratory units is far superior to that in process units. In contrast, in column system and pneumatic technology, the reverse is true. Rapid advances in detector technology are required because of the increased demands for tighter process control, for compliance with OSHA and EPA regulations, and for accurate and precise measurement of trace components in process streams. This development has forced the use of the inherently highly sensitive ionization detectors.

Currently, many manufacturers do offer flame ionization detectors and many users do operate them successfully. The benefits obtained by using such a sensitive detector have, in the past—with the notable exception of atmospheric monitoring—usually been outweighed by a number of mitigating factors. These are (a) the higher capital cost, (b) the need for two additional pure gas supplies (hydrogen and air), (c), the intolerance of the detector to column imbalance, (d) the effect of contamination from stationary phase bleed, and (e) the greatly increased complexity of the associated electronics. These aspects add to the amount of maintenance required.

However, as mentioned earlier, it is becoming essential that process analyzers provide sensitive and selective analyses. Trace analyses often have to be performed at the part per million level or even less, with long-term reliability and accuracy. In order to fulfill this need, the hydrogen flame ionization detector will probably become the detector of choice owing to its sensitivity, linearity, and relative simplicity. Over the next few years, other ionization detectors such as the electron capture and flame photometric detectors can be expected to appear in onstream applications in which their unique selectivity is important. Other detectors, such as the photoionization detector and the helium ionization detector, are difficult enough to use in a closely monitored, relatively clean environment in the laboratory. They are obviously less likely to be used in onstream analyzers to any significant extent.

The thermal conductivity detectors used in process analyzer applications are basically not different from those used in laboratory situations. The only significant difference is that they are usually operated at a lower filament temperature and, therefore, lower sensitivity, in order to prolong filament life. A case in which a unique thermal conductivity detector is used is the single thermistor detector employed by Applied Automation as an intercolumn detector [8].

A flame ionization detector showing some unique features is manufactured by Arcas (Figure 10). In this detector the flame is inverted. The manufacturer claims that this design is less noisy than others because any particulates or droplets formed fall away from the flame and thus do not contribute to thermionic noise in the system. The Model 2G-039 analyzer made by Process Analyzers is constructed of glass and Teflon. It is, therefore, not as susceptible to corrosion as the more conventional units that commonly use stainless steels as materials of construction.

One detector that is radically different is the pneumatic detector used in the Foxboro PCT [9, 10]. Detection is based on the change in pressure drop across an orifice (Figure 11). The differential pressure is a function of density and viscosity changes in the carrier gas (helium). Since this variable also reflects the square of any flow changes a capillary restrictor is used to effectively eliminate the flow rate sensitivity over a narrow range of flow rates. The differential pressure drop across the capillary, which is proportional to flow, is scaled such that $\Delta p \approx q^2$; in which q represents flow. This value is subtracted from the orifice signal. This allows the detector to be independent of flow rate changes over a reasonable range of carrier gas flow rates.

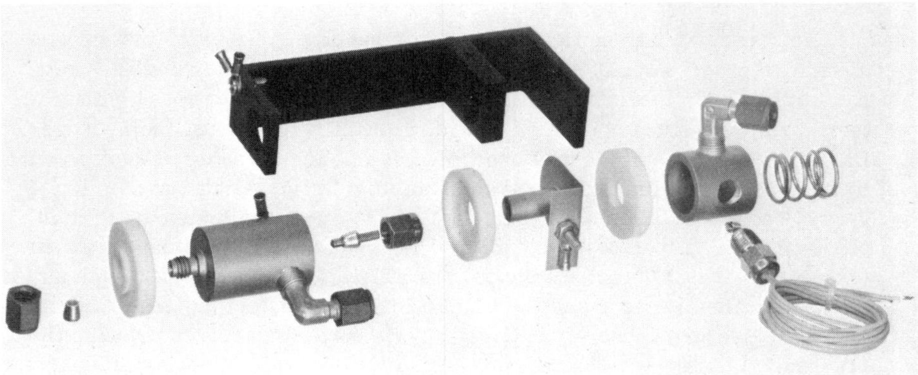

FIG. 10. ARCAS flame ionization detector (exploded parts).

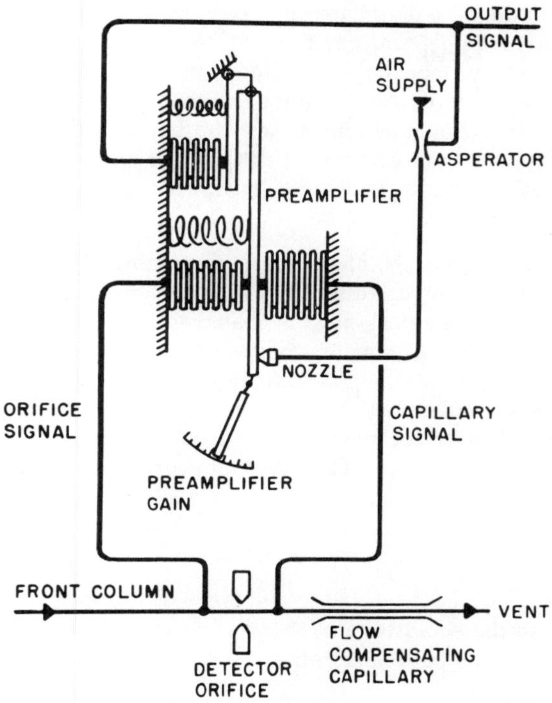

FIG. 11. Pneumatic detector (Foxboro).

This analyzer is also unique in that the total system is pneumatic and therefore can be field-mounted in hazardous areas. It is also designed to be particularly efficient for the monitoring of dual or single components in a stream. The Honeywell KCA 500 is another suitable analyzer for this purpose.

VI. DATA PRESENTATION

One traditional mode of data presentation is in the form of a bar graph output of either the peak height or the peak area of discrete components. In this output the amount of each component of interest can be read off directly from the height of the corresponding bar. Another mode is to display the trend of one or two discrete components in the mixture, where the amounts are plotted as a function of time.

Usually, chromatographic data are gated to the data collection system such that only selected peaks are processed. This approach has the advantage of reducing the data handling needs of the system to a minimum. In a typical control-loop situation, this can be as little as one peak per chromatograph. This imposes several constraints on the operation of the system. The sample size, temperature, carrier gas flow rate, and other factors must be maintained within narrow limits in order to obtain good precision. Second, this mode requires the use of an external standard, or standards, for calibration. These standards have to be reliable and accurately known.

Using these methods, the short-term repeatability of a process chromatograph has been reported to be ±1/4 to ±1/2% of full scale when in the bar graph mode and using peak height measurement [6]. The long-term repeatability, over several months, is reported by Osborne [11] to be much worse, ±8.2%. This is not as good as can be obtained with a laboratory instrument. We should expect much better results since computer data acquisition and reduction techniques were used. Advances in electronics systems and especially the advent of the microprocessor have enabled the manufacturers of process analyzers to offer sophisticated calculation and control techniques. These greatly improve the accuracy and precision of the analyses performed. Data reduction techniques can be utilized to perform several tasks, such as

1. Minimize the sensitivity of the system to external events by normalizing a complete analysis.
2. Combine and ratio component concentrations in any combination.
3. Adjust for retention time changes.
4. Provide calculated results such as BTU content and specific gravity.

A typical process gas chromatograph programmer with these features is shown in Figure 12.

These trends in data processing and reduction capabilities reinforce the impact that process gas chromatographs have on the complete analysis needs of petroleum and chemical process plants. It is clear that modern process analyzers can use the laboratory man's tools of area integration, relative retention time, multipoint calibration, and so on; it is also clear that the errors inherent in taking single grab samples with the associated problems of phase changes, leaks, and the difficulty of withdrawing an analytical sample from the portable sample container impose a handicap on the laboratory analysts. Therefore, we will probably see an increasing trend toward onstream chromatographs not only for loop control but also for quality control.

In addition to the data-reduction capabilities, the hardware available can also perform all of the functions previously carried out with

FIG. 12. Optichrom 2100 Microprocessor Programmer (Applied Automation).

electromechanical timers. It can vary from the single processor type shown in Figure 12 to large systems that can control as many as 32 chromatographs from the same minicomputer. Over the years, the requirements for stream selection valve switching, column valve switching, and data transmission back to the control room have become more demanding as analyses have become faster. An added benefit from the more sophisticated programmers—chromatograph controllers, which they really are nowadays—is that self-checking and alarm features can be built in. Variables which can be monitored are, for example, temperature, flow, carrier gas pressure, electronic functions, and calculated sample alarm limits.

The choices offered are broad and advantages in performance and promised reliability are always possible at the expense of increased electronic sophistication. The value of these features will depend on the purchaser's confidence in the manufacturer's ability to produce the necessary reliability in hardware and software.

The choice between single unit microprocessor and large integrated systems is another one that can only be resolved by considering all aspects of the systems and the plant system and maintenance philosophy of the user. The obvious major trade-offs are cost and risk of downtime. With the totally independent, more costly systems, failure of any part of the unit puts only one single analyzer out of operation. On the other hand, the minicomputer-based systems can cost less. But if the minicomputer fails, all analyzers are disabled. It should be pointed out, however, that the modern minicomputer system hardware is much less likely to fail than the sampling system and analyzer hardware, particularly the valves and filters.

VII. PROCESS LIQUID CHROMATOGRAPHS

In spite of the explosive growth of high-pressure liquid chromatography over the past 6 to 8 years, the use of this technology in the process field is best described as imminent rather than having arrived. Two manufacturers, Du Pont Process Instruments and Applied Automation, offer process liquid chromatographs, but the number of units sold is small, probably 25 per year and can best be described as experimental. The reasons for the slow acceptance of HPLC units is somewhat difficult to determine but cost and reliability obviously have an effect. Another factor is that HPLC units are extremely susceptible to particulates and contamination by "foreign" substances. In process gas chromatography, a wealth of experience has been developed in column switching techniques in order to preserve the analytical column and the detector. Similar technology has not yet been developed in HPLC, although there are some signs of research work in this area [12, 13]. Another factor to be borne in mind with HPLC systems is the cost of the analytical columns, ca. $300, and the short life, ca. 2 to 3 months.

This is only 20% of the minimum acceptable life for a process gas chromatograph column system.

VIII. SELECTED APPLICATIONS

The diversity of the applications for process gas chromatographs is, as mentioned earlier, one of the major reasons for its widespread use.

In the petroleum industry typical applications are on the following processes:

Alkylation

Isomerization

Thermal cracking

Catalytic cracking

Reforming

Distillation

Fluid catalytic cracking

Sulfur recovery

Each of the manufacturers has proprietary methods of analysis for these and other applications.

A typical example of an application chromatogram, but one that also demonstrates what is possible with microcolumns is shown in Figure 13. Illustrated is the separation of C_1 to C_3 hydrocarbons in 12 sec; this is much faster than the response time of most processes.

Figure 14 shows the separation of saturated compounds through C_5 with a backflush of C_{6+}. This analysis, similar to ASTM D 2427, takes 25 sec.

IX. SUMMARY

Modern process gas chromatographs are reliable, cost-effective for analyses and increasingly sophisticated electronically. They are compact, efficient, and increasingly being designed for ease of maintenance by the use of plug-in cards, built-in diagnostics, and careful design and placement of mechanical parts. Figures 15 and 16 show two of the many units currently available.

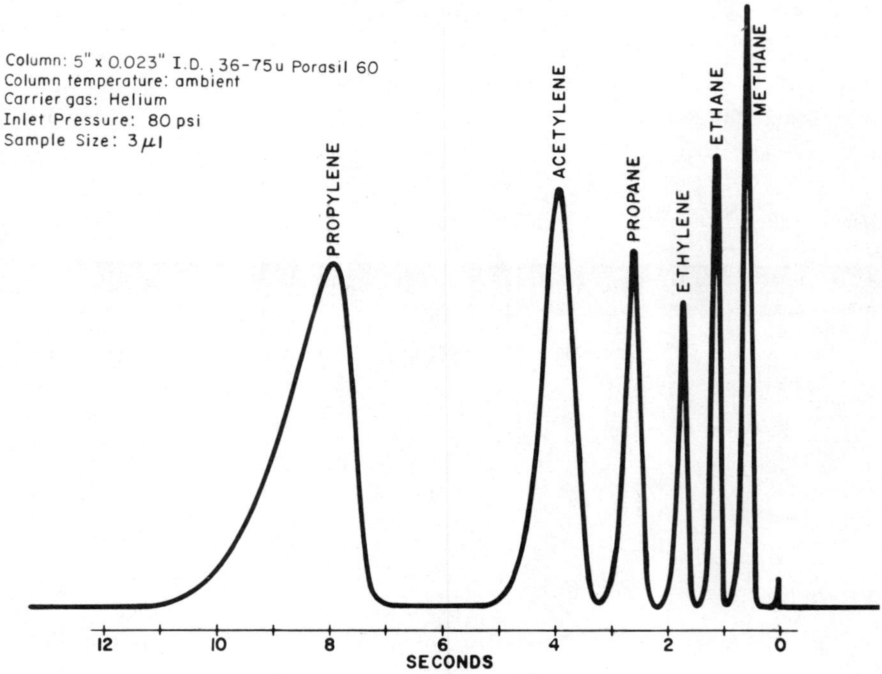

FIG. 13. Separation of C_1 to C_3 hydrocarbons (courtesy Honeywell).

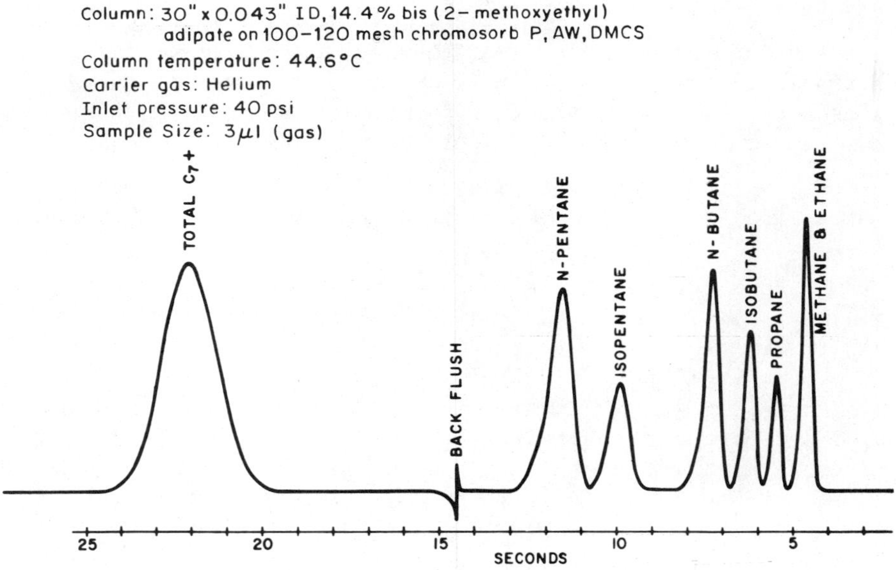

FIG. 14. C_5 and lighter analysis (courtesy Honeywell).

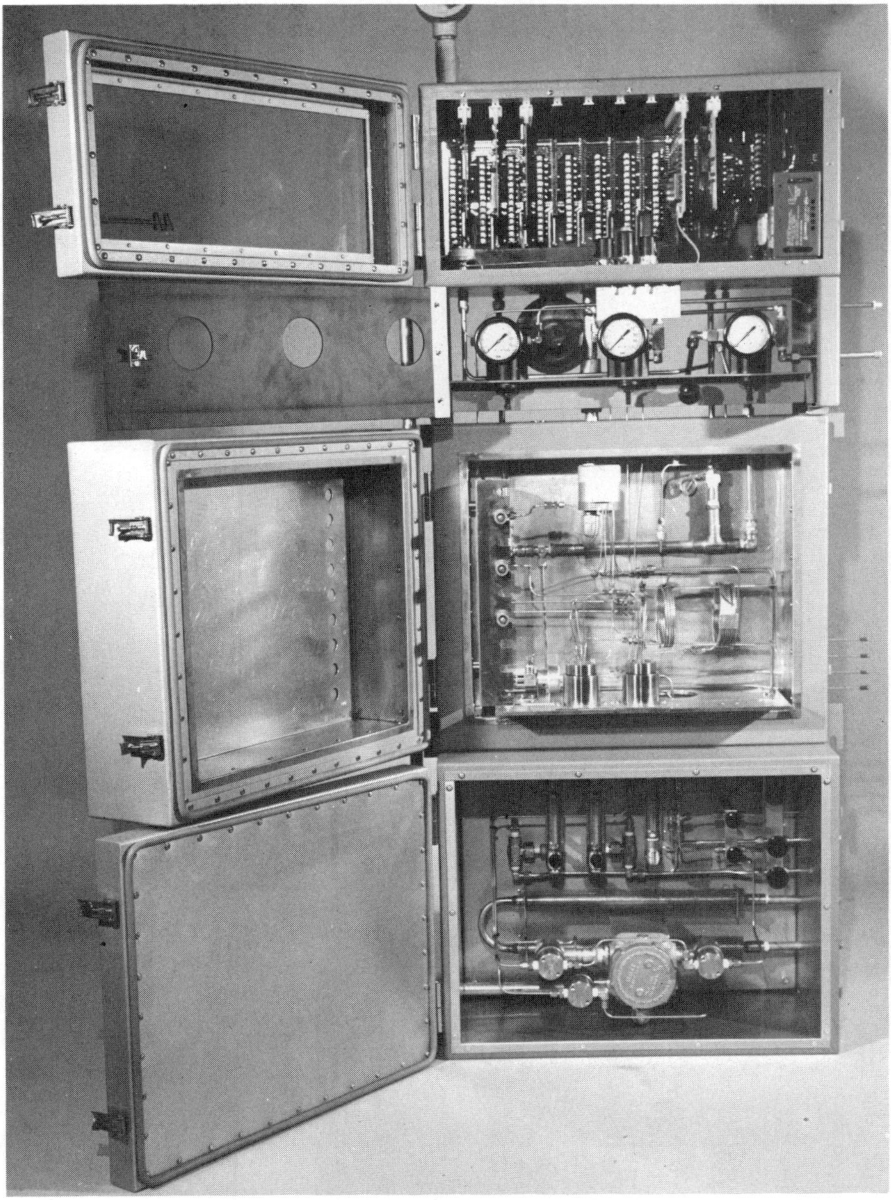

FIG. 15. Applied Automation Optichrom 2100.

FIG. 16. Honeywell Model M-1000 equipped with FPD detector for sulfur analysis (electronics and flow controllers to left of main analyzer).

ACKNOWLEDGMENTS

The author gratefully acknowledges the courtesy of the Instrument Society of America for permission to publish Figs. 3 and 6. Thanks are also due to the various instrument manufacturers who supplied him with the figures of the products shown in this chapter.

REFERENCES

1. A. T. James and A. J. P. Martin, Biochem. J. 50:679 (1952).
2. P. R. Carl and J. G. Kerley, Hydrocarbon Process. 47(5):127 (1968).
3. E. A. Houser, "Principles of Sample Handling and Sample System Design for Process Analysis," Instrument Society of America, Pittsburgh, 1972.
4. O. L. Hollis, "Advances in Chromatography 1965 (A. Zlatkis and L. S. Ettre, eds.), Preston Technical Abstracts Co., Evanston, Illinois, 1966, p. 56.
5. O. L. Hollis and W. V. Hayes, J. Gas Chrom. 4:235 (1966).
6. R. Villalobos, Anal. Chem. 47(11):983A (1975).
7. D. E. Durbin and J. Fruge, Anal. Chem. 44(8):1502 (1972).
8. Optichrom 102 Process G.C. Manual, Applied Automation Inc., Bartlesville, Oklahoma.
9. R. Annino, J. Curren, Jr., R. Kalinowski, E. Karas, R. Lindquist, and R. Prescott, J. Chromatog. 126:301 (1976).
10. Pneumatic Composition Transmitter Manual, Foxboro Instruments, Foxboro, Massachusetts, GS 6-3A1A (November, 1975).
11. J. L. Osborne, Chem. Eng. Prog. 80(11):76 (1974).
12. J. F. K. Huber, R. van der Linden, E. Ecker, and M. Oreans, J. Chromatog. 83:267 (1973).
13. R. J. Dolphin and F. W. Willmatt, J. Chromatog. Sci. 14:584 (1976).

AUTHOR INDEX

Numbers in brackets are reference numbers and indicate that an author's work is referred to although his name is not cited in the text. Underlined numbers give the page on which the complete reference is listed.

E

F

G

N

T

Y

Z

SUBJECT INDEX

T

U